Modern Methods of Plant Analysis

New Series Volume 7

Editors

H.F. Linskens, Erlangen/Nijmegen

J.F. Jackson, Adelaide

Volumes Already Published in this Series:

Volume 1: Cell Components
1985, ISBN 3-540-15822-7

Volume 2: Nuclear Magnetic Resonance
1986, ISBN 3-540-15910-X

Volume 3: Gas Chromatography/
Mass Spectrometry
1986, ISBN 3-540-15911-8

Volume 4: Immunology in Plant Sciences
1986, ISBN 3-540-16842-7

Volume 5: High Performance Liquid Chromatography
in Plant Sciences
1987, ISBN 3-540-17243-2

Volume 7: Beer Analysis
1988, ISBN 3-540-18308-6

Forthcoming:

Volume 6: Wine Analysis
1988, ISBN 3-540-18819-3

Volume 8: Analytical Method for Non-alcoholic Beverages
ISBN 3-540-18820-7

Volume 9: Gas Analysis in Plants
ISBN 3-540-18821-5

Volume 10: Plant Fiber Analysis
ISBN 3-540-18822-3

Beer Analysis

Edited by
H. F. Linskens and J. F. Jackson

Contributors

S. Aastrup M. J. Allison M. Bohl Pedersen B. E. Christensen
J. A. Delcour S. Donhauser H. Eriksen C. Gjermansen
R. van Heeswijck B. Jende-Strid L. P. Jepsen K. G. Jørgensen
A. K. Keiding G. C. J. Muts T. Peppard U. Rasmussen
M. I. G. Roncero P. Sigsgaard J. Strating L. C. Verhagen

With 97 Figures

Springer-Verlag
Berlin Heidelberg New York
London Paris Tokyo

Professor Dr. Hans-Ferdinand Linskens
Goldberglein 7
D-8520 Erlangen

Professor Dr. John F. Jackson
Department of Biochemistry
Waite Agricultural Research Institute
University of Adelaide
Glen Osmond, S.A. 5064
Australia

Library of Congress Cataloging-in-Publication Data. Beer analysis/edited by H.F. Linskens and J.F. Jackson; contributors, S. Aastrup ... [et al.]. p. cm. – (Modern methods in plant analysis; new ser., v. 7) Includes bibliographies and index.
ISBN-13: 978-3-642-83212-3 e-ISBN-13: 978-3-642-83210-9
DOI: 10.1007/ 978-3-642-83210-9
1. Beer–Analysis. I. Linskens, H.F. (Hans F.), 1921– . II. Jackson, J.F. (John F.), 1935– . III. Aastrup, S. IV. Series. TP577.B38 1988 663'.42–dc19 87-28644

Softcover reprint of the hardcover 1st edition 1988

Typesetting: Brühlsche Universitätsdruckerei, Giessen
2131/3130-543210

Introduction

Modern Methods of Plant Analysis

When the handbook *Modern Methods of Plant Analysis* was first introduced in 1954 the considerations were:

1. the dependence of scientific progress in biology on the improvement of existing and the introduction of new methods;
2. the difficulty in finding many new analytical methods in specialized journals which are normally not accessible to experimental plant biologists;
3. the fact that in the methods sections of papers the description of methods is frequently so compact, or even sometimes so incomplete that it is difficult to reproduce experiments.

These considerations still stand today.

The series was highly successful, seven volumes appearing between 1956 and 1964. Since there is still today a demand for the old series, the publisher has decided to resume publication of *Modern Methods of Plant Analysis*. It is hoped that the New Series will be just as acceptable to those working in plant sciences and related fields as the early volumes undoubtedly were. It is difficult to single out the major reasons for success of any publication, but we believe that the methods published in the first series were up-to-date at the time and presented in a way that made description, as applied to plant material, complete in itself with little need to consult other publications.

Contributing authors have attempted to follow these guidelines in this New Series of volumes.

Editorial

The earlier series *Modern Methods of Plant Analysis* was initiated by Michel V. Tracey, at that time in Rothamsted, later in Sydney, and by the late Karl Paech (1910–1955), at that time at Tübingen. The New Series will be edited by Paech's successor H. F. Linskens (Nijmegen, The Netherlands) and John F. Jackson (Adelaide, South Australia). As were the earlier editors, we are convinced "that there is a real need for a collection of reliable up-to-date methods for plant analysis in large areas of applied biology ranging from agriculture and horticultural experiment stations to pharmaceutical and technical institutes concerned with raw material of plant origin". The recent developments in the fields of plant biotechnology and genetic engineering make it even more important for workers in the plant sciences to become acquainted with the more sophisticated methods, which sometimes come from biochemistry and biophysics, but which also have been developed in commercial firms, space science laboratories, non-university research institutes, and medical establishments.

Concept of the New Series

Many methods described in the biochemical, biophysical, and medical literature cannot be applied directly to plant material because of the special cell structure, surrounded by a tough cell wall, and the general lack of knowledge of the specific behavior of plant raw material during extraction procedures. Therefore all authors of this New Series have been chosen because of their special experience with handling plant material, resulting in the adaptation of methods to problems of plant metabolism. Nevertheless, each particular material from a plant species may require some modification of described methods and usual techniques. The methods are described critically, with hints as to their limitations. In general it will be possible to adapt the methods described to the specific needs of the users of this series, but nevertheless references have been made to the original papers and authors. While the editors have worked to plan in this New Series and made efforts to ensure that the aims and general layout of the contributions are within the general guidelines indicated above, we have tried not to interfere too much with the personal style of each author.

There are several ways of classifying the methods used in modern plant analysis. The first is according to the technological and instrumental progress made over recent years. These aspects were used for the first five volumes in this series describing methods in a systematic way according to the basic principles of the methods.

A second classification is according to the plant material that has to undergo analysis. The specific application of the analytical method is determined by the special anatomical, physiological, and biochemical properties of the raw material and the technology used in processing. This classification will be used in Volumes 6 to 8, and for some later volumes in the series. A third way of arranging a description of methods is according to the classes of substances present in the plant material and the subject of analytic methods. The latter will be used for later volumes of the series, which will describe modern analytical methods for alkaloids, drugs, hormones, etc.

Naturally, these three approaches to developments in analytical techniques for plant materials cannot exclude some small overlap and repetition; but careful selection of the authors of individual chapters, according to their expertise and experience with the specific methodological technique, the group of substances to be analyzed, or the plant material which is the subject of chemical and physical analysis, guarantees that recent developments in analytical methodology are described in an optimal way.

Volume Seven – Beer Analysis

Beer is one of the oldest beverages of man. In its classical form it is the product of interaction between three different plant products: a specially prepared carbohydrate substrate, an active yeast, and additional flavour substances. A definition of beer is that it is derived from the anaerobic alcoholic fermentation of aqueous extracts of plant origin. In classical beer, the substrate for activity of the yeast cell is malt, produced from germinated caryopses of barley.

Besides barley, other substrates such as rice, corn and millet are used as raw material for the fermentation process. The presence of water-soluble sugar is important; it is generated by the activation of enzymes during the germination process. Normally, barley seedlings after starting germination transform 60–70% of the dry weight of starch, which is stored in the endosperm of the caryopse to soluble sugars. Additional components of the barley, such as enzymes, fat, cellulose, proteins and ash are important elements which characterize the quality of the beer. For taste and preservation the fermentation product is blended with extracts from hops. The bitter principles, resins and etheric oils of the hop flowers characterize the innumerable types of beer. Finally, the water quality is decisive for the specific taste of a beer type; it has not only to be clean and to contain as little salt as possible, but should also be soft. In some cases some gypsum is needed.

As long as man has manufactured beer, he has maintained a continuing investigation of the materials used, and of the control of the production process and the final product. This quality control of a pure plant product becomes of ever-increasing importance with the world-wide growth of beer production. The consumers must therefore demand a product which is not only of high quality, but in all cases comes under the definition of the word beer. Worldwide, and especially in Europe, there is much discussion on the purity of beer. This means that analytical methods are of increasing importance.

The progress made in biotechnology includes manipulation of the yeast, so that many new strains, characteristic for certain types of beer, have been developed. Breeding has long been involved in the development of barley strains, especially destined and suitable for beer making. The inclusion of substrates from plants other than barley make it even more important to have the constant control of the materials included in the beer-making processes.

Quality control of the material used for beer making as the final product has a long tradition. Analytical methods have been used for more than 100 years for checking the fermentation process.

The recent progress in plant analysis and the inclusion of modern biochemical methods in beer making are the prerequisite for the high standard of a beer industry. This volume is not reporting on the standard methods of water, sugar, protein and alcohol control, but is rather restricted to modern methods of analysis of the plant material involved in the production process, including raw material analysis of the fermentation process and the final product. These special methods are an excellent example of the rapid progress that plant analytical methods have made over the last decade. The application of fundamental methods in beer analysis have contributed to the vast extension of world beer production, to the

Front page of the first monograph on beer analysis. It was written by Josef Kraeger from Pilsen. The title reads (translated) as follows: "The investigation and judgement of beer and of the raw materials used in beer brewery – methods for the chemical and technical examination of beer and the raw materials used in brewery, compiled for use in the laboratory", with 30 illustrations. The publisher was A. Hartleben, who edited a Chemical–Technical Library, of which Kraeger's book was a part. The monograph was published in Vienna and Leipzig in the year 1906.

The front page, a wood cut, shows various aspects of chemical technology from the beginning of the 20th century, clockwise: gas–lighting, boxes with starch, laundry blue (a mixture of ferri and ferro salts of ironcyanide used for whitening of linen and laundry), varnish, a pair of scales as used by pharmacists, barrels and tubes for wine and beer, wine bottles in a basket, a tankard, stamp and ink-pad, a brewery installation, in the foreground a fermentation bottle and instruments from a chemical laboratory, installation for preparing vinegar, a camera with darkening cloth, a set of tobacco pipes, a petrol lamp, various bottles and wood preparations, and at the top: a display of fireworks.

large spectrum of specific beer characters and to the high quality of an industry based exclusively on plant products.

Acknowledgements. The editors express their thanks to all contributors for their efforts in keeping to production schedules, and to Dr. Dieter Czeschlik, Ms. K. Gödel and Ms. I. Samide of Springer-Verlag for their co-operation with this and other volumes in Modern Methods of Plant Analysis. The constant help of José Broekmans is gratefully acknowledged. The editors would like to thank Professor D. von Wettstein (Carlsberg Laboratory, Copenhagen) for his advice in the preparation of the volume.

Erlangen/Nijmegen and Adelaide, Autumn 1987

H. F. LINSKENS
J. F. JACKSON

Contents

Raw Material

Barley and Malt Analysis
G. C.J. Muts

Reversed Phase HPLC as an Aid for the Identification of Barley Cultivars
M. J. Allison (With 5 Figures)

The Application of Immunofluorescence for Detecting Specific Proteins in Barley Seeds
U. Rasmussen (With 7 Figures)

Fluorimetric Methods for the Analysis of Malt Modification and Pre-Harvest Sprouting
S. Aastrup and K. G. Jørgensen (With 6 Figures)

Automated Testing of Brewer's Yeast Strains for Fermentation Characteristics (Multiferm, System Carlsberg)
P. SIGSGAARD (With 12 Figures)

The Use of Nucleotide Sequence Polymorphisms and DNA Karyotyping in the Identification of Brewer's Yeast Strains and in Microbiological Control
M. Bohl Pedersen (With 8 Figures)

Screening and Testing New Distillers' Yeasts for Their Potential in Molasses Ethanol Fermentations
B. E. Christensen and H. Eriksen (With 3 Figures)

Genetic Analysis and Manipulation of *Mucor* Species by DNA-Mediated Transformation
R. Van Heeswijck, M. I. G. Roncero and L. P. Jepsen (With 1 Figure)

Final Product

Malt and Hop Flavanoids in Pilsner Beer
J. A. Delcour (With 5 Figures)

Short Introduction to Headspace Analysis
J. Strating

The Use of Principal Components Analysis in Monitoring the Quality of Beer
T. Peppard (With 5 Figures)

German Beer Purity Law and its Influences on the Properties and Analysis of Beer
S. Donhauser (With 14 Figures)

List of Contributors

AASTRUP, STEN, Carlsberg Research Center, Department of Brewing Chemistry, Gamle Carlsberg Vej 10, 2500 Valby, Copenhagen, Denmark

ALLISON, MICHAEL J., Scottish Crop Research Institute, Invergowrie, Dondee, DD2 5DA, Scotland

BOHL PEDERSEN, MOGENS, Carlsberg Research Center, Department of Brewing Chemistry, Gamle Carlsberg Vej 10, 2500 Valby, Copenhagen, Denmark

CHRISTENSEN, BJØRN E., Novo Industri, Novo Alle'I, 2880 Bagsvoerd, Denmark

DELCOUR, JAN A., Katholieke Universiteit Leuven, Faculteit Landbouwwetenschappen, Kard. Mercierlaan 92, 3030 Heverlee, Belgium

DONHAUSER, SIEGFRIED, Hittostr. 4, 8050 Freising, FRG

ERIKSEN, HENRIK, Aaloekkevej 13, 5800 Nyborg, Denmark

GJERMANSEN, CLAES, Carlsberg Laboratory, Department of Physiology, Gamle Carlsberg Vej 10, 2500 Valby, Copenhagen, Denmark

VAN HEESWIJCK, ROBYN, Department of Genetics, Melbourne University, Parkville 3052, Australia

JENDE-STRID, BARBRO, Department of Physiology, Carlsberg Laboratory, Gamle Carlsberg Vej 10, 2500 Valby, Copenhagen, Denmark

JEPSEN, LARS P., Department of Physiology, Carlsberg Laboratory, Gamle Carlsberg Vej 10, 2500 Valby, Copenhagen, Denmark

JØRGENSEN, KIM G., Carlsberg Research Center, Department of Biotechnology, Gamle Carlsberg Vej 10, 2500 Valby, Copenhagen, Denmark

KEIDING, ANNE K., Carlsberg Laboratory, Department of Physiology, Gamle Carlsberg Vej 10, 2500 Valby, Copenhagen, Denmark

MUTS, G. C. J., Heineken Technisch Beheer B. V., P.O. Box 510, 2380 BB Zoeterwoude, The Netherlands

PEPPARD, TERRY, The Stroh Brewery Co., 100 River Place, Detroit, MI 48207, USA

RASMUSSEN, U., Carlsberg Research Laboratory, Department of Biotechnology, Gamle Carlsberg Vej 10, 2500 Valby, Copenhagen, Denmark

RONCERO, M. ISABEL G., Departamento de Genetica, Facultad de Ciencias, Universidad de Cordoba, 14071 Cordoba, Spain

SIGSGAARD, POUL, Carlsberg Research Center, Department of Brewing Chemistry, Gamle Carlsberg Vej 10, 2500 Valby, Copenhagen, Denmark

STRATING, J., Heineken Technisch Beheer B. V., P.O. Box 510, 2380 BB Zoeterwoude, The Netherlands

VERHAGEN, L. C., Heineken Technisch Beheer B. V., P.O. Box 510, 2380 BB Zoeterwoude, The Netherlands

List of Contributors

Raw Material

Barley and Malt Analysis

G. C. J. Muts

1 General Introduction

Barley (*Hordeum vulgare* sp.) has been cultivated for many centuries. Nowadays, its use is mainly restricted to cattle feed and brewing purposes.

Several groups of people have different interests in the analysis of barley. Apart from the feed application of barley, which will not be further discussed here, breeders, maltsters and brewers hold different viewpoints.

The breeder is interested in methods of analysis that are suitable for daily screenings of large numbers of small samples in a breeding programme. The properties he tries to give the barley should serve the farmer (for instance, high yield), the maltster (for instance, quick and even germination) and the brewer (for instance, high extract yield). The maltster needs methods that enable him to monitor the malting procedure and to control the quality of barley and the produced malt. The brewer wants analysis methods that provide proper entrance control in terms of extract yield, processability and beer quality. In other words, the malt analysis should predict the behaviour of the malt through the brewery until in the final product.

Most analysis methods in use in breweries were developed years ago. The scope of this chapter will be limited to giving an insight into modern developments in the analysis of barley and malt.

The analysis methods for barley and malt that are in common use are regularly updated and extended in four publications:

- Recommended Methods of Analysis of the Institute of Brewing (IOB)
- Brautechnischen Analysenmethoden, Methodensammlung der Mitteleuropäischen Brautechnischen Analysenkommission (MEBAK)
- Analytica EBC, Analysis Committee of the European Brewery Convention (EBC)
- Recommended Methods of Analysis of the American Society of Brewing Chemists (ASBC).

These publications will be referred to as "Recommended Methods".

All methods and analyses that are common, but are not discussed in this chapter, such as sampling, moisture determination, Kjeldahl nitrogen determination, etc., can be found in these publications.

Many of the methods mentioned in this chapter apply to both barley and malt.

2 Barley Analysis

Freshly harvested barley needs to be analysed before it is stored for further trading and processing. The moisture content should be sufficiently low to prevent mould growth and deterioration, the value in terms of extract should be known in order to predict the trading prices and a germination test should be performed to inform the maltster about the processability.

The breeder again analyses barley in order to predict its value for malting and brewing properties. This analysis has a completely different approach.

2.1 Visual Inspection

Methods for visual inspection are always subjective and are only powerful in the hands of skilled people. Visual inspection is therefore often somewhat underestimated. Nevertheless, it is usually a very quick way to obtain a bird's-eye view of a barley or a malt. Visual inspection should comprise smell, moisture, colour, unwanted material (stones, straw, grains), size and form of the kernels, etc.

The publications on Recommended Methods of Analysis contain descriptions of how each of these items should be judged.

2.2 Mechanical Analysis

Because large kernels yield more extract than small ones, the analysis of corn size distribution is very common practice. It is performed on sieves with a 2.8, 2.5, and 2.2 mm screen. Very important for the maltster is a narrow corn size distribution. Small kernels are modified more quickly than large ones. The weight of 1000 corns also gives information on the expected extract content of barley.

There have been some publications on the measurement of barley kernel hardness, homogeneity and steeliness (Chapon 1961 a, b, 1963 a, b). Industrial application, however, has been limited mainly to malt and will be discussed there. Several instruments, such as murbidimeter and cut half kernels (Michel and Michel 1910), have been used for the measurement of kernel hardness.

2.3 Germination Tests

When barley is processed by the maltster it must be germinating for more than 97%. Dormancy, i.e. a "sleeping" status of barley after harvesting, can be a serious problem. The grains then are viable, but do not germinate and need rest or temperature shocks in order to overcome dormancy (Palmer and Taylor 1980).

Two germination tests are therefore in common use. The first test gives information on the actual number of germinating grains, i.e. it measures the germinative energy (Pollock 1962).

Germinative Energy (EBC)
Two lots of 500 corns are taken by sample divider with rejection of half corns and of foreign corns and other foreign matter. 500 corns are placed in a large beaker and covered with tap water at 18–20 °C. They are steeped for 1 h, the water poured off and the grains washed five times with tap water. They are again covered with water and steeped for a further 6 h. After this the corns are strained off on a stainless steel gauze and transferred to a funnel, 85 mm in diameter, and equipped with a small piece of flat-ended glass rod in the neck.

The funnel is covered with a Petri dish to avoid evaporation. At the commencement of work next day, at midday and at the end of work that day, the corns are tipped out on the gauze, moistened slightly with water using an atomizer and returned to the funnel. At 72 h and 120 h from the beginning of steeping the numbers of non-germinated (n) corns are counted and the result is reported as:

$$\text{Germinative energy } \% = \frac{500 - n}{10}.$$

This method has been automated recently (Leysen 1984). A germination test that overcomes dormancy with the use of H_2O_2 (Thunaeus 1938) determines the maximal possible number of germinating grains, which is called germinative capacity.

Germinative Capacity
Two samples of 500 corns are obtained using a sample divider with rejection of half corns and foreign matter. Each lot of 500 corns is steeped for 2 days in a beaker at with 200 ml of 0.75% hydrogen peroxide at 18–21 °C, freshly prepared by dilution of 30% H_2O_2 with tap water. After this the solution is poured off, 200 ml of fresh 0.75% H_2O_2 is added and left for 2 more days at 18–21 °C. The corns are then strained off on a stainless steel gauze and the number of ungerminated corns is counted (n)

$$\text{Germinative capacity } \% = \frac{500 - n}{5}.$$

More details can be found in the EBC-Analytica.

An alternative to this test is the staining of the viable germ with 2-p-iodophenyl-3-p-nitrophenyl-5-phenyltetrazolium chloride (Bishop 1957; de Clerck 1962a).

In order to measure water sensitivity of barley, which can also be a cause of delayed germination, several tests have been proposed (Essery et al. 1955; Macey and Stowell 1959; Scriban 1963) which are used mainly by maltsters because the test indicates how a barley should be steeped (De Becker 1974; Krüger and Bielig 1976).

The opposite of dormancy, sprouting in the ear, can also be a problem. The kernels then already germinate in the field. Three methods for determining this condition have recently been developed. In one, methylene blue is used to stain the already modified parts of the kernel blue. This method is outlined in the section on malt. A second method was developed by Palmer and Taylor (1981), who use acetic orcein staining as detection method. The third approach is the measurement of lipase activity, which develops early in germination. This is done on half kernels that are stained with fluoresceine dibutyrate which produces a fluorescent signal after hydrolysis by lipase (Jensen and Heltved 1982, 1983). The accuracy of this method has been questioned (Henry and McLean 1986).

Pregermination Analysis
A sample of barley seeds is fixed into clay plates using a perforated steel plate to align the seeds and pressure to sink them in the clay. The kernels are then mechanically sanded down half way using sand paper. The half kernels are stained with a 1 m*M* solution of fluoresceine dibutyrate in 80% (v/v) aqueous ethanol. Lipase action hydrolyses the fluorochrome and in due time a yellow fluorescence can be detected under ultraviolet light.

The Recommended Methods include an amylase detection method using a starch solution in agar that is stained blue, after incubation of half kernels on the gel, with iodine.

2.4 Chemical Analysis

Husk Content. The quality of the husk is very important when the brewer uses a lautertun for wort filtration. The husk should be firm and flexible on visual inspection. Too high a content of husk, however, reduces the extract content by weight of the barley. Many husk determination tests have been developed. The most recent are those by Essery et al. (1956), Whitmore (1960), and Seeleitner and Püspök (1983). Below an example is given of a husk determination test.

Husk Content
A quantity (g) of about 4 g of barley, with moisture content v, from which half corns and foreign material are removed, is added to 100 ml of 5% potassium carbonate in water in a flask equipped with a reflux cooler. The mixture is boiled for 25 min, decanted and 10 ml of 2 N hydrochloric acid is added. The corns are then washed three times with tap water. By rubbing, the husks are removed from the corns, transferred to a Jena no. G1 glass filter and washed with distilled water of 80–100 °C. The husks are dried in an oven at 105 °C for 3 h, and weighed after cooling to room temperature in a desiccator (a)

$$\text{Husk content } \% = \frac{1.31 \times a \times 10{,}000}{g\,(100 - v)}.$$

Nitrogen Content and Proteins. The nitrogen content of barley is usually determined by Kjeldahl analysis. The protein content is calculated from that using $6.25 \times N$. For determination of the price of barley a formula is in use that relates protein content of barley directly to extract yield: extract % = varietal constant $-0.75 \times$ protein $+ 0.1 \times 1000$ corn weight.

The varietal constant can be 84.4 for two-row barleys and 80.0 for six-row barleys (Bishop 1930, 1948).

Recently, methods using near infrared reflectance analyses (NIR) have come into use for the simultaneous determination of nitrogen and moisture content of barley (Grandclerc et al. 1982; Donhauser and Linsenmann 1983; Smith 1984). The method is extremely rapid and convenient but implies the purchase of expensive equipment.

There is very little literature and certainly no analysis method available that determines the *quality* of the protein or the protein composition for control purposes in maltings and breweries. An attempt has been made by Van den Berg et al. (1981), who isolated gel protein from barley as an important factor that influences lautertun filtration (Muts et al. 1984). The analysis of gel protein content of barley is as follows (Muts et al. 1984).

Gel Protein
4.0 g of barley flour, sieved through a 150- μm sieve, is extracted with petroleum ether 40–60 to remove fatty material and is dried in the air at room temperature. The defatted flour is suspended in 90 ml of a 1.5% sodium dodecyl sulphate (SDS) solution at 10 °C under vigorous stirring, avoiding lumping, and as quickly as possible. The suspension is then ultracentrifuged at 80,000 *g* for 30 min, decanted and the upper gel layer is scraped off. The remaining pallet is resuspended in 60 ml of 1.5% SDS solution and again ultracentrifuged at 80,000 *g* for 30 min. The second gel layer is scraped off after decantation of the supernatant and added to the first one.

20 ml 80% aqueous ethanol (v/v) is added and the suspension is mixed in a Vortex. After centrifugation, the Kjeldahl nitrogen of the pellet is determined.The result is given as mg gel protein (Nkj × 6.25) per gram of barley flour.

This determination can also be used for malt, and can also be done using HPLC or SDS-PAGE (Moonen et al. 1987).

Some attempts have been made to relate the protein distribution pattern of barley in an electrophoresis method directly to brewing quality (Baxter and Wainwright 1979). Such a relationship has, however, not been confirmed by other workers (Shewry et al. 1980; Riggs et al. 1983; Smith and Simpson 1983). The methods that are used for these electrophoresis analyses are discussed in the section on varietal identification.

Proanthocyanidins. The Carlsberg laboratories have developed new barley mutants that lack certain polyphenols, the proanthocyanidins, also called anthocyanogens in the brewing literature. These mutants enable the production of a beer that is physically stable, i.e. does not develop a haze on storage, without any use of stabilizing agents.

In order to detect quickly if barley grains contain anthocyanogens, a test has been developed in which these compounds can be stained and visualised in half kernels using HCl/vanillic acid (Aastrup 1985; Aastrup et al. 1984).

Proanthocyanidins
A sample of 100 barley or malt grains is fixed in a block of thermoplastic clay or glued on a glass plate using two-component glue. The grains are then sanded half way down, using sand paper, and stained for 30 min with a freshly prepared 1% vanillin, 6-M hydrochloric acid solution in water. A distinct red colour that is stable for at least 30 min develops in the testa of the proanthocyanidin-containing grains which is absent in proanthocyanidin free grains.

Betaglucans. The cell walls of the barley endosperm for 75% consist of betaglucan, (Fincher 1975; Thompson and Laberge 1981), a mixed linkage beta (1–3)(1–4)-bonded block polymer of glucose. During malting the cell walls should be sufficiently broken down (Bourne et al. 1982). Inadequately degraded betaglucans present problems in brewing because of enhanced viscosities in the wort, which cause a slow beer filtration (Muts et al. 1984).

Furthermore, the extract yield remains too low and lautertun filtration will be slow. An excellent review has been published by Bamforth (1982). Although brewers have always been very interested in betaglucans, it was the breeders who promoted the rapid development of analyses in order to be able to screen large quantities of barley seed for low betaglucan content.

The analysis of barley betaglucan focused mainly on two difficulties. The first was the extraction of the betaglucan which was done by 4% sodium hydroxide (Palmer 1975), boiling water (Fleming and Kawakami 1977), sodium carbonate at 80 °C (Prentice et al. 1980), autoclaving at 130 °C (Anderson et al. 1976), by hydrazine at 105 °C (Forrest and Wainwright 1977) or by perchloric acid (Ahluwalia and Ellis 1984).

The other problem has been the specificity of the enzyme to degrade the solubilized betaglucan. This has been circumvented by differential measurement using total acid hydrolysis and alphaglucan digestion with amyloglucosidase (Fleming et al. 1974). Other approaches use direct betaglucan hydrolases from *Trichoderma reesei* (Forrest and Wainwright 1977; Martin and Bamforth 1981). An improvement was the cellulose preparation from *Penicillium funiculosum* (Bamforth 1983), which is more stable during the heat inactivation of the inferring activities, mainly amyloglucosidase. Recently, a method was described using purified lichenase according to Anderson (1976) and betaglucosidase to degrade the betaglucan which is solubilized in the flour without previous extraction (McCleary and Glennie-Holmes 1985).

Betaglucan Determination in Barley (Mc Cleary)
Approximately 0.5 g of barley flour is weighed in screw-capped tubes (weighed and oven-dried at 80 °C for 20 h) after drying at 80 °C for 20 h and cooling in a desiccator.

1 ml of 50% phosphate buffer pH 6.5 is added to the flour and the mixture stirred in a Vortex mixer. The tubes are incubated for 2 min in a steam bath, removed for vigorous stirring in a Vortex mixer and again heated for 3 min.

After cooling, 0.2 ml lichenase[1] solution of 10 U in 0.1 M sodium phosphate buffer pH 6.5 is added, the solution stirred and incubated at 40 °C for 1 h.

The volume of the tube is adjusted to 30 ml with distilled water, mixed and filtered through Whatman no. 42 filter paper. 0.1 ml of the filtrate is transferred to test tubes. To one of these (the blank) 0.2 M acetate buffer (0.1 ml) pH 4.0 is added, and to two others 0.1 ml, 0.2 U betaglucosidase in 0.2 M acetate buffer pH 4.0. The tubes are incubated at 40 °C for 15 min. Glucose oxidase/peroxidase[2] is then added (3.0 ml) and after exactly 20 min at 40 °C, the absorbance at 510 nm is measured using a 1 cm flow cell. Standards containing 50 μg and 100 μg of glucose in duplicate are also measured

$$\text{Betaglucan (\% w/w)} = \frac{\text{E (test-blank)} \times \text{F} \times 27}{\text{sample weight (mg)}}$$

More non-specific methods have been reviewed by Bamforth (1982). When betaglucan determinations are to be performed in malt, the presence of in situ betaglucanase can give rise to underestimations. Usually these determinations are preceded by enzyme inactivation by boiling with 80% (Preece and MacKenzie, 1952) or 85% ethanol (Meredith et al. 1951), also adapted and employed by McCleary (1986) in an application of his method to malt. Soluble betaglucans in malt have been determined using many different extraction methods (Preece and Mackenzie 1952; Schuster et al. 1967; Meredith et al. 1951; Bourne and Pierce

[1] Lichenase, betraglucosidase can be purchased from Biocon.
[2] Glucose oxidase/peroxidase can be purchased from Merck.

1970; Luchsinger et al. 1958). Subsequent precipitation with 30% ammonium sulphate, or 50% acetone, is very common, after which the precipitate is hydrolysed and the reducing sugars are determined.

The indirect determination of betaglucans by the viscosity of a wort from malt has been described in the Recommended Methods. Because the viscosity of wort is only partially caused by betaglucans (Jørgensen and Aastrup 1986), deviating results are possible. The viscosity of the wort is a good indication for the ease of beer filtration (Muts et al. 1984).

2.5 Varietal Identification

Following the development of new electrophoretic analysis methods in biochemistry, the application of this analysis for varietal identification of barley and also of malt has become increasingly important. Barley can be malted much more evenly and to a narrower specification when pure varieties are used instead of mixtures.

Different methods have been tested, ranging from isoelectric focusing (IEF) to sodium dodecyl sulphate poly acrylamid electrophoresis (SDS-PAGE) (Van Lonkhuysen and Marseille 1978; Günzel and Fischbeck 1979; Scriban et al. 1979; Schildbach and Burbidge 1979; Scriban and Strobbel 1979; Montembault et al. 1983; Fleschurz et al. 1983; Smith and Payne 1984; Nielsen and Johansen 1985; Schildbach and Burbidge 1985). An electrophoresis method on acrylamide slab gels with urea at low pH seems to have become generally accepted (EBC):

Varietal Identification of Barley by Electrophoresis

The following solutions are used in this analysis:

Extractant solution		Gel stock solution	
Urea	180 g	Acrylamide	100 g
2-mercaptoethanol	10 ml	Bis-acrylamide	4 g
2-chloroethanol	200 ml	Urea	60 g
Methylgreen	0.1 g	Glacial acetic acid	20 ml
Distilled water upto	1 l	Glycine	1 g
		Ferrous sulphate (5 g/l)	3 ml
		Ascorbic acid	1 g
		Distilled water up to	1 l

Buffer solution		Staining solution	
Glacial acetic acid	20 ml	Trichloro acetic acid	100 g/l (aqueous)
Glycine	2 g	Serva Blau R	1 g/100 ml (ethanolic)
Distilled water up to	1 l	For use: add 5 ml serva blau solution to 200 ml trichloro acetic acid solution.	

A vertical slab gel electrophoresis apparatus is used.

Fifty grains of the sample to be examined are crushed individually and each is mixed with 400 µl of extractant solution. The hordeins are left to dissolve for at least 16 h at room temperature, after which the mixture is centrifuged.

In the meantime, gels of 1.5 mm thick and 10–15 cm long are casted. To 100 ml of gel stock solution 150 µl of freshly prepared hydrogen peroxide solution (2 ml 30% H_2O_2 and 100 ml distilled water) is added after deaeration of the gel stock solution and cooling to 4 °C. The mixture is poured between the glass plates and allowed to polymerize with a comb to form slots on top of the gel. After removal of the comb, the slots are filled with 10–20 µl of the centrifuged extract, using a pipette. The glass plates with gel are then mounted vertically in the electrophoresis apparatus with buffer solution. Optionally the gel is cooled with tap water at a temperature of 10–20 °C by using the cooling device of the apparatus. Run the gel at 200 V for 10 min and then at 500 V for twice the time required for the coloured band (methyl green) to pass through the gel.

Stain the gel directly after the run in the staining solution, in which it can be left overnight.

The obtained band patterns are characteristic of (classes of) barley varieties. The varietal composition of the barley sample can be calculated by counting the number of identical patterns, taking into account the statistical levels of significance regarding the number of kernels analysed from a sample.

2.6 Micro-Malting

The most direct way to determine the properties of a barley for malting and brewing is of course to malt it and brew with it. In a breeding programme, however, only small quantities of barley are available in an early stage. In order to overcome these problems, micromaltings have been developed that enable the breeder to evaluate small samples of barley (Gothard et al. 1980). Micromaltings have been in use for several years now (Draper 1979; Gromus et al. 1980; Hall 1981; Smeaton 1985).

Every year numerous reports are published all over the world to evaluate the barley crop. Usually pilot maltings and pilot breweries are used for that purpose and the analysis does not go beyond the normal barley and malt analysis that is discussed in this chapter.

3 Malt Analysis

3.1 Wort Production and Analysis

The usual way to analyse malt in the brewery is to make a small brew and to determine the parameters of interest in the wort. The methods described in the Recommended Method collections differ mainly in the way in which the brew is made. There are two principal differences. The IOB recommends the determination of the Hot Water Extract (HWE) using a mashing scheme of 65 °C that is maintained for 1 h, whereas the EBC, for instance, refers to a scheme with temperatures from 45 °C up to 75 °C.

The results obtained with both methods will therefore be different. Furthermore, the British brewers use the analysis of the cold water extract, obtained by mashing at 20 °C for 3 h.

The milling, the mashing equipment and the wort filtrations are all described in the Recommended Methods and are not essentially different from each other. Recently a new mill has been developed and tested (Schur et al. 1979; Moll 1979 b), the so-called DLFU-mill, which is now in general use.

Occasionally there have been suggestions to change the mashing scheme to obtain results closer to brewing practice (Moll et al. 1980; Moll and Flayeux 1975).

Below an example is given of the EBC extract analysis of malt.

Preparation of an EBC-Wort from Malt

Duplicate quantities of approximately 55 g of malt are ground in the EBC-mill, using a 1-mm sieve. The ground malt is transferred to a mash beaker and well mixed. A portion of 50.0 g is weighed in the beaker. Warm distilled water (200 ml) is stirred into the mash with a glass rod, so that a temperature of 45 °C is reached in the mash.

The beaker is placed in the mash bath at 45 °C and the content is stirred. After exactly 30 min the temperature is raised 1 °C per minute during 25 min. Then, at 70 °C, another 100 ml of water at 70 °C is added. The temperature is maintained at 70 °C for 1 h, then cooled to room temperature in 10–15 min. The stirrer is rinsed off and the contents of the beaker is made up to a weight of 450.0 g with distilled water.

After stirring, the contents of the beaker is at once completely emptied on to a 30–32 mm diameter fluted filter (for instance S & S. no. 597½) in a funnel. The first 100 ml of the filtrate is returned to the funnel. The filtration is continued until dry or up to 2 h.

In the obtained wort, several parameters can be determined that give information on the quality of the malt. The extract content has always been determined with pycnometers as specific gravity at a specific temperature (20 °C). This specific gravity is then calculated to extract content using the official Plato tables (Schild 1961).

Today this method can be replaced by automated density measurements with instruments based on the principle of damping of vibration (Krueger and Oliver-Daumen 1980).

Total soluble nitrogen can be determined by Kjeldahl analysis, whereas reagents giving a spectro-photometrically measurable extinction, such as ninhydrin or trinitrobenzene sulfonic acid are usually used for free amino nitrogen determinations. An example of the use of ninhydrin in wort is given below. This determination is the best reproducible of the two mentioned.

Free Amino Nitrogen (FAN)

The wort is diluted with distilled water 100 times to a concentration of 1–3 mg/l FAN. In a test tube, 2 ml of this dilution and 1 ml colour reagent (100 g $Na_2HPO_4 \cdot 12H_2O$, 60 g KH_2PO_4, 5 g ninhydrin, 3 g fructose, made up to 1 l, pH = 6.6–6.8) are mixed, the tube stoppered with a glass ball and heated for 16 min in a boiling water bath. The tube is cooled to 20 °C in 20 min in a water bath, 5 ml of diluting solution (2 g KIO_3 in 600 ml distilled water and 400 ml 96% ethanol) is added, mixed, and the extinction is read within 30 min at 570 nm. The blank contains 2 ml of distilled water instead of diluted wort. A standard in triplicate is made by an analysis as described above of 2 ml of a standard solution that is 100-fold diluted. The standard solution is prepared by dissolving 107.2 mg glycine in 100 ml of distilled water, giving 200 mg/l FAN

$$\text{FAN (mg/l)} = \frac{\text{E 570 (test solution)}}{\text{E 570 (standard, mean)}} \times 2 \times \text{dilution.}$$

The fermentability of the wort is determined as the attenuation limit, which is the quantity of unfermentable dextrines left over after a wort has been fermented by yeast measured by density. All Recommended Methods have a test for this important parameter.

Furthermore, pH, colour and viscosity (discussed already in the section on betaglucans) can be determined.

During the preparation of the wort, the saccharification time is determined. This is the time from the point in the mashing scheme that 75 °C is reached until the mash no longer stains blue with iodine. A correlation exists with diastatic power (Zangrando 1974).

3.2 Modification Analysis

The degree to which cell walls and storage proteins in the barley endosperm are dissolved enzymatically during germination of the barley is called the modification degree.

An undermodified malt gives lower extract yields, because the starch is shielded from amylase attack, but also impairs wort and beer filtration (Muts et al. 1984). An overmodified malt, however, also gives rise to lower extract yield because of higher malting losses through starch digestion, and can lead to high-coloured, badly foaming beers with a specific flavour.

Analysis methods that enable the brewer to control the malt and the maltster to monitor the modification process during germination are of great importance.

The methods available have been updated lately to modern standards, although several principles were already known at the beginning of this century. Two approaches exist: the ones using physical properties such as kernel hardness or milling fractions, and those that use the staining of kernels cut in half.

Physical Methods. Determination of the hardness of the malt grain gives information on its degree of modification. Many methods have been developed in the course of time, with varying success.

The Brabender hardness tester was introduced in 1939 (Enders 1939a) and perfected by Paukner (1951), Kleber et al. (1952), Diedering (1958), and Blum et al. (1960, 1961). The application has been tested by Reiner et al. (1967) and Steiner (1963).

The sclerometer was developed by Beck and studied by Chapon and Kretschmer (1964a).

The murbimeter has replaced the sclerometer and has been described extensively. The most relevant references are Ullman (1973), Bellmer (1974), de Becker (1971) and Chapon (1961 b, 1964), Isebaert and Rotti (1964).

Nowadays the friabilimeter, first described by Chapon (1979), is the most commonly used instrument. It crushes the malt between a rubber roller and a screen on which the unmodified parts are retained. The flour is weighed and the weight is an indication of the degree of modification. (Kretschmer 1982; Burbidge 1984; Thomas 1986).

Methods Using Half Kernels. A method to assess the degree of modification of malt was described by Palmer (1975 b). He cutted grains longitudinally and extracted the grains with water at 65 °C. The non-solubilized part of the endosperm is unmodified (Palmer 1975 a).

Other methods employ half grains that are stained. At the Carlsberg laboratories, a staining method for betaglucans was developed for this purpose using calcofluor white.

In this case the unmodified part of the endosperm, where the betaglucans are still present, is stained (Gibbons 1980; Aastrup et al. 1980; Gibbons 1981; Gibbons et al. 1983). At Heineken, the methylene blue method was developed, in which the modified endosperm parts are penetrated by an alcoholic methylene blue solution (Aalbers 1980; Van Eerde 1984). Both methods give essentially the same results (Carnielo et al. 1982).

Malt Modification

A sample of 100 malt kernels is placed in a matrix to bring them into alignment and is then glued onto a glass plate with two-component glue. After drying, the grains are sanded off for one third, using a mechanical sanding apparatus. After removing the dust the plate is immersed in a solution of 1% methylene blue in 96% ethanol for 30 min. The grains are then dried in the air and again sanded down to one third of their diameter.

The modified part of the endosperm has a blue colour, the unmodified part is white or glassy.

The degree of modification can then be calculated using a classification system: kernels modified for <5%, 5–25%, 35–50%, 50–75%, 75–95%, and >95%. From these figures the homegeneity of the modification can also be calculated (Carnielo et al. 1982).

The Heineken method does not need any additional equipment, whereas the Carlsberg method needs at least an ultraviolet lamp. The latter, however, is more convenient to screen wet green malt samples during malting. This method is essentially outlined in the section on germination tests.

An indirect way of determining the degree of modification of a malt sample is the estimation of the fine-coarse difference. In the usual extract determination method, the mill setting is a fine one. When a second, carefully chosen, mill setting is used, a coarse flour is obtained, producing a lower extract in the wort. The difference between these two is called the fine-coarse difference and is an indication of the degree of modification of the malt.

The method is rather laborious and not very accurate, but is used very commonly. Its description can be found in the Recommended Methods of Analysis.

3.3 Chemical Analysis

Many of the analysis methods mentioned in the section on barley can also be used for malt. Examples are varietal identification, betaglucan analysis, determination of gel protein content, proanthocyanidins and protein by Kjeldahl analysis.

Specific analyses on malt have evolved since the detection of possibly mutagenic N-dimethyl nitrosamines (NDMA) in malt. The development of these compounds has since then been controlled by using indirectly fired kilns or by sulphuring. NDMA analysis involves an extraction from the malt and gas chromatographic separation of the mixture, using a Thermal Energy Analyser (TEA) as detection device. A review of the chemistry of NDMA formation was recently published by Wainwright (1986).

The polyphenols in barley and malt have always inspired many people to develop analysis methods. Especially the development of proanthocyanidin free barley has caused much research (Von Wettstein et al. 1984). Nowadays HPLC methods are the ones most commonly, but certainly not routinely, used (Jerumanis 1985; Jende-Strid and Lindberg-Moller 1981). Other methods are less specific and deal with colour development in acid and adsorption on polyvinylpyrrolydone (PVP) (Chapon and Kretschmer 1970). Most polyphenol analyses are used for wort and beer and are not applied to the raw materials.

3.4 Enzyme Analysis

The activity of the various enzymes in the malt is, of course, very important for the brewer because these are partly the instruments he has to achieve the right wort composition. Among the many enzymes that a malt contains only the most important ones will be dealt with, as the others are mainly of research importance.

In general, the practical significance of enzyme analysis in malting and brewing is limited. In the assays mentioned below deviations from practice can be found in: pH, temperature, water activity (malt!), concentration, substrate structure and enzyme availability. The assays should only be used for the purpose of comparison, if at all.

Alpha Amylase. Although malt alpha amylase usually is not the limiting enzyme activity in the brewery (except where more than 50% adjunct is used), there are several methods in use to determine this activity. The most common one uses beta limit dextrin (Recommended Methods). Anderegg et al. (1976) reviewed several of these methods. One of the most convenient methods uses dyed amylose that is solubilized by alpha amylase action after which a blue colour can be measured spectrophotometrically.

Alpha Amylase Activity
2.0 g of finely ground malt is extracted with 100 ml 0.5% NaCl at 20 °C for 60 min with continuous stirring. Filter the suspension through a fluted filter. Dilute 1 ml of the filtrate to 100 ml with 0.5% NaCl solution.

Add a Phadebas-tablet (Pharmacia Diagnostics) to a test tube, equipped with a magnetic bar and immersed in a water bath at 45 °C on top of a magnetic stirrer. Add 5.0 ml buffer pH 5.50 (0.25 g $CaCl_2 . 2H_2O$ in 100 ml distilled water and 100 ml 0.2 M sodium acetate) and preincubate for 10 min at 45 °C.

Add 1 ml of diluted sodium hydroxide with immediate mixing. Let the tubes stand for 15 min at 20 °C in water and filter through Whatmann no. 1.

Read the extinction at 630 nm using 1.0 cm cuvettes. The blank, consisting of 1 ml 0.5% NaCl instead of enzyme extract in the above sequence, is also read and substracted giving A

$$\text{EBC Alpha Amylase (U/mg malt)} = \frac{A \times 100}{100 - M},$$

in which M is the moisture content (%) of the malt.

This method, which uses a dyed substrate, resembles those used for the determination of other enzymes, such as endoprotease and beta glucanase. These assays will therefore not be described.

Beta Amylase. Analysis of beta amylase has usually been carried out by determination of diastatic power. In this analysis the diastatic power is expressed as Windish Kolbach Units, which is a well-known unit among brewers and maltsters. The maltose which is generated from starch by beta amylase in a malt extract is measured iodometrically. A more specific assay was developed by Dellweg et al. (1977). In this assay maltotetraose is used as a substrate, which is hydrolysed to maltose by beta amylase. The maltose is subsequently transformed by maltose phosphatase to glucose and glucose-1-phosphate, which can be transformed to glucose-6-phosphate with betaphosphoglucomutase. This product is in its turn transformed to 6-phosphogluconate and NADH by glucose-6-phosphate dehydrogenase and NAD^+. The NADH can be measured spectrophotometrically at 340 nm (Krüger and Nordmann 1983).

The enzyme complex is commercially available as Enzamyl K, and the assay is performed by measuring the reaction kinetics of the NADH formation.

Beta Amylase

Add 2.0 g milled malt flour to 100 ml extraction solution (5 g NaCl and 1.25 g $CaCl_2 \cdot 2H_2O$ in distilled water) in a 400-ml beaker that was incubated for 10 min at 20 °C, and stir for 30 min at 20 °C. Centrifuge at 3000 rpm during 3 min and filter the supernatant over S & S no. 597 filter paper. Dilute 1 ml of the filtrate to 10 ml with extraction solution. Attemperate the spectrophotometer and the solution Enzamyl K to 25 °C. Pour 1.0 ml Enzamyl K and 0.10 ml diluted extract into a cuvette (1 cm), mix, and measure the extinction at 340 nm exactly 8 min after addition and every minute thereafter until 15 min after addition. Measure a blank with 0,10 ml distilled water instead of extract.

Calculate a ΔE/min from the linear part of the plot of extinction versus time

Beta amylase activity (U/g malt) = $\Delta E \times 4421.2$.

Enzamyl K contains maltotetraose and enzymes and can be obtained from Goedecke.

This method shows a high correlation with diastatic power.

Endoprotease. Many assays for endoprotease activity in malt are performed using substrates such as haemoglobin (Bhatty 1969), casein (Burger et al. 1970), gliadin (Jacobsen and Varner 1967), gelatin (Enari and Mikola 1983) or synthetic peptides. As their composition and structure differ from those of hordein, the main protein to be broken down during malting, the results of these assays might not reflect the actual activity of the enzyme. Endo-activity is the most important and most rate-limiting factor (Mikola and Kolehmainen 1972) in the proteolysis during malting and brewing. The use of hordein as a substrate was described by Baxter (1976) who developed an assay in which exo- and endo-activity could be measured simultaneously. The method has not attracted many users, however. Another assay procedure is the use of a dyed hide powder substrate (collagen). It does not meet the above requirements, but is very suitable for routine use (John et al. 1975).

Exoprotease. The release of amino acids from proteins by either amino or carboxy peptidases is conveniently monitored by using assays employing synthetic substrates or by simply determining quantities of amino acids by trinitrobenzene sulphonic acid (TNBS) or ninhydrin.

Since hordein contains 20% proline, it should be borne in mind that these determinations will underestimate the true quantity. Proline does not react with TNBS or ninhydrin, although the latter can be used in an *ethanolic* assay (Schweet 1954).

Betaglucanase. The activity of betaglucanase that is important for brewing purposes is endo-activity. The (1–3), (1–4) mixed linkage betaglucan is rapidly reduced in size by this activity.

Several assays have been proposed, of which the viscometric procedure is the most common one (Analysis Committee Institute of Brewing 1982). It has been criticised because of its unnatural substrate (Bamforth 1982), variations between betaglucan batches, and non-linearity of reciprocal specific viscosity plots against time (Bathgate 1979). A radial gel diffusion method has been proposed by Martin and Bamforth (1983). It is convenient, but did not prove sufficiently precise in an informal collaborative trial (Buckee 1985).

Betaglucanase by Radial Diffusion
Add 0.05 g barley betaglucan and 0.004 g Congo Red to 100 ml of 0.01 M sodium acetate solution pH 5.5 in a 250-ml beaker. The solution is boiled under stirring on a magnetic stirrer until the whole has dissolved (cover the beaker with a watch glass). Cool to about 50–60 °C and adjust the volume to 100 ml with distilled water. Reboil the solution and immediately pour it into Petri dishes, 9 cm in diameter, to a depth of 5 mm. After cooling, cut out five wells, 9 mm in diameter, using a stainless steel tube, add 100 µl of the malt extract to each of four wells and 100 µl of 0.1 M sodium chloride to the fifth. (The malt extract is obtained by milling 25 g of malt, adding 200 ml 0.1 M sodium chloride and stirring at 4 °C for 1 h. The mash is then centrifuged at 25,000 g for 20 min at 4 °C and the supernatant decanted through a plug of glass wool). Cover the plates and leave them for 16.5 h at 25.0 °C. Add sufficient 1.0 M hydrochloric acid to cover the gel surface to stop the reaction and decant it after 3 min. Measure the total diameter of clearing around each well, using dividers.

Betaglucanase activity (U/100 g malt) $= 0.8 \times A \times F$

$F =$ dilution factor malt extract

$A =$ calculated as betaglucanase units (A) from the clearing diameter (C) in mm using $\log A = 0.17\ C - 1.56$.

A promising method could be the use of a dyed betaglucan as a substrate enabling the measurement of the stained solubilised fragments after enzymic hydrolysis (McCleary and Nurthen 1986).

4 Further Reading

The literature on the analysis of barley and malt, not to speak of cereals in general, is a vast and disordered field. In this chapter only a few of the methods have been given, usually as an example. Good reviews have been written by Wainwright and Buckee (1977) and Moll (1979a).

References

Aalbers VJ (1980) Verbesserte Methode für die Bestimmung der Glasigkeit von Malz. Monatsschr Brau 33:212–213

Aastrup S (1985) A test for the presence or absence of proanthocyanidins in barley and malt. Carlsberg Res Commun 50:37–42

Aastrup S, Gibbons GC, Munck L (1980) A rapid method for estimating the degree of modification in malt by measurement of cell wall breakdown. Carlsberg Res Commun 49:77–86

Aastrup S, Outtrup H, Erdal K (1984) Location of the proanthocyanidins in the barley grain. Carlsberg Res Commun 49:105–109

Ahluwalia B, Ellis EE (1984) A rapid and simple method for the determination of starch and betaglucan in barley and malt

Anderegg P, Schur F, Pfenninger H (1976) Bestimmung der Alphaamylaseaktivität in Cerealien und microbiellen Enzympräparaten. Schweiz Brau Rundsch 87:239–250

Anderson RL, Anderson MA, Stone BA (1976) Some structural and enzymological aspects of non-starchy polysaccharides in cereal grains. Proc Conv Inst Brew (Aust N Z Sect) 14:153–160

Bamforth CW (1982) Barley betaglucans. Their role in malting and brewing. Brew dig 57:22–27, 35

Bamforth CW (1983) *Penicillium funiculosum* as a source of beta glucanase for the estimation of barley betaglucan. J Inst Brew 89:391–392

Bathgate GN (1979) The determination of endo-beta-glucanase activity in malt. J Inst Brew 85:92–94

Baxter ED (1976) The use of hordein fractions to estimate proteolytic activity in barley and malt. J Inst Brew 82:203–208

Baxter ED, Wainwright T (1979) Hordein and malting quality. J Am Soc Brew Chem 37:8–12

Bellmer HG (1974) Studie über die Beziehungen zwischen der Viskosität der 65 °- und 80 °-Würze nach Hartong-Kretschmer und verschiedenen Malzanalysendaten. Brauwissenschaft 27:215–222

Bhatty RS (1969) Note on the development of proteolytic enzymes in germinating barley. Cereal Chem 47:74–77

Bishop LR (1930) The prediction of extract, part 1. J Inst Brew 36:421–434

Bishop LR (1948) The prediction of extract. IV. The adjustment of prediction to give true extract in malt. J Inst Brew 54:330–333

Bishop LR (1957) Ultra-rapid method for measurement of the germinative capacity of barley grain. J Inst Brew 63:516–530

Blum PH, Panos G, Smith RJ (1960) Measurement of hardness of barley and malt with the Brabender hardness tester. II. Modification during malting. Proc Annu Meet Am Soc Brew Chem 95–100

Blum PH, Frederickson BL, Smith RJ (1961) Measurement of hardness of barley and malt with the Brabender hardness tester. III. Evaluation of hardness graphes for analysis control. Proc Annu Meet Am Soc Brew Chem 119–122

Bourne DT, Pierce JS (1970) Betaglucan and betaglucanase in brewing. J Inst Brew 76:328–335

Bourne DT, Powlesland T, Wheeler RE (1982) The relationship between total betaglucan of malt and malt quality. J Inst Brew 88:371–375
Buckee GK (1985) Estimation of betaglucanase. J Inst Brew 91:264–266
Burbidge M (1984) Homogenität und Malzanalyse. Monatsschr Brauwiss 37:4–9
Burger WC, Prentice N, Moeller M, Robbins GS (1970) Stabilisation, partial purification and characterisation of peptidyl peptide hydrolases from germinated barley. Phytochemistry 9:49–58
Carnielo M, Foucault M-A, Moll M (1982) Application of a test for modification in green and kilned malts. Brauwissenschaft 35:168–170
Chapon L (1961 a) Le murbimètre. Un nouvel appareil pour la messure objective de la friabilité du malt. Brasserie 16:190–201
Chapon L (1961 b) Das Mürbimeter. Ein neuer Apparat für die objektive Messung der Mürbigkeit der Malzes. Brauwelt 101:752–756
Chapon L (1963 a) Anwendung des Mürbimeters, zum Studium der Härte und der Homogenität der Gersten. Brauwelt 103:746–750
Chapon L (1963 b) Neue Betrachtungen über die Rolle der Gerbstoffe beim Bierbrauen. Brauwissenschaft 16:225–231
Chapon L (1964) Cohesion de l'endosperme du grain d'orge et friabilité du malt. Bull Assoc Econ Brass Louvain 60:49–70
Chapon L (1979) Das Friabilimeter, ein Gerät für die objektive Bestimmung der Malzmürbigkeit. Monatsschr Brau 160–164, 166, 167
Chapon L, Kretschmer KF (1964 a) Malzmürbigkeit als Qualitätsfaktor für die Bierbereitung. Brauwissenschaft 17:1–10
Chapon L, Kretschmer KF (1970) Beziehungen zwischen der Braugerstenqualität und dem Brauwert des Kleinmalzes. Brauwelt 110:395–405
Dadic M (1971) Analytical methods for polyphenols in brewing. I. Polyphenols in brewing and their estimation. Proc Am Soc Brew Chem 29:149–158
Dadic M, van Gheluwe JEA, Valyi Z (1970) Chemical constituents of Nylon 66 adsorbat. Preliminary investigations and anthocyanogen contents. J Inst Brew 76:267–280
De Becker J (1971) Etude du murbimètre. Bios 2:3–12
De Becker J (1974) Methode pour la determination de la sensibilité a l'eau de l'orge. Bios 5:23–32
De Clerck J (1962) Cours de brasserie I, 2nd éd, Université de Louvain
Dellweg H, John M, Schmidt J, Flindt E (1977) Amylasen in gekeimter Gerste. Eine neue Methode zur getrennten Bestimmung der Alpha- und Beta-Amylaseaktivitäten in Gerstenmalz. Proc Eur Brew Conv 16:77–91
Diedering P (1958) Die Bedeutung des Brabenderschen Härteprüfers für die Malzuntersuchungen. Monatschr Brau 11:59–65
Donhauser S, Linsenmann O (1983) Bisherige Erfahrungen mit der NIR Spektroskopie zur Analyse von Brauereirohstoffen und Zwischenprodukten. GIT Fachz Lab 27:1063–1067
Draper SR (1979) The performance in micromalting tests of twenty varieties of spring barley. J Natl Inst Agric Bot 15:76–87
Enari TM, Mikola J (1983) Peptidases in germinating barley grain: properties, localisation and possible functions. J Chromatogr Libr:335–352
Enders C (1939 a) Neuere Methoden zur Beurteilung von Gerste und Malz. Wochenschr Brau 56:1–8
Enders C (1939 b) Die Malzanalyse und die Aufschlüsse, die sie über die Güte des Malzes liefern kann. Wochenschr Brau 56:309–312
Essery RE, Kirsop BH, Pollock JRA (1956) Studies in barley and malt. II. Tests for germination and water sensitivity. J Inst Brew 61:25–28
Fincher GB (1975) Morphology and chemical composition of barley endosperm cell walls. J Inst Brew 81:116–122
Fleming M, Kawakami (1977) Studies of the fine structure of beta-D-glucans of barleys extracted at different temperatures. Carbohydr Res 57:15–23
Fleming M, Manners DJ, Jackson RM, Cooke SC (1974) The estimation of betaglucan in barley. J Inst Brew 80:399–404

Fleschurz W, Seeleitner G, Liebl M, Püspök J (1983) Die Identifizierung von Gerstensorten durch die PAGE-Elektrophorese. Mitt Versuchsst Gärungsgewerbe Wien 37:9–14

Forrest IS, Wainwright T (1977) Differentiation between desirable and troublesome betaglucans. Proc Eur Brew Conv 16:401–413

Gibbons GC (1980) On the sequential determination of alfa amylase transport and cell wall breakdown in germinating seeds of *Hordeum vulgare*. Carlsberg Res Commun 45:177–184

Gibbons GC (1981) Visualization of alpha amylase movement and cell wall breakdown during barley malting. Practical application of current research. J Am Soc Brew Chem 39:55–59

Gibbons GC, Janssen SA, Aastrup S, Guildal T, Olsen G, Damstrup J (1983) Practical application of fluorescent seed-analysis systems in a commercial malthouse. J Am Soc Brew Chem 41:117–120

Gothard PG, Morgan AG, Smith DB (1980) Evaluation of a micromalting procedure used to aid a plant breeding programme. J Inst Brew 86:69–73

Grandclerc J, Lebordais B, Carnielo M (1982) Utilisation de l'Infralyzer 400 R à l'analyse de l'orge, du malt et de la bière. Brauerei Rundsch 93:177–179

Gromus J, Baecker M, Baunas M (1980) Zur Beurteilung der Braugerstenqualität. Die Bremer Kleinmälzung im Brauereilaboratorium. Brauwelt 120:1074–1080

Günzel G, Fischbeck G (1979) Die Sortendiagnose am Gerstenkorn. Brauwissenschaft 32:226–232

Hall SC (1981) Micro-malting trials 1979/1980. J Inst Brew 87:282–285

Henry RJ, McLean BT (1986) Re-evaluation of fluorescein dibutyrate staining as a method for assessment of pre-harvest sprouting in wheat and barley. J Cereal Sci 4:51–55

Isebaert L, Rotti K (1964) Etude comparative de différent résultats d'analyse d'orge et de malt en fonction de leur teneur en proteine. Petit J Brass 72:757–761

Jacobsen JV, Varner JE (1967) Gibberellic acid-induced synthesis of protease by isolated aleurone layers of barley. Plant Physiol 42:1596–1600

Jende-Strid B, Lindberg-Moller B (1981) Analysis of proanthocyanidins in wild type and mutant barley (*Hordeum vulgare* L). Carlsberg Res Commun 46:53–63

Jensen SA, Heltved F (1982) Visualisation of enzyme activity in germinating cereal seeds using a lipase-sensitive fluorochrome. Carlsberg Res Comm 47:297–303

Jensen SA, Heltved F (1983) An improved method for the determination of pregerminated grains in malted barley. Carlsberg Res Comm 48:1–8

Jerumanis J (1985) Quantitative analysis of flavonoids in barley, hops and beer by high perfomance liquid chromatography. J Inst Brew 91:250–252

Jørgensen KG (1983) An improved method for determining betaglucan in wort and beer by the use of calcofluor. Carlsberg Res Comm 48:505–516.

Jørgensen KG, Aastrup S (1986) Analysis of betaglucan in wort. XI. EBC Monograph on wort production, Maffliers, pp 262–273

John M, Schmidt J, Dellweg H (1975) Eine neue Methode zur Bestimmung der proteolytischen Aktivität im Gerstenmalz. Monatsschr Brau 28:14–19

Kleber W, Paukner E, Heyer W (1952) Auflösung 1952er Mälze nach der Viskositätsmethode. Brauwelt 92:1273

Kretschmer H (1982) Friabillimeterwert: Bedeutung für die Braupraxis. Brauwelt 122:54–58

Krüger E, Bielig HJ (1976) Betriebs- und Qualitätskontrolle in Brauerei und alkoholfreier Getränkeindustrie. Paul Parey, Berlin

Krüger E, Nordmann A (1983) Enzymatische Analysen im Brauereilaboratorium. II. Modifizierte Analyse zur Bestimmung der Betaamylasenaktivität. Monatsschr Brauwiss 36:30–36

Krüger E, Oliver-Daumen B (1980) Aktueller Stand der konventionellen Bieranalytik. Monatsschr Brau 33:12–21

Leysen B (1984) Automatisierte Schoenfeld-Methode. Analyse zur Bestimmung der Keimenergie von Braugerste. Monatsschr Brauwiss 37:516–518

Luchsinger WW, Cochrane DG, Kneen E (1958) Enzyme properties and kinetics. Proc Am Soc Brew Chem 46–53

Macey A, Stowell KC (1959) Practical studies on scientific control in commercial maltings. Proc Eur Brew Conv 7:105–112
Martin H, Bamforth CW (1981) An enzymic method for the measurement of total and water-soluble betaglucan in barley. J Inst Brew 87:88–91
Martin HL, Bamforth CW (1983) Application of radial difusion assay for the measurement of beta-glucanase in malt. J Inst Brew 89:34–37
McCleary BV, Glennie-Holmes M (1985) Enzymic quantification of (1–3) (1–4) beta-D-glucan in barley and malt. J Inst Brew 91:285–295
McCleary BV, Nurthen E (1986) Measurement of (1–3) (1–4) beta-D-glucan in malt, wort and beer. J Inst Brew 92:168–173
Meredith WOS, Bass EJ, Anderson JA (1951) Some characteristics of barley, malt and wort gums. Cereal Chem 28:176–188
Michel K, Michel O (1910) Betriebskontrolle für Brauerei und Mälzerei, 3rd edn. Brautech Verl Allg Anz Brauereien, Mannheim, p 206
Mikola J, Kolehmainen (1972) Localization and activity of various peptidases in germinating barley. Planta 104:167–177
Moll M (1979 a) Analysis and composition of barley and malt. In: Pollock JRA (ed) Brewing science, vol I, Academic, London
Moll M (1979 b) Auswertung der DLFU Bühler Miag Mühle. Brauwissenschaft 32:346–348
Moll M, Flayeux R (1975) Usefulness of malt analyses. II. EBC monograph on barley and malt, Zeist, pp 236–270
Moll M, Flayeux R, Jeandemange JM (1980) Analyse von industriellen Mälzen mit der Tepral-Methode. Monatsschr Brau 33:170–174
Moll M, Flayeux R, Carnielo M (1982) New methods for the evaluation of barley and malt. J Am Soc Brew Chem 40:155–158
Montembault JC, Autran JC, Joudrier P (1983) Varietal identification of barley and malt. J Inst Brew 89:299–302
Moonen JHE, Graveland A, Muts GCJ (1987) The molecular structure of gelprotein from barley, its behavior in wort filtration and analysis. J Inst Brew (in press)
Muts GCJ, Kakebeeke M, Pesman L, Van de Berg R (1984) Malt characteristics and filtration. Proc Conv Inst Brew (Aust N Z Sect) 18:155–120
Nielsen G, Johansen H (1985) Identification of barley varieties by isoenzyme and hordein patterns. 2nd ISTA symposium, 5–9 May, Braunschweig
Palmer GH (1975a) Influence of endosperm structure on extract development. Proc Am Soc Brew Chem 33:174–180
Palmer GH (1975 b) A method for direct assessment of malt modification. J Inst Brew 81:408–409
Palmer GH, Taylor JA (1980) Dormancy drying and pregermination of malting barleys. The Brewer 66:105–110
Palmer GH, Taylor JA (1981) Acetic orcein staining test for assessment of pregermination in barley. J Inst Brew 87:286–288
Paukner E (1951) Die objektive Mürbigkeitsmessung zur Bestimmung des Malzlösunggrades I, II, III, Brauwissenschaft 4:138–142, 171–174, 187–190
Pollock JRA (1962) In: Cook AH (ed) Barley and malt. Academic, New York, pp 303–427
Preece IA, Mackenzie KG (1952) Non-starchy polysaccharides of cereal grains. II. Distribution of water-soluble gumlike materials in cereals. J Inst Brew 58:457–464
Prentice N, Bobler S, Faber S (1980) Enzymic analysis of betaglucans in cereal grains. Cereal Chem 57:198–202
Reiner L, Hanus H, Piendl H (1967) Zum Mikromalzungsverfahren. Die Bestimmung der Viskosität an geringer Würzemenge als Selektionsmerkmal für die praktischen Braugerstenerzüchtung. Brauwissenschaft 18:328–338
Riggs TJ, Sanada M, Morgan AG, Smith DB (1983) Use of acid gel electrophoresis in the characterisation of B hordein in relation to malting quality and mildew resistance of barley. J Sci Food Agric 34:576–586
Schild E (ed) (1961) Die brautechnischen Untersuchungsmethoden, 8. Aufl. Hans Carl, Nürnberg

Schildbach R, Burbidge M (1979) Identifizierung von Gerstensorten an Einzelkörnern durch Flachgel-Elektrophorese der Proteine und Aleuronfärbung. Monatsschr Brau 32:470–480

Schildbach R, Burbidge M (1985) The identification of barley and malted barley varieties. Brauwelt Int 1:23–30

Schur F, Ullmann F, Pfenninger H (1979) Eignungsprüfung der neuen DLFU-Scheiben-Mühle. Brauerei Rundsch 90:161–167

Schuster K, Narziss L, Kumada J (1967) Über die Gummistoffe der Gerste und ihr Verhalten während der Malz und Bierbereitung. I, II. Brauwissenschaft 20:125–135, 185–206

Scriban R (1963) Die Empfindlichkeit der Gerste beim Weichen. Brauwissenschaft 16:141–146

Scriban R, Strobbel B (1979) Chemotaxonomic study of barley and malt by gel electrofocalization. MBAA Tech Q 1:28–32

Scriban R, Autran JC, Strobbel B, Nicolaidis M (1979) Synthèse préliminaire de différents recherches analytiques sur la chemotaxonomie des orges et des malts. Bios 10:80, 83–87

Seeleitner G, Püspök J (1983) Die Bestimmung des Spelzengehaltes der Braugerste. Mitt Versuchsst Gärungsgew Wien 37:134–135

Shewry PR, Faulk SAJ, Parmar S, Miflin BJ (1980) Hordein polypeptide pattern in relation to malting quality and the varietal identification of malted barley grain. J Inst Brew 86:138–141

Smeaton RH (1985) Evaluating new barley varieties. Brew Guardian 44:20, 21, 24, 25–31

Smith DB (1984) Near infrared analysis of barley and malt. The Brewer 20:246–248

Smith DB, Lister PR (1983) Gel-forming proteins in barley grain and their relationships with malting quality. J Cereal Sci 1:229–239

Smith DB, Payne PI (1984) A procedure for the routine determination of electrophoretic band patterns of barley and malt endosperm proteins. J Natl Inst Agric Bot 16:487–498

Smith DB, Simpson PA (1983) Relationships of barley proteins soluble in sodium dodecylsulphate to malting quality and varietal identification. J Cereal Sci 1:185–197

Steiner K (1963) Zur statistischen Beurteilung von Untersuchungsergebnissen unter besonderer Berücksichtigung der Malzanalyse. Schweiz Brau Rundsch 74:131–135

Thomas DA (1986) A novel result of malt friabilimeter analysis: case hardened malt. J Inst Brew 92:65–68

Thompson RG, Laberge DE (1981) Barley endosperm cell walls: morphological characteristics and chemical composition. MBAA Tech Q 18:116–121

Tunaeus H (1938) Über die Bestimmung der endgültigen Keimfähigkeit nicht keimreifer Braugerste mittels Wasserstoffsuperoxyd. Wochenschr Brau 55:129–132

Ullmann F (1973) Moderne Interpretation der Malzanalyse. Schweiz Brau Rundsch 84:168–171

Van den Berg R, Muts GCJ, Drost BW, Graveland A (1981) Proteins from barley to wort. Proc Eur Brew Conv 18:461–469

Van Eerde P (1984) Simplified method for the determination of the degree of modification of malt. Brew Dig 59:12–14

Van Lonkhuyzen HJ, Marseille JP (1978) Identifizierung von Gerstensorten durch Stärkegelelektrophorese. Getreide Mehl Brot 32:288–291

Wettstein D von, Nilan TA, Ahrenst-Larsen B, Erdal K, Ingversen J, Jendestrid B, Kristiansen KN, Larsen J, Outtrup H, Ullrich SE (1984) Proanthocyanidin free barley in brewing: Progress in breeding for high yield and research tool in polyphenol chemistry. Am Soc Brew Chem Congr

Wainwright T (1986) Nitrosamines in malt and beer. J Inst Brew 92:73–80

Wainwright T, Buckee GK (1977) Barley malt analysis – a review. J Inst Brew 83:325–347

Whitmore ET (1960) Rapid method for determination of the husk content of barley and oats

Zangrando T (1974) Kostenbeeinflussung durch Wahl der Rohstoffqualität und technologische Maßnahmen. Brauwelt 114:519–526, 659–662, 733–738, 748–750

Reversed Phase HPLC as an Aid for the Identification of Barley Cultivars

M. J. Allison

1 Introduction

At the different generation stages of plant breeding programmes, it is advantageous to have a means of identifying cultivars uniquely. This is especially so at the later stages of breeding, when potentially new cultivars are tested rigorously for phenotypic distinctness from existing cultivars. In addition, any new cultivar must show homogeneity over different field plots, and genetic stability when multiplied, before acceptance and recommendation are possible.

In recent years SDS-polyacrylamide gel electrophoresis (SDS-PAGE) has proved to be a powerful tool for the resolution of proteins from crop plants (Matta et al. 1981; Shewry et al. 1978c). The resolving power of this method, together with a wide range of different forms observed for the storage proteins of cereal and leguminous crops, has led to fruitful studies on the genetics of storage proteins and how their electrophoretic patterns may be used as markers in crossing and for cultivar identification (Doll and Brown 1970; Oliver and Martinez-Zapiter 1985; Wrigley and McCousland 1975).

Proteins from the legumes and cereals have been defined according to their solubility in a range of solvents (Osborne 1924). In barley, for example, the alcohol-soluble storage proteins are known as hordeins (they have similar properties to the gliadins in wheat and the zein proteins in maize). The water- and salt-soluble proteins are the albumins and globulins, respectively. A fourth group of proteins soluble in dilute acid or alkali comprises the glutenins, although this group may be subdivided further on the basis of solubility in urea into the glutelins versus glutenins. Despite the usefulness of this division into groups, the classification is not clear-cut. For instance, one of the barley hordeins, hordein D, exhibits some of the properties of the high molecular weight glutenins (Smith and Lister 1983). Other proteins which aggregate strongly may appear in more than one fraction. Nevertheless, the broad solubility divisions are generally useful and have been adopted for the present investigation.

Studies (Shewry et al. 1979; Miflin et al. 1982) on SDS-PAGE electrophoresis of the barley hordeins (soluble in 55% iso-propanol) have revealed that the hordeins can be separated into four groups, the A, B, C, and D hordeins in order of increasing molecular weight (SDS-PAGE separations depend on molecular weight rather than electrical charge). No variation for the low molecular weight A proteins has yet been observed, so their genetic control and the chromosomal location of the control is still unknown. Genetic studies on the B and C hordeins, both of which show extensive polymorphism, have established that these groups of proteins are governed by two loci, *Hor1* and *Hor2*, sited approximately 8 units apart on the short arm of chromosome 5 (Doll and Brown 1970). Genetic investigation (Blake et al. 1982) has also shown that there is a locus controlling the production of D hordein on the long arm of chromosome 5 in barley.

In addition, surveys have been made of barley collections (Shewry et al. 1978a, 1979) using SDS-PAGE to resolve B and C hordeins with the aim of demonstrating variation for these proteins which could be of use in cultivar identification. The results of one survey (Shewry et al. 1979) showed that 30 distinct hordein patterns were observed in a total of 160 spring and winter barleys tested.

Although some of the above advances were made because of the resolving power of SDS-PAGE, this method has been largely unsuccessful in providing a means for unique cultivar identification of European barleys (Shewry et al. 1979). Treatment with SDS denatures protein and dissociates multimers into their constituent subunits. One consequence is that the subunit electrical charges are swamped by the SDS negative charge, so that differences in migration rates of proteins in SDS-PAGE are due to variation in molecular weights of the subunits. Thus qualitative differences in amino acid composition are not exploited by this technique. Moreover the electrophoretic bands resolved using SDS-PAGE may be difficult to quantify by densitometry as the bands may be diffuse or uneven across the gel track. It is also difficult to extract the resolved proteins from the gel matrix.

1.1 Reversed-Phase High Performance Liquid Chromatography

In this chapter the use of reversed-phase high-performance liquid chromatography (RP-HPLC) is considered as a potential method for an efficient resolution of the proteins of European barley.

The possible applications of RP-HPLC analysis have widened recently, particularly with reference to biological polymers of high molecular weight, such as polysaccharides, DNA fragments and proteins. This increase in versatility of RP-HPLC is due in part to the development of silica beads with wide pore sizes ($\simeq$ 300 Å pore size) such that large polymers can be exposed to chemical interaction with the stationary phase over a large surface area. A second main improvement in RP-HPLC of less recent origin arose from the use of gradients for column elution. For example, good separations of proteins are now possible by using an increasing ratio of acetonitrile to water (both with trifluoroacetic acid) and this gradient will elute proteins at different rates from the stationary phase of the column according to the surface hydrophobicity of the proteins.

The above improvements in RP-HPLC technology have been exploited by Bietz et al. (1984) to effect good separations of wheat gliadins and glutenins, which were markedly superior to those achieved by SDS-PAGE.

A similar technique was employed by Marchylo and Kruger (1984) to separate the B and C hordeins of Canadian barleys of fairly diverse origins (both two-rowed and six-rowed barleys). This raises the possibility that the RP-HPLC method may also be a useful aid to the identification of European barleys and also that the application may be widened to include salt-soluble proteins. It seems likely, however, that the European barleys are more closely related than the Canadian cultivars. Pedigree analyses (Lein 1963; Allison 1986) of European barleys have indicated that there have been very few inputs of malting quality components into plant breeding programmes. Similarly, there have been few inputs of disease resistance, one example being the mlv locus which is crossed into breeding material as a source of mildew resistance. This fact, together with the practice of intensive backcrossing, has ensured that the genomes of European barley cultivars do not vary greatly, but differ only in a few genes. However, even part of the success of RP-HPLC in separating the proteins from Canadian barleys would be a welcome aid to the identification of European barleys, especially in the large groups unresolved by SDS-PAGE.

1.1.1 The RP-HPLC System

The RP-HPLC system used was a Gilson gradient chromatograph consisting of 2 single piston pumps, an Apple 2C with software for sophisticated changes of gradient, a variable UV detector and a Shimadzu integrator. The main reagents used to create the gradient were acetonitrile and water (both Fisons HPLC grade). Solvent A contained 15% acetonitrile plus 0.1% Trifluoroacetic acid (TFA from Sigma London Co. Ltd.) dissolved in water (care was taken to introduce the TFA below the solution surface). Solvent B was 80% acetonitrile plus 0.1% TFA in water. For glutelin separations, Solvent A was 100% water plus 0.1% TFA whereas Solvent B was 100% acetonitrile plus 0.1% TFA.

Before use in any of the separations, both solvents were filtered and extensively degassed. The gradients were created with the aid of the Apple 2C gradient programme. Twenty µl of sample was injected at time zero. Any precipitation appearing in the sample was re-dissolved by gentle heating and the elution rate through the system was set at 1 ml per minute.

Injected sample was first pumped through a precolumn 60 mm by 4.1 mm (internal diameter) packed with Synchroprep RP-P beads, 30 µm in diameter. The sample then entered the main column which was a 250 mm by 4.1 mm (internal diameter) Synchropak PP C18 column (reversed-phase C18 support with a 300 Å pore size). The column was obtained from Anachem Ltd. The column was held in a thermo-regulated jacket with the chromatography run at 25 °C. Elution of the proteins was monitored at 210 nm by a variable UV detector and the peaks were plotted out with retention times and integrated areas on a Shimadzu integrator. The column was flushed between samples by a steep linear gradient from 25% to 80% B in 20 min, and then back to 25% B in 20 min.

1.1.2 Sample Preparation

In most of the RP-HPLC separations reported here, a few grains of barley were finely ground in a Glen Creston Retsch mill fitted with 0.2 mm sieve and the resulting flour was extracted in the appropriate solvent (Osborne 1924). Details are given under the headings for individual protein groups.

After extraction, the solutions were centrifuged at 2000*g* at room temperature and 20 µl of the clear supernatant was taken for HPLC column injection.

1.1.3 Hordeins

Flour from the milling of several barley grains was obtained as described earlier. The hordein storage proteins were extracted from 0.5 g milled flour in 2 ml of an extracting solution comprising 55% isopropanol containing 1% dithiothreitol. The extraction mixture was initially vortexed in a whirlimixer and then incubated in a water bath at 60 °C for 30 min with periodic vortexing. After centrifugation at 2000 *g* for 10 min at room temperature, the clear supernatant was used for column injection. The separation of the hordeins was achieved by a long gradient from 25% to 60% B over a period of 120 min, the detector setting was at sensitivity 0.5 AUFS.

1.1.4 Albumins and Globulins

Preliminary experiments on the albumins and globulins of barley established that these protein fractions from the embryo can cause the disappearance of peaks from the endosperm salt-soluble proteins in the same grain. Thus it may be that a fast-acting proteinase, which can digest specific endosperm proteins and which retains enzymic activity in salt solution, is produced by the embryo. To avoid this occurrence, samples used for albumin and globulin separations were taken from de-embryonated endosperms (grains sliced transversely). Individual de-embryonated endosperms were ground in a mortar and pestle and the flour was extracted in 0.2 ml of a 0.1 molar sodium chloride solution for 30 min at room temperature. After centrifugation (10 min) at 2000 *g* at room temperature the clear supernatant was loaded onto the HPLC column as described above except that the gradient was held at 15% solvent B for 60 min, raised to 80% B between 60 and 75 min, held at 80% B for 5 min and then decreased to 15% B from 80 to 95 min. On the detector the AUFS setting was 0.2. The separations achieved are similar to those reported in previous investigations (Bietz 1983).

1.1.5 Glutelins

This protein fraction was obtained from a residue left after sequential extraction of the albumins, globulins and hordeins. In this procedure, 5 g of flour was extracted in 20 ml 0.1 molar sodium chloride for 30 min at room temperature. The extraction mixture was centrifuged at 2000 *g* for 10 min at room temperature and the resulting supernatant was discarded. The pellet was taken up in 20 ml 0.1 molar sodium chloride to repeat the extraction step. After the second salt extraction, the pellet was taken up in 20 ml 55% isopropanol containing 1% dithiothreitol. This was spun as before and the supernatant discarded. This hordein extraction step was also repeated. After discarding the supernatant, the pellet was taken up in 0.05 molar acetic acid. After 30 min extraction at room temperature, the glutelins were precipitated by adjusting the pH to 6.6 with sodium hydroxide and allowing the precipitate to fall out of solution over night at 4 °C.

The glutelins were dissolved in 1 ml of a buffer containing 0.05 molar Tris and 8 molar urea adjusted to pH 5.0 with normal nitric acid. The proteins were then reduced for 2 h at room temperature in 0.1% dithiothreitol. Reduced proteins were alkylated in 1.5% 4 vinylpyridine for a further 2 h at room temperature. Alkylation was terminated by the addition of 0.5 ml glacial acetic acid. The solution was centrifuged at 2000 *g* at room temperature and the clear supernatant was injected into the HPLC column.

The gradient created to separate the glutelins consisted of 15% solvent B for 15 min (Solvent B was 100% acetonitrile plus 0.1% TFA). From 15 min to 80 min B was increased from 15% to 60%. The gradient was then decreased from 60% back to 15% for 90 min. The detector setting was 0.2 AUFS.

1.2 Hordein Identification

In the previous survey made of barley cultivars from Canada (Marchylo and Kruger 1984), two main groups of hordeins from the cultivar Betzes were separated on RP-HPLC. Those protein groups which eluted at 25 to 48 min and from 50 to 75 min were identified as C and B hordeins, respectively, from comparisons made of electrophoretic migration rates of eluted fractions on SDS gels. A sample of Betzes was run on the chromatographic system reported here and two protein groups each with the same range of retention time as in the Canadian survey were

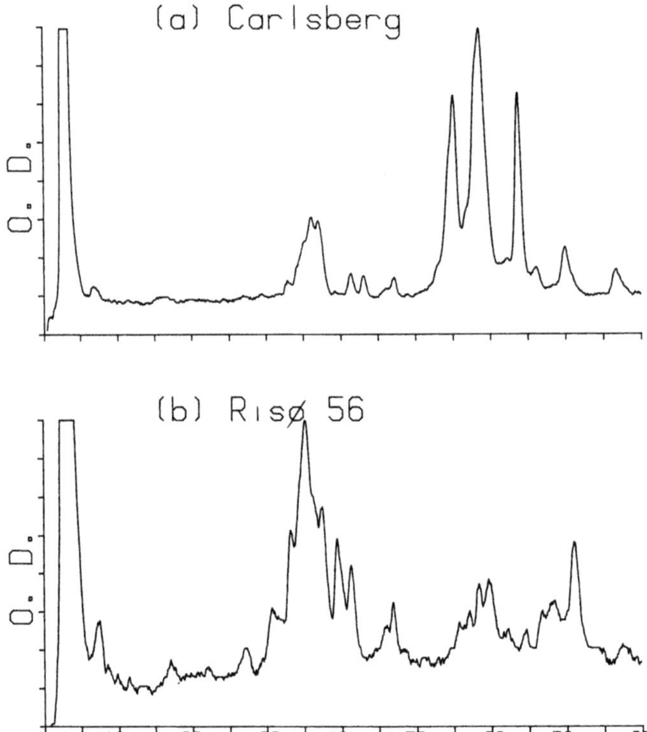

Fig. 1 a, b. RP-HPLC separations of the hordein proteins from the cultivars **a** Carlsberg and **b** Risø 56. The optical density axis is in arbitrary units and the x-axis is column retention time in minutes; and the axes are the same units in all the figures

observed with similar patterns to the Canadian sample. That the two main protein groups are composed of C and B proteins is supported by the separations shown in Fig. 1 a and b which show the hordeins from the cultivar Carlsberg and a mutant derived from it, Risø 56, respectively. It has been established from electrophoretic studies that Risø 56 has a severely diminished B hordein fraction Køie and Doll 1979; Doll et al. 1974) so that this cultivar has a very high C to B hordein ratio, whereas most cultivars have a high B to C ratio. The separations in Fig. 1 provide confirmatory evidence for the conclusion from the Betzes comparison, that the slower eluting proteins on RP-HPLC (50 to 75 min are B hordeins, while the relatively fast eluting group (25 to 48 min retention time) are C hordeins:

It has not been established whether the A and D hordeins are among the proteins retained on the column (column cleaning between runs removes some protein) or if they are eluted as minor peaks on the HPLC chromatograms. Whichever is the case, it seems that the A and D hordeins exhibit little significant variation after RP-HPLC, as is the case on electrophoretic separation (Shewry et al. 1978 b; Blake and Nilan 1982) and are thus of less consequence in cultivar identification.

1.3 Repeatability of Hordein Elution

The cultivar Golden Promise produces a range of C hordein and B hordein peaks, so samples of this cultivar were chosen to check repeatability of protein retention time both within and between days. Five samples are included from within one day (Table 1) plus five samples run on consecutive days (Table 2). It is evident from the results in these tables that for most peaks, and particularly the B hordeins, there is less variation for samples run in the same day. However, the repeatability reported here is not as good as that reported in the RP-HPLC study of Canadian cultivars (Marchylo and Kruger 1984). It is possible that this may be due to small differences in acetonitrile concentration which could lead to more

Table 1. Repeatability of retention times for hordein proteins of the cultivar Golden Promise. Separated by RP-HPLC, for five samples run within 1 day

Peak no.	Mean retention time for five samples (min)	Standard deviation of the mean (min)	Peak no.	Mean retention time for five samples (min)	Standard deviation of the mean (min)
1	12.2	0.2	9	53.3	0.4
2	23.0	0.9	10	56.3	0.1
3	31.0	0.3	11	58.2	0.1
4	33.3	0.2	12	60.3	0.2
5	35.5	0.2	13	63.1	0.1
6	39.0	0.6	14	64.7	0.1
7	46.3	0.3	15	70.2	0.3
8	48.7	0.2	16	77.9	0.2

The hordeins (separated by RP-HPLC) from the cultivar Golden Promise show similar patterns to those from the cultivar Proctor (see Fig. 2a).

Table 2. Repeatability of retention times for hordein proteins of the cultivar Golden Promise. Separated by RP-HPLC for five samples, each run on a different day

Peak no.	Mean retention time for five samples (min)	Standard deviation of the mean (min)	Peak no.	Mean retention time for five samples (min)	Standard deviation of the mean (min)
1	11.3	0.2	9	53.8	0.2
2	25.3	0.8	10	56.2	0.4
3	30.0	0.3	11	58.1	0.4
4	32.5	0.4	12	60.1	0.3
5	34.8	0.5	13	63.5	0.5
6	40.0	0.4	14	65.8	0.4
7	47.1	0.4	15	69.4	0.4
8	48.3	0.4	16	79.5	0.3

variable retention times in proteins that have a large number of hydrophobic groups. This may also explain why variation in peak retention time is uneven, with some peaks having an exceptionally high variation in retention time. For example, peak 2 (between 23 min and 26 min retention time) may have a relatively high number of hydrophobic groups. In addition to specific examples, the retention times are generally more variable than in previous studies, although variants in peak profile can be readily detected (see Fig. 2).

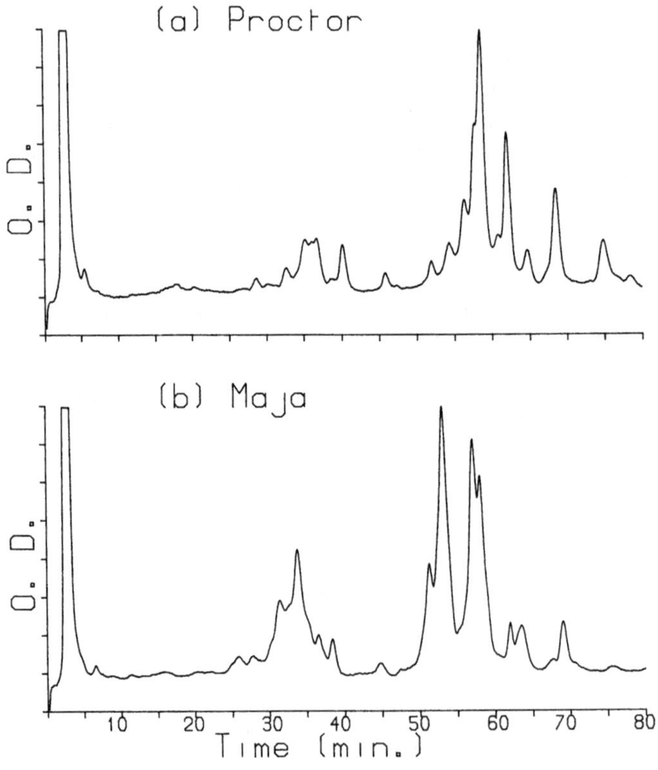

Fig. 2 a, b. RP-HPLC separations of the hordein proteins from the cultivars **a** Proctor and **b** Maja

1.3.1 RP-HPLC Separations of Hordeins

In electrophoretic studies (Shewry et al. 1978b) on the hordeins from European barleys, the cultivars were grouped according to distinctness of the B hordein or C hordein band patterns. After electrophoretic typing of spring and winter European cultivars using numbers for distinct B hordein patterns and a letter for C hordeins distinguishable after SDS PAGE, it was observed that large numbers of cultivars fell into only three or four groups (Shewry et al. 1979). For example, Shewry typed 15 cultivars as Group 1A and 9 cultivars as 3B. Thus the hordein patterns differed between groups 1A and 3B but were classified as indistinguishable within these groups using the SDS-PAGE unidirectional system (Shewry et al. 1979).

In order to test how RP-HPLC separations compared to those obtained after SDS-PAGE, a number of cultivars within each of the large groups were subjected to RP-HPLC analysis. Protein patterns after RP-HPLC are shown for cultivars belonging to electrophoretic groups 1A and 3B in Fig. 2 and 3.

Two kinds of polymorphism were observed after RP-HPLC. In one type, there are different numbers of peaks in the patterns and individual peaks have significantly different retention times compared to their counterparts in the other cultivars. An example of this is shown in Fig. 2. In a second type, the peak appears in a similar place in the pattern (i.e. roughly the same retention time) but there is a marked difference in the concentration of protein present. An example of this type of "gene regulation" variant is shown in Fig. 3 and has been observed in previous electrophoretic studies (Allison 1973; Schwartz 1962) on isoenzymes with the same electrophoretic migration rate but markedly different amounts of protein.

In Fig. 2a and b the hordein peaks of the cultivars, Proctor and Maja (both typed in group 1A after SDS-PAGE) are shown after resolution on RP-HPLC. In the Proctor chromatograph there are a number of incompletely resolved B hordein peaks which appear as "shoulders" at approximate retention times 55, 57, and 58 min before the main peak at 59 min.

The profile then falls away to smaller peaks at 64 and 69 min. This is in marked contrast to the Maja chromatograph which has one "shoulder" at approximately 51 min to a main peak at 53 min. This is followed by a second main B hordein peak at 57 min.

Although the C hordeins are less well resolved than the B hordeins, again the chromatographs differ in that Proctor has a cluster of C peaks from 34 to 36 min with a clear peak at 40 min. In contrast, Maja has unresolved peaks from 30 to 34 min with a main one at 34 min, followed by smaller peaks at 37 to 38 min.

These cultivar patterns are markedly different; it is worth noting that differences between patterns could occur because a cultivar has been mis-named, or there may have been an outcross during stock maintenance so that two small samples from the same cultivar may differ. However, RP-HPLC separations of the hordeins from 13 cultivars electrophoretically typed as 1A revealed five different hordein patterns, and within one of the these groups each one of six cultivars within the sub-group showed quantitative differences in the B hordeins. This subgroup included a number of current barley cultivars. Although other electro-

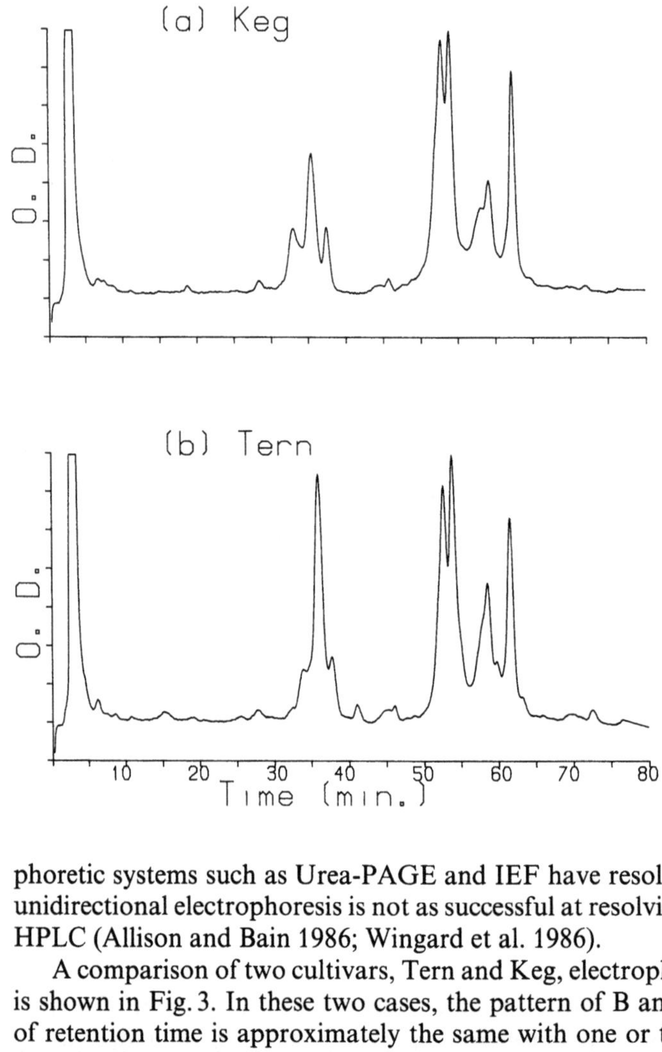

Fig. 3 a, b. RP-HPLC separations of the hordein proteins from the cultivars **a** Keg and **b** Tern

phoretic systems such as Urea-PAGE and IEF have resolved groups within 1A, unidirectional electrophoresis is not as successful at resolving the hordeins as RP-HPLC (Allison and Bain 1986; Wingard et al. 1986).

A comparison of two cultivars, Tern and Keg, electrophoretically typed as 3B is shown in Fig. 3. In these two cases, the pattern of B and C hordeins in terms of retention time is approximately the same with one or two minor differences. One significant difference, however, is in the amount of C hordein at 36 min retention time. In the cultivar Tern there is a relatively large peak, approaching the B hordeins in the height of the peak, whereas in Keg this peak is a smaller proportion of the C or B proteins, but has a similar retention time, 35.5 min to its counterpart in Tern. This is an example of a difference in the amount of proteins produced, i.e. rather like a "control gene" rather than a "structural gene" variant. In a study of ten samples of Keg grown in different UK environments, the main C protein was approximately 50% of the C hordeins $\pm 2.5\%$, whereas in two samples of Tern tested the main C hordein was over 60% of the C hordeins in both examples.

Further investigations of the 3B group revealed the presence of four different groups resolved after RP-HPLC out of eight cultivars tested. Three of the differences related to amount of protein and for one cultivar, Terrax, there was a different number of proteins in the separation.

An investigation of hordein types in electrophoretic group 15A, however, revealed only two cultivars with different groups after RP-HPLC among eight cultivars tested. Taken together, the results indicate that RP-HPLC can achieve a superior resolution of hordeins compared to SDS-PAGE, particularly in those cases where proportions of the different hordeins are diagnostic for the cultivar.

1.3.2 RP-HPLC Separation for Albumins and Globulins

Electrophoretic separations of extracted total protein (hordeins, glutelins, albumins and globulins) have revealed differences between cultivars from the 1A group (Smith and Payne 1984), but did not distinguish between the hordein patterns within 1A. RP-HPLC of the salt-soluble proteins (albumins and globulins) from European barley cultivars also revealed a number of different proteins separable on the basis of differences in surface hydrophobicity. Preliminary studies indicate that there is less polymorphism for the albumins and globulins than that observed for the hordeins, but detection of polymorphism may depend on a careful comparison of protein amounts. Two cultivars, Bomi Abed and Risø 1508, known to differ (Rhodes and Gill 1980) in soluble protein composition, are shown in Fig. 4a and b, respectively. It has been estimated that Risø 1508 (Fig. 4b) de-

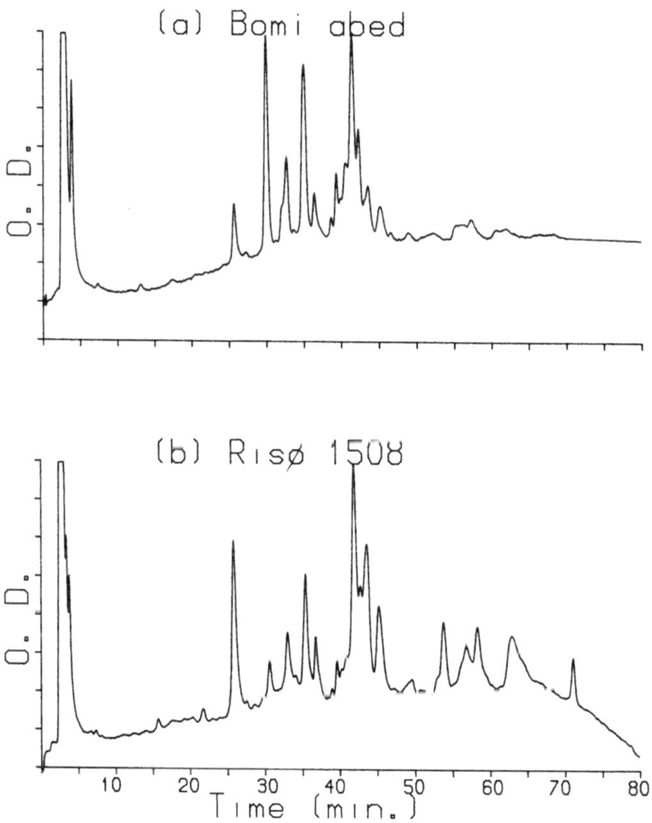

Fig. 4a, b. RP-HPLC separations of the albumin and globulin proteins of the cultivars **a** Bomi Abed and **b** Risø 1508

rived by mutation from the cultivar Bomi Abed (Fig. 4a) (Doll et al. 1974) has 35% more lysine-rich protein than the parent and that this is a net increase (Køie and Doll 1979; Kreft et al. 1975).

It is evident from Fig. 4a and b that there are a number of small differences in the amount of protein between Risø 1508 and Bomi Abed in the earlier eluting proteins with the protein being relatively less in Risø 1508 (the peak at 35 min retention time), or substantially more in Risø 1508 compared to Bomi Abed (the peak at 26 min retention time). However, there are major differences between the two chromatographs in the more hydrophobic (later eluting) proteins, with a series of distinct peaks between 49 to 73 min in Risø 1508, which are scarcely present in Bomi Abed. These results may indicate that the more lysine-rich salt-soluble proteins tend to be more hydrophobic, therefore selection for hydrophobicity may effectively select for nutritional quality.

1.3.3 RP-HPLC Separation of Glutelins

The glutelins separated on RP-HPLC were derived from residues left after sequential extraction of the other main proteins. The glutelins derived in this way from the cultivar Keg are shown in Fig. 5. Preliminary investigations indicate that there is less polymorphism for the glutelins than for the hordeins. However, calculations of gradient changes show that the glutelins are much less hydrophobic than the hordeins and would appear at about 20 min retention time on a gradient designed for hordein separation. Thus they may be studied in addition to the hordein pattern, as they do not overlap with C or B hordeins.

The results of these investigations of the proteins of European barleys separable by RP-HPLC show that it was possible to distinguish between hordein patterns which had been previously classified as indistinguishable after SDS-PAGE. Thus RP-HPLC can be regarded as a very useful addition to SDS-PAGE in cultivar identification. It was also possible to obtain a good resolution of the albumins, globulins and glutenins in barley, and these may be important in further work on barley protein identification, particularly the nutritionally important proteins.

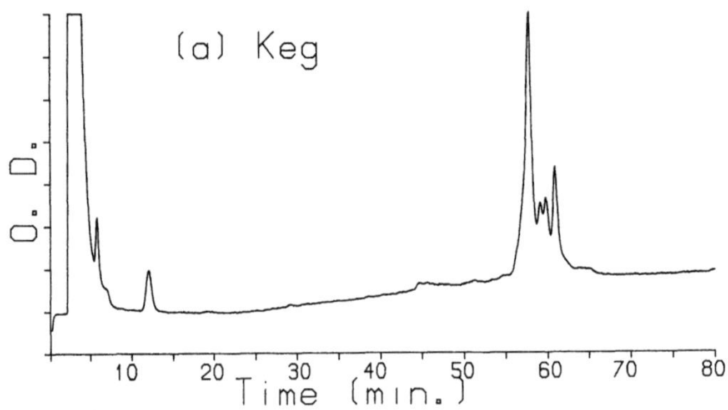

Fig. 5. RP-HPLC separation of the glutelins from the cultivar Keg

References

Allison MJ (1973) Genetic studies on the β-amylase isozymes of barley malt. Genetika 44:1–15

Allison MJ (1986) Relationships between milling energy and hot water extract values of malts from some modern barleys and their parental cultivars. J Inst Brew (in press)

Allison MJ, Bain H (1986) The use of reversed-phase high performance liquid chromatography as as aid to the identification of European barley cultivars. Euphytica 35:345–353

Bietz JA (1983) Separation of cereal proteins by reversed-phase high-performance liquid chromatography. J Chromatogr 255:219–222

Bietz JA, Burnouf T, Cobb LA, Wall JS (1984) Wheat varietal identification and genetic analysis by reversed-phase high-performance liquid chromatography. Cereal Chem 61:129–136

Blake TK, Ullrich SE, Nilan RA (1982) Mapping of the Hor-3 locus encoding D hordein in barley. Theor Appl Genet 63:367–371

Doll H, Brown AHD (1970) Hordein variation in wild *(Hordeum spontaneum)* and cultivated *(H. vulgare)* barley. Can J Genet Cytol 21:391–404

Doll H, Køie B, Eggum BO (1974) Induced high lysine mutants in barley. Radiat Bot 14:73–80

Køie B, Doll H (1979) Protein and carbohydrate components in the Risø high-lysine barley mutants. In: Seed protein improvement in cereals and grain legumes. IAEA, Vienna, pp 205–215

Kreft I, Javornik B, Milkovic C (1975) Studies on the gene control of electrophoresis protein patterns in barley. In: Gaul H (ed) Proc 3rd Barley Genetics Symposium, Wageningen, pp 310–325

Lein A (1963) Breeding for malting quality. In: Brochnizen S et al. (eds) Proc first barley genetics symposium Wageningen, pp 310–325

Marchylo BA, Kruger JE (1984) Identification of barley cultivars by reversed-phase high-performance liquid chromatography. Cereal Chem 61:295–301

Matta NK, Gatehouse JA, Boulter D (1981) The structure of legumin of *Vicia faba* L–a reappraisal. J Exp Bot 32:185–197

Miflin BK, Field JM, Shewry PR (1982) Cereal storage proteins and their effects on technological properties. In: Mosse J, Daussant J (eds) Seed proteins, Academic, London

Oliver JL, Martinez-Zapiter JM (1985) A genetic classification of potato cultivars based in allozyme patterns. Theor Appl Genet 69:305–313

Osborne TB (1924) The vegetable proteins. Longmans Green, London, p 154

Rhodes AP, Gill AA (1980) Fractionations and amino acids analysis of the salt-soluble fraction of normal and high lysine barleys. J Sci Food Agric 31:467–473

Schwartz D (1962) Genetic studies on mutant enzymes in maize 11 control of gene action in the synthesis of pH 7.5 esterase. Genetics 47:1609–1615

Shewry RP, Pratt NM, Miflin BJ (1978a) Varietal identification of single seed of barley by analysis of the hordein polypeptides. J Sci Food Agric 29:387–396

Shewry RP, Ellis JRS, Pratt HM, Miflin BJ (1978b) A comparison of methods for the extraction and separation of hordein factions from 29 barley varieties. J Sci Food Agric 29:433–441

Shewry RP, Faulks AJ, Pratt HM, Miflin BJ (1978c) The varietal identification of single seeds of wheat by sodium dodecyl sulphate electrophoresis of gliadins. J Sci Food Agric 29:847–849

Shewry RP, Pratt HM, Faulks AJ, Parmar S, Miflin BJ (1979) The storage protein (hordein) polypeptide pattern of barley *(Hordeum vulgare)* in relation to varietal identification and disease resistance. J Natl Inst Agric Bot 15:34–50

Smith DB, Lister PR (1983) Gel-forming proteins in barley grain and their relationships with malting quality. J Cereal Sci 1:229–239

Smith DB, Payne PI (1984) A procedure for the routine determination of electrophoretic band patterns of barley and malt endosperm proteins. J Natl Inst Agric Bot 16:487–499

Wingard CE, Iqbal M, Griffin M (1986) Separation of hordein proteins from European barley by high performance liquid chromatography: its application to the identification of barley cultivars. Chromatographia 21:49–55

Wrigley CW, McCousland J (1975) Instruction manual for laboratory procedures for identifying grain samples of Australian wheat varieties. Tech Publ No 3 Wheat Res Unit CSIRO

The Application of Immunofluorescence for Detecting Specific Proteins in Barley Seeds

U. RASMUSSEN

1 Introduction

Antibodies are precise molecular recognition tools in biochemical research and analytical procedures. The property of antibodies to recognize and react with specific sites (epitopes) on the antigen has been utilized in different immunological methods such as immunoprecipitation (in solutions and in gels), immuneblot assay and nephelometry. Detection of the antigen-antibody complex includes labelling of the complex with radioactivity (RIA, p. 48) and coupling of the antibodies with enzymes (ELISA), or fluorochromes (IF). The use of immunofluorescence to trace antigen was pioneered by Albert H. Coons (Coons et al. 1941, 1942; Coons and Kaplan 1950), who showed that antibodies could be coupled with fluorescein isocyanate with retention of the antigen-binding properties, and that these fluorescent antibodies could be used as very sensitive and specific probes to detect and localize antigens.

The potential and specificity of immunological techniques have increased considerably with the introduction of the hybridoma technique (Köhler and Milstein 1975). Hybridoma cell lines produce monoclonal antibodies in unlimited amounts and these recognize one single epitope.

Although an early report of the use of fluorescent antibodies in plant analysis was given in 1964 by Hagmann (1964), detailed immunological studies were not published until 1970, when the immunofluorescence method was used to localize pollen grain antigen (Knox et al. 1970) and antigens in bean cotyledons (Graham and Gunning 1970). Since then, the method has been used extensively in plant analysis and is now a well established and highly specific tool for the localization of different antigens in plant tissue.

With cereals the immunofluorescence has been used on rice seeds for the localization of β-amylase in germinating seeds (Okamoto and Akazawa 1979). In maize it has been employed for the demonstration that zein, the major storage protein in maize, is confined to the endosperm tissue (Dierks-Ventling and Ventling 1982). In wheat the lectin, called wheat germ agglutinin, has been localized in the surface layer of the radicle, roots, coleoptile and scutellum (Mishkind et al. 1982). In oat, the distribution of globulin during germination has been analysed (Zee et al. 1983), while in barley α-amylase was located in aleuron cells (Jacobsen and Knox 1973; Jones and Chen 1976) and the transport of α-amylase through the endosperm followed during germination (Gibbons 1979, 1980 a, b). The time of deposition of chymotrypsin inhibitor 2 (CI-2) during grain development was determined in barley (Rasmussen 1985).

My own work has focussed on immunofluorescence methods for detection of specific proteins in barley endosperm. One field of application would be the breeding for improved nutritional value of barley, e.g. by increasing the content of lysine-rich proteins in this cereal. A protein, which contains 11% lysine, CI-2 (Svendsen et al. 1980), has been shown to be present in elevated amounts in the high lysine barley variety Hiproly (Jonassen 1980a). The hordeins are, conversely, low in lysine. Therefore development of a non-destructive screening method using immunofluorescence for detection of an elevated content of CI-2 and a reduced content of individual polypeptides of the B-hordein family in single kernels is the ultimate goal.

For malt and beer quality, the content of B-hordeins may be of interest (Baxter and Wainwright 1979; Marchylo et al. 1986; Shewry et al. 1980) as is β-amylase or protein Z (Hejgaard and Boisen 1980) and various β-glucans and starch-degrading enzymes.

The purpose of this chapter is to give an outline of different immunofluorescence methods, some practical aspects concerning the conjugation of antibodies with fluorochromes or biotin and the application of fluorescing antibodies for screening of seeds for specific protein contents.

2 Immunofluorescence

2.1 Fluorescent Dyes

Fluorochromes are dyes which attain a different electronic distribution (become excited) when illuminated by light of an appropriate wavelength. If the excited molecule is sufficiently stable, it may return to its ground electronic state by emitting radiation. This effect is called fluorescence. The wavelengths which are capable of causing a molecule to fluoresce are known as the excitation spectrum, and the wavelengths of emitted fluorescent light are termed the emission spectrum. The emitted light is normally at a longer wavelength than that of the light which excited the fluorochrome. Many fluorochromes have been used successfully as fluorescent histological stains but only a few of these can be conjugated to proteins, because the modification of the dye molecule which may be necessary to make a stable link to the protein often results in loss of fluorescence.

In considering conjugation of fluorochromes to antibodies, the choice of fluorochrome is determined by a number of factors, but the colour of the emitted light should contrast strongly with any background fluorescence emitted by the specimen.

Three of the most popular fluorochromes for immunofluorescence microscopy are: fluorescein isothiocyanate (FITC), tetramethylrodamine isothiocyanate (TRITC) and Texas Red, composed of a derivative of sulforhodamine 101 (Fig. 1).

FITC is the most widely used fluorochrome and has an apple-green fluorescence. It is now commercially available in a rather pure state, can easily be conjugated to antibodies and is relatively hydrophilic. The conjugation of antibodies with FITC will be described in Sect. 3.2.1.

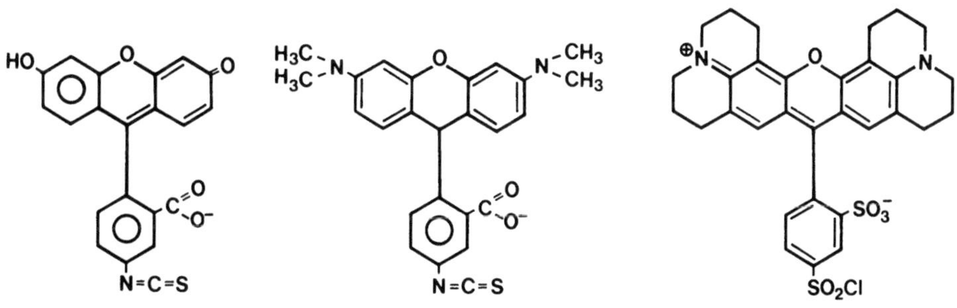

Fig. 1. Structure of commonly used fluorochromes *FITC* fluorescein isothiocyanate; *TRITC* tetramethylrhodamine isothiocyanate and Texas Red

Table 1. Exitation and emission wavelengths of conjugated FITC, TRITC, and Texas Red

Fluorochrome	Maximum exitation (nm)	Maximum emission (nm)	Range of emitted light (nm)
FITC	495	525	500–580
TRITC	554	573	550–650
Texas Red	596	615	600–615

TRITC and Texas Red are two rhodamine derivatives with orange-red fluorescence. TRITC has been used for many years, both alone and in combination with FITC. It is a rather hydrophobic fluorochrome and has therefore a tendency to bind non-specifically to the specimen, which requires careful controls. When used in combination with FITC, the emitted light from either fluorochrome should, by use of appropriate filters, be measured at wavelengths which are as widely separated as possible, because the emission spectra of FITC and TRITC overlap (Table 1).

Some of the disadvantages of TRITC are overcome in the newer fluorochrome Texas Red. It is a more hydrophilic compound and the emission spectrum is much separated from that of FITC.

For a detailed description of other fluorochromes see Nairn (1976) and Hansen (1967).

2.2 Staining Methods

Immunofluorescence analysis entails the use of a fluorochrome conjugated to antibodies or conjugated to a compound that can combine with the antibodies.

Four main methods of immunofluorescent staining can be used to localize antigens in tissue. Two of these involve fluorescence-labelled antibodies – the direct and indirect methods – while in the other two procedures, the biotin-(strept)

avidin system is inserted between the antibodies and the label. Streptavidin is a protein isolated from Streptomyces avidinii, capable of binding four molecules of biotin very strongly (Chaiet and Wolf 1964). Biotin can easily be conjugated to proteins, and up to 90 biotin molecules can theoretically be attached to each antibody molecule (Kendall et al. 1983). By adding fluorochrome-labelled streptavidin to biotin-labelled antibodies, each antibody and thus also each antigen in the tissue fluoresces extensively. Avidin may be used instead of streptavidin, but is, in contrast to streptavidin, a glycoprotein with a very high isoelectric point (Green 1975) and should therefore be more prone to non-specific adsorption than streptavidin. The very large signal amplification that can be achieved allows detection of very small amounts of antigen, use of high dilutions of the primary antibody and a strong signal is obtained even if only one monoclonal antibody is available for detection of small amounts of antigens.

2.2.1 The Direct Method with Fluorochrome-Labelled Primary Antibodies

The simplest way to localize a certain antigen is by using a fluorescence-labelled antigen specific antibody (primary antibody). The conjugated antibody is applied to the tissue and will react with the antigen ((Fig. 2a). Excess antibodies are removed by washing, and the antigen is visualized upon excitation, e.g. in a fluorescence microscope.

The direct method can be performed very quickly and with low probability of non-specific reactions.

2.2.2 The Indirect Method with Fluorochrome-Labelled Secondary Antibodies

In this double layer method, the unlabelled antigen-specific antibody is applied to the specimen and binds to the antigen. To localize this antibody-antigen complex, a fluorescent species-specific antibody (secondary antibody) is added to bind to the first one. The secondary antibody must be raised in animal species other than the primary antibody. For example, if the primary antibody was made in rabbit, then the conjugated secondary antibody could be a swine antibody which specifically recognizes the rabbit antibody ((Fig. 2b). The advantage of this method is the increase in fluorescence, since several secondary antibodies can bind to each of the primary antibodies. However, as more antibodies are applied than in the direct method, background fluorescence may increase due to non-specific adsorption, and the procedure takes approximately twice as long as the direct method.

2.2.3 Use of Biotin-Labelled Primary Antibodies

The biotin-labelled antigen-specific antibodies are added to the tissue and react with the antigens. Non-reacted antibodies are removed by washing. The fluorochrome-labelled streptavidin is added and binds with high affinity to biotin (Fig. 2c). By washing, the excess of streptavidin is removed. The antigens are visualized by excitation of the fluorochromes in the microscope.

2.2.4 Use of Biotin-Labelled Secondary Antibodies

In this method, three reagents are used. The first is the primary antibody specific for the antigen to be localized. The secondary antibody, capable of binding to the primary antibody, is conjugated with biotin. The third reagent is the fluorochrome-labelled streptavidin which binds to biotin, whereby the whole complex is visualized ((Fig. 2d).

This method is more sensitive than the direct streptavidin-biotin method but may be less specific, because there is greater risk of non-specific binding.

The principle in the above-described methods is comparable to enzyme-linked immunosorbant assay (ELISA) (Vaag 1988), except that a fluorochrome is used as label instead of an enzyme and the antigen to be detected is immobilized in the

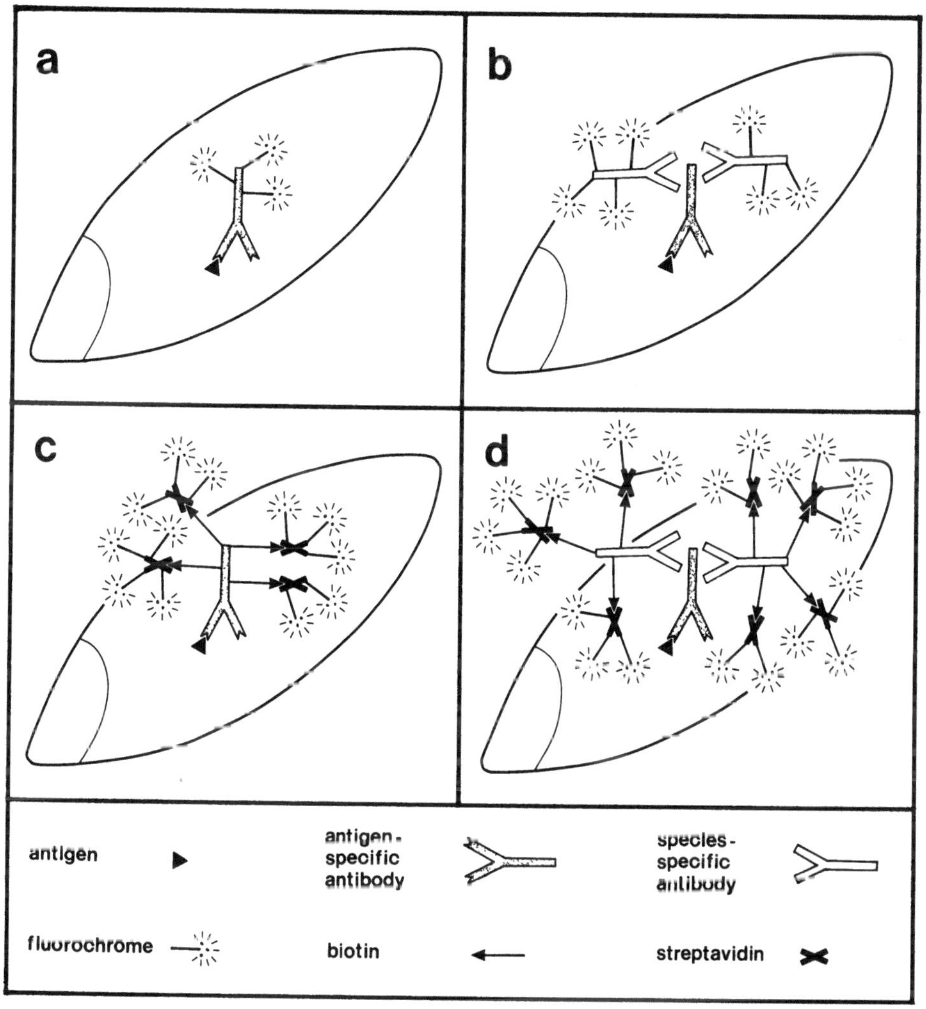

Fig. 2a–d. Immunofluorescence methods based on fluorescence-labelled antibodies **a** and **b**, and fluorescence labelled streptavidin combined with biotin labelled antibodies **c** and **d**

tissue, in situ. Depending upon the solubility of the antigen, an immobilization of the antigen by a fixation of the tissue can be necessary. This will be discussed in Sect. 3.4.

3 Materials, Protocols and Comments

3.1 Production and Purification of Antibodies

For production of polyclonal antibodies, goat, sheep and rabbit are the most commonly used animals. In order to obtain monospecific antibodies, the antigen for immunization should be as pure as possible. Molecules of low molecular weight are generally very poor immunogens, but may elicit a stronger immune response when coupled to larger, more immunogenic molecules (carriers) or polymerized before injection. For production of antibodies against the protein chymotrypsin inhibitor 2 [MW 9100 (Jonassen 1980a)], the antigen was polymerized with glutaraldehyde as described by Bollum (1975).

Many different immunization protocols have been developed. In rabbit, one normally immunizes with a mixture of the antigen and Freund's complete Adjuvant for the first injection and a mixture of antigen and Freund's incomplete Adjuvant for the following injections. The first five injections are given at 2-week intervals, and the first bleeding is taken 1 week after this. The following injections are given at 1-month intervals and bleedings are taken 1 week after each injection. Because rabbits vary in their immune response at least two rabbits should be injected with the antigen.

For some applications, production of monoclonal antibodies specific for a single epitope on the antigen are desirable. Production of monoclonal antibodies is not discussed here. A procedure for production of monoclonal antibodies to B-hordeins has been described by (Ullrich et al. 1986), and general methodologies are presented in Kennett et al. (1980).

The purification of both polyclonal and monoclonal antibodies can be performed as described by Warr (1982). A crude purification may be obtained by ammonium sulphate precipitation, where a saturated ammonium sulphate solution is added to the serum. Most immunoglobulins will precipitate by 45% of saturation (Heide and Schwick 1978). Ion exchange chromatography on a DEAE-Sephadex column is another possibility for purification (Fahey and Terry 1978). A third possibility is affinity chromatography on protein A-Sepharose columns, where the property of protein A, a protein from *Staphylococcus aureus,* to bind many, but not all sub-classes of IgG with high affinity is utilized (Kornvall et al. 1970; Goding 1978). The procedure for purification is according to the manufacturer's direction.

The purified antibody or antiserum is to be tested carefully for specificity before use. This can be done for instance by crossed-immunoelectrophoresis (Weeke 1973) or by immuneblot assay (Hønberg 1984). In the case where the antibodies cross-react with other antigens in the sample, the antibody can be further purified by affinity chromatography on immobilized antigen, as far as the antigen is available in large amounts.

3.2 Conjugation of Antibodies with Fluorochromes

The conjugation of antibodies with fluorochromes is technically very simple to perform but bright, specific fluorescence can only be obtained if the antibodies are optimally conjugated.

Conjugation of antibodies with the acid chlorides and isothiocyanate derivatives of rhodamine and fluorescein is performed by covalent binding to the ε-amino group of lysine on the antibodies. Conjugation is most effective under alkaline conditions, where most of the lysine residues are unprotonated.

3.2.1 Protocol for FITC Conjugation

The molar ratio of FITC to protein (F/P) in the conjugate is one of the most important factors in immunofluorescence. The net charge of heavily conjugated antibodies ($F/P > 5$) will be different from that of their native state, and because of this additional negative charge, they often stain in a non-specific manner by binding to positively charged components on the specimen (Goldstein et al. 1962; Herbert 1967; The and Feltkamp 1970b; Brandtzaag 1973a, b). If the antibodies are underconjugated ($F/P < 1$) the sensitivity of the fluorescence assay will decrease. Factors affecting the F/P ratio will be discussed in Sect. 3.2.2. Our procedure to obtain an optimal F/P ratio is as follow:

1. The antibodies are dialysed against 7.5 m*M* PBS (7.5 m*M*-sodium phosphate containing 112.5 m*M* sodium chloride at pH 7.3) for 2 × 12 h at 4 °C. The concentration is adjusted to 6 mg per ml, if necessary by pressure dialysis.
2. The necessary amount of FITC (50 μg FITC per mg IgG) is dissolved in 150 m*M* Na_2HPO_4 (pH 9.2), to 1 mg per ml.
3. The FITC solution is added dropwise to the antibody solution during stirring.
4. pH is adjusted to 9.5 with 30 m*M* K_3PO_4 (pH 11).
5. The mixture is left at room temperature for 2 h under continuous stirring.
6. The reaction is stopped by adjusting the pH to 7.3 with 500 m*M* potassium photostate buffer (pH 5.3).
7. The solution is applied to a Sephadex G-25 column pre-equlibrated with 7.5 m*M* PBS (pH 7.3). The first emerging coloured band is collected.
8. The absorbances of the conjugate at 280 nm and 495 nm are measured. From these values, the F/P ratio and the antibody concentration can be calculated according to:

$$F/P = \frac{2.87 \times OD_{495}}{OD_{280} - 0.35 \times OD_{495}} \quad \text{(The and Feltkamp 1970a)},$$

and

$$\text{IgG (in mg/ml)} = \frac{OD_{280} - 0.35 \times OD_{495}}{1.4} \quad \text{(Wood et al. 1965)}.$$

9. The conjugate is stored, protected from light, at 4 °C in the presence of 15 m*M* sodium azide.

3.2.2 Comments to the FITC Conjugation Procedure

FITC is available in different purities but only the purest preparations can be recommended for conjugation for instance FITC isomer I obtained from Research Organic. The purity of FITC can be evaluated by paper chromatography (Felton and McMillion 1961; Corey and McKinney 1962, 1965) or by thin layer chromatography (Steinbach 1974). FITC is very susceptible to hydrolysis, and the bottle must be warmed to room temperature before opening to avoid condensation.

Numerous factors affect the number of FITC molecules that actually bind to the antibodies, among them the initial ratio of FITC to antibodies, temperature, reaction time and pH.

If the initial antibody concentration is low (1–2 mg per ml) more FITC per mg antibody must be added in order to compensate for the competing hydrolysis of the fluorochrome. Conversely, less FITC is added if the antibody concentration is high (The and Feltkamp 1970a). A guideline for the appropriate amount of FITC to be added to a given protein concentration can be found in Goding (1983).

If the conjugation is performed at room temperature, a plateau level of labelling is attained after 1–2 h, depending on the protein concentration. At low concentration a longer incubation time can be necessary (The and Feltkamp 1970a). Under the conditions described in the protocol above, we have measured the molar F/P ratio after ½, 1, 1½, 2, and 2½ h of conjugation with an antibody against chymotrypsin inhibitor 2 of barley (Table 2). The various preparations were used for immunoreactions on thin sections of barley grain, and the amount of fluorescence was measured by use of a scanning spectrofluorometer, as shown in Fig. 3. The results indicate that a bright fluorescence and a F/P ratio between 1–5 is only obtained after 1½–2 h of conjugation.

The pH of the solution must be measured a few times during the reaction and adjusted to 9.5 with K_3PO_4 in order to avoid protonation of the lysine groups.

Unconjugated and conjugated dye can be separated by gel filtration on a Sephadex G-25 column. If the column with an initial diameter of 1.5 cm is used, the height should be approximately 2 cm per ml reaction mixture. Unreacted dye may also be removed from the preparation by dialysis.

The conjugate obtained consists of a mixture of antibodies with different F/P ratio. If the procedure is followed precisely, all complexes will have an F/P ratio in the range 1–5, which is adequate for immunofluorescence technique. A sepa-

Table 2. The effect of reaction time on conjugation of FITC with antibodies (F/P ratio)

h	F/P
½	1.3
1	2.2
1½	2.9
2	3.9
2½	5.6

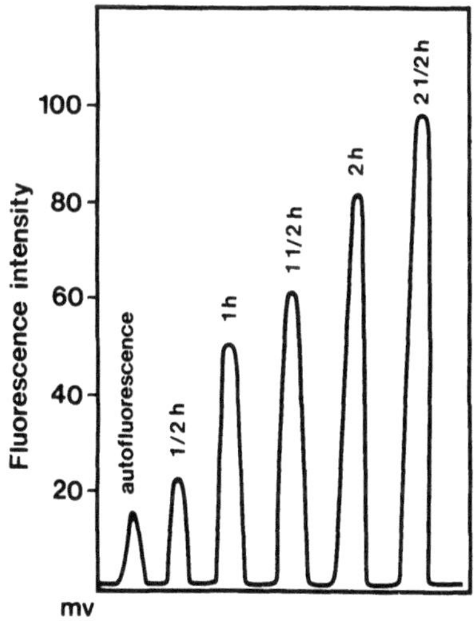

Fig. 3. The effect of reaction time on conjugation of FITC with antibodies against chymotrypsin inhibitor 2 of barley. Thin sections of barley were treated with antibodies which had been conjugated with FITC for various periods of time. The amount of fluorescent was measured by a scanning spectrofluorometer at 520 nm

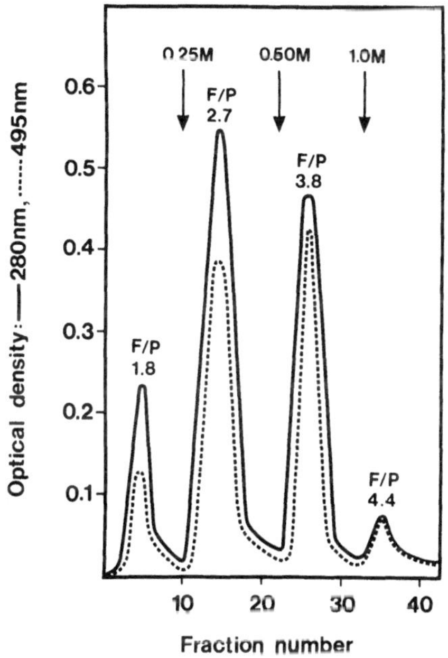

Fig. 4. Chromatography on DEAE-cellulose of polyclonal antibodies against barley chymotrypsin inhibitor 2 conjugated with FITC. Fractions were eluted with steps of NaCl as indicated. F/P ratio of each peak was calculated

ration according to the F/P ratio can be obtained by ion exchange chromatography on DEAE-Sepharose or DEAE-Sephacel (Cebra and Goldstein 1965; Goding 1976). In Fig. 4, the F/P ratios of the conjugates eluted at various salt concentrations are indicated.

The conjugates are stored at 4 °C in the presence of 15 m*M* sodium azide. Under these conditions, they can still be used after 3–4 months, although the fluorescence intensity may decrease somewhat. Conjugates may also be stored in aliquots at −20 °C or −70 °C for up to 2 years (Nairn 1976), but repeated cycles of freezing and thawing should be avoided.

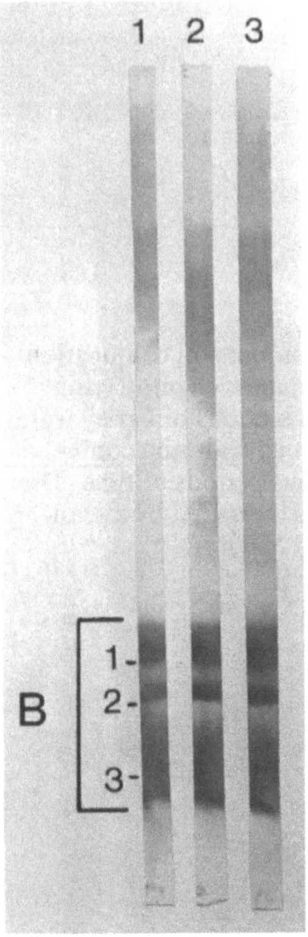

Fig. 5. Immuneblot assay for testing possible inactivation of monoclonal antibodies by conjugation with FITC or biotin. The test was performed on monoclonal antibodies to B-hordeins. *Strip 1* unconjugated antibodies; *2* FITC conjugated antibodies; *3* biotin conjugated. The same concentration of antibody was used in each lane

Inactivation of the immunological activity of polyclonal antibodies by conjugation with fluorochromes is highly unlikely, because it would require the simultaneous destruction of many different binding sites. In contrast, inactivation of particular clones of monoclonal antibodies by conjugation may occur if the epitopes contain lysine, and conjugated monoclonal antibodies should always be tested for activity by an immuneblot assay (Fig. 5) (Hønberg 1984; Ullrich et al. 1986).

3.2.3 Conjugation with Other Fluorochromes

The procedures for conjugation of antibodies with other fluorochromes are similar to the procedure described for FITC conjugation but for each fluorochrome, different precautions are to be observed.

Some fluorochromes, notably the rhodamine derivatives, are rather insoluble in water and should therefore be dissolved in dimethyl sulphoxide (DMSO) immediately prior to conjugation (Bergquist and Nilsson 1974; Goding 1976). Sulphonyl chlorides react with DMSO, and this solvent cannot be used with Texas Red. Instead, Texas Red can be added as a powder or dissolved in acetonitrile (Titus et al. 1982).

Different protocols for conjugation of antibodies with TRITC have been suggested (Goding 1976, 1983; Amante et al. 1972). For Texas Red, a conjugation procedure has been developed by Titus et al. (1982). Conjugation procedures for other rhodamine derivates and their properties are described by Brandtzaeg (1973 a, b).

3.3 Biotinylation

Biotin can be coupled to antibodies by means of various derivatives. The most convenient method for coupling makes use of the N-hydroxysuccinimide ester of biotin (BNHS), which couples to antibodies by reaction between the ester group and unprotonated ε-amino groups of lysine on the antibodies. When BNHS is directly coupled to the antibody, the binding ability of avidin or streptavidin for biotin is more or less hindered by steric constraints. This effect can be alleviated by insertion of an appropriate spacer arm between the biotin and the ester group (BXNHS), allowing the subsequent binding between biotin and avidin or streptavidin to proceed unhindered.

3.3.1 Protocol for Conjugation with Biotin (from Guesdon et al. 1979)

1. The antibodies are dialysed overnight in 0.1 *M* $NaHCO_3$, pH 9.5, and adjusted to a concentration of 10 mg per ml.
2. 57 µl of a freshly made 0.1 *M* solution of BXNHS dissolved in DMSO are added to the antibody solution during stirring.
3. Leave for 1 h at room temperature.
4. Unreacted biotin is separated from the conjugate by gel filtration on a Sephadex G-25 column, pre-equilibrated with 10 m*M* PBS, pH 7.3, or by dialysis.
5. The conjugate is stored at −20 °C.

3.3.2 Comments to the Biotin Conjugation

According to Guesdon et al. (1979), about one half of the lysine residues are biotinylated by the procedure described above. N-hydroxy succinimide esters are in general very prone to hydrolysis so the bottle should not be opened before warmed to room temperature in order to avoid condensation. During coupling there is a competing hydrolysis reaction, as in the case with FITC. The extent to which coupling and hydrolysis compete depends on the protein concentration. At low antibody concentration, the hydrolysis reaction dominates. According to Goding (1983), this can to some extent be balanced by addition of larger amounts of the N-hydroxy succinimide ester.

The immunological activity of polyclonal antibodies is unaffected by biotinylation (Guesdon et al. 1979). However, in the case of monoclonal antibodies this should be checked for each clone as described in Sect. 3.2.2 (Fig. 5). If inactivation is a problem, biotinylation procedures employing other functional groups than amino groups can be tried (Bayer et al. 1985; Forster et al. 1985).

3.4 Fixation

Chemical fixation of tissue by various means is used extensively in different fields of microscopy. For immunohistochemistry in particular, a fixation procedure which retains both the morphology of the tissue and the immunoreactivity of the antigens must be selected.

The sensitivity of protein detection by immunolabelling depends upon retention of the antigenicity of the protein in question during tissue processing, and any loss in the antigenicity will affect the sensivity of the methods. It is well known, that especially tissue fixation is a critical step in this context (Tokuyasu and Singer 1976; Sexton et al. 1981; Craig and Goodchill 1982).

For plant tissue, numerous fixation procedures have been developed. Some of the most commonly used fixatives are ethanol, glutaraldehyde and formaldehyde. Ethanol is known to preserve the antigenicity quite well but on the other hand it does not preserve the morphology of the tissue as well as cross-linking fixatives such as glutaraldehyde or formaldehyde. Glutaraldehyde is the most effective cross-linking fixative but may at times, even in low concentrations, permit very little specific staining of the tissue, either because it inactivates certain protein antigens or because it renders them inaccessible to the antibody stain (Tokuyasu and Singer 1976; Craig and Goodchill 1982). Formaldehyde alone is also commonly used for fixation of plant tissue (Jacobsen and Knox 1973, Manen and Pusztai 1982; Reed and Chollet 1985). It cross-links less strongly than glutaraldehyde, but preserves the antigenicity better. This has led to the combined use of glutaraldehyde and formaldehyde as introduced by Karnovsky (1967).

The fact that cereal seeds contain large amounts of carbohydrate is taken advantage of in a special fixative developed by McLean and Nakane (1974) containing periodate, lysine, and paraformaldehyde (PLP). Using this fixative, the carbohydrate moieties in the tissue are oxidized by the periodate to form aldehyde groups. Lysine, a divalent amine, then cross-links the carbohydrate by reacting with these aldehyde groups. Paraformaldehyde is added to crosslink the proteins. This fixative preserves the ultrastructure of the tissue equally well as glutaraldehyde and retains the antigenicity to the same extent as formaldehyde. For immunofluorescent studies, the fixative has been used sucessfully on seeds of rice (Okamoto and Akazawa 1979), maize (Dierks-Ventling and Ventling 1982) and barley (Gibbons 1979, 1980a, b; Rasmussen 1985).

3.4.1 Protocol for PLP Fixation of Seeds to be Sectioned

1. The seeds are slightly abraded with a scalpel blade to facilitate entry of the fixative.
2. The periodate-lysine-paraformaldehyde (PLP) solution is prepared fresh as described by McLean and Nakane (1974), and contains 10 m*M* periodate, 80 m*M* lysine-HCl and 4% paraformaldehyde (w/v).
3. The seeds are covered with fixative, (30 ml per 10 seeds), and incubated at 4 °C for 12 h. Subsequently, the seeds are washed for 8 h in PBS (10 m*M*-sodium phosphate, 150 m*M*-sodium chloride) pH 7.4. The seeds are frozen at 20 °C until use.

3.4.2 Surface Fixation of Sanded Seeds for Retention of Germination Capacity

The fixative, prepared as described above, is applied lightly to the sanded surface of the seed with a brush. Application of too much fixative may impair the germination of the seed. After 2 h at 4 °C, the seeds are washed for 30 min by immersion in PBS. The seeds are stored at 5 °C until use. The fixed seeds should not be stored more than 3–4 days before immuno-labelling.

3.4.3 Comments to the Fixation

The optimal choice of fixative will depend on the stability of the antigen in question and the ability of the fixative to preserve the localization of the antigen. For each given antigen and tissue, experimental determination of the optimal fixative and its concentration is required.

For soluble proteins in barley, the minimal concentration of paraformaldehyde which immobilizes the protein in tissue may be found by an enzyme-linked immunosorbant assay (ELISA). In an investigation of chymotrypsin inhibitor 2 (CI-2), PLP fixation of seeds was performed as described under Sect. 3.4.1 at different paraformaldehyde concentrations. The seeds were milled and the proteins extracted according to Jonassen (1980b). The amount of extracted CI-2 was measured by a double antibody sandwich ELISA using biotin coupled antibody and streptavidin linked to peroxidase for detection (Vaag 1986). The effect of paraformaldehyde on the extractability of the protein from the tissue is shown in Fig. 6. The results indicate that a 4% paraformaldehyde is sufficient to imobilize CI-2 in the tissue.

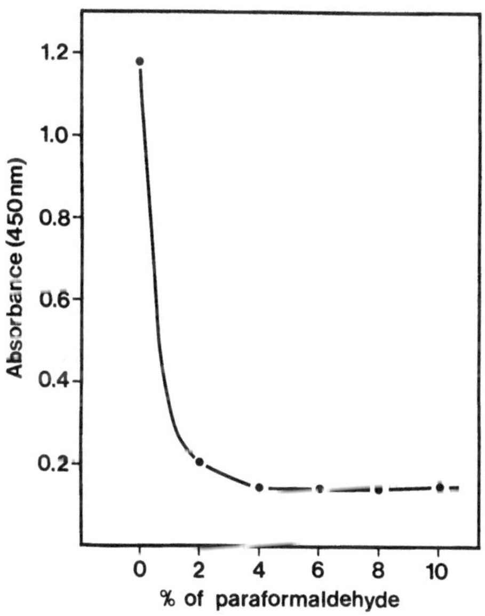

Fig. 6. The effect of different paraformaldehyde concentrations on retention of the water soluble protein CI-2 in the tissue, measured by an ELISA procedure

The effect of fixation on antigenicity can also be investigated by ELISA (Sexton et al. 1981) or other immunological methods such as rocket immunoelectrophoresis or radioactive immunoassay, RIA (Gibbons 1979; Craig and Goodchill 1982). Such assays can be useful in estimating the loss of antigenicity in solution, but it should be kept in mind that the reactivity of the antigen in these assays may differ from its reactivity in situ, where it is surrounded by a multitude of components. After an appropriate fixative or an adequate concentration has been found, the results need to be reevaluated by applying an immunofluorescence method on the fixed tissue. For the protein CI-2 a concentration of 4% paraformaldehyde was shown to give the optimal fluorescence, Rasmussen (1985).

3.5 Preparation of Tissue

3.5.1 Thin Sections

Prior to sectioning the seeds were mounted in Tissue Tek II [an embedding medium for frozen tissue specimes (Miles Laboratories, Naperville, Illinois, USA)]. Sections of 12 μm were cut in a cryostat and the sections placed on 1% gelatine-coated microscope glass slides. Other methods for tissue sectioning are described by Knox (1970) and Nairn (1976).

3.5.2 Sanded Seeds

Seeds wre embedded in a Cernit-plate (Heltved et al. 1982), using the Carlsberg seed fixation system, which is described in detail elsewere in this volume by Aastrup (1987). The seeds are sanded obliquely with a disc grinder in such a way that the embryos remain undamaged if the viability of the kernels is to be preserved.

3.6 Staining Procedures

3.6.1 General Considerations

All steps in the staining procedure are carried out in a moist chamber in order to prevent evaporation. It is important that the tissue area on the slide does not become dry at any stage during the entire procedure, as this results in a local concentration of salt which may produce artefacts.

In order to block non-specific sites and thereby reduce non-specific binding, the tissue should be treated with pre-immune serum or a solution of bovin serum albumin prior to the application of the antibody. Application of the antibody preparation for 30 min at room temperature is in general adequate. Longer exposure does not enhance specific staining, but may increase non-specific staining. Non-specifically bound reactants are removed after each step by washing the slides in a bath of PBS (pH 7.4). The stained preparations are finally mounted in 80% glycerol adjusted to pH 7.4 with PBS. In order to reduce the rapid fading of fluorescence during microscopy, which is especially pronounced for FITC, different retarding agents can be added to the mounting medium (Johnson and Aravjo 1981; Giloh and Sedat 1982; Valnes and Brandtzaeg 1985).

The reagent concentrations and time-course in the procedures outlined in this section should be taken as examples. When immunofluorescence is applied to other antigens and other antibody preparations, optimal dilutions must be determined in each case.

3.6.2 Direct Staining of Thin Sections

1. Wash in PBS for 30 min.
2. Apply 30 μl PBS containing 1% BSA or 30 μl pre-immune serum diluted 1:20. Incubate for 30 min at room temperature.
3. Suck off the solutions.
4. Incubate with 30 μl of a solution of FITC conjugated primary antibody for 30 min. To retain a uniform concentration of antibody the solution is removed and a new aliquot reapplied after 15 min.
5. Wash by a dip in distilled water and immersion in PBS for 15 min.
6. Mount in glycerol.

Controls
In order to distinguish staining attributable to the specific immunological reaction from non-specific reactions, the following controls must be included:

A. A tissue section treated with fluorochrome conjugated pre-immune serum.
B. An untreated section showing the autofluorescence.

This procedure has been used for localization of the protein CI-2 in the endosperm of barley Fig. 7a, b. The sections shown in the figure were incubated with polyclonal rabbit antibodies at a concentration of 0.7 mg per ml.

3.6.3 Indirect Staining of Thin Sections

As described above, but point (4) is substituted by:

4a. Treat with the appropriate dilution of the primary antibody for 30 min.
4b. Wash in PBS for 30 min.
4c. Incubate with fluorochrome-conjugated secondary antibody for 30 min.

Controls
It is advisable to run the same controls as described for the direct test and additionally:

C. A tissue section treated only with the fluorochrome-conjugated secondary antibody in order to ensure that this preparation reacts only with the primary antibody and not with any components in the tissue.

3.6.4 Use of Biotin-Labelled Primary Antibodies

This staining procedure is also performed as described for the direct staining (Sect. 3.6.2) with the exception of point (4), which is substituted by:
4a′. Incubate with biotin conjugated primary antibody for 30 min.
4b′. Wash in PBS for 30 min.
4c′. Treat with FITC conjugated streptavidin (from Amersham, diluted 1:50) for 20 min.

Controls
The same controls as for the direct staining and additionally:

D. A tissue section treated only with FITC-conjugated streptavidin.

An application of this method is seen in Fig. 7c, where biotinylated monoclonal antibodies against B-hordeins at a concentration of 0.3 mg per ml have been used for localization of the proteins. In order to illustrate the amplification effect of this system, the monoclonal antibodies were also FITC-conjugated and used in a direct test at the same concentration, Fig. 7d. Evidently, the biotin method provides a brighter fluorescence.

3.6.5 Use of Biotin-Labelled Secondary-Antibodies

This method is performed as the direct staining procedure except for point 4:

4a″. Incubate with primary antibodies for 30 min.
4b″. Wash in PBS for 30 min.
4c″. Incubate with biotin conjugated secondary antibodies for 30 min.
4d″. Wash in PBS for 30 min.
4e″. Incubate with FITC conjugated streptavidin (1:50) for 20 min.

Controls
It is advisable to run the controls described in Sect. 3.6.4 and additionally:
E. A tissue section treated with the biotinylated secondary antibody and FITC-conjugated streptavidin.

3.6.6 Two-Colour Staining of Thin Section

The same tissue section may be examined simultaneously for two different antigens in various ways. Thus, the two antibody preparations can be directly conjugated with different fluorochromes, and the staining procedure is in this case exactly as described in Sect. 3.6.2 with incubation of a mixture of the two antibodies. The indirect method can also be used, for one of the antigens or both, but in this case, the primary antibodies must be obtained from two different animals and the labelled secondary antibodies must be chosen in such a way that they will react only with the appropriate primary antibodies.

Another strategy is to couple one antibody directly to a fluorochrome, and couple the other with biotin, or as illustrated in Fig. 7g and h, where a mixture of unlabelled polyclonal antibodies, raised in rabbits and specific for the protein barley α-amylase/subtilisin inhibitor (BASI) (Mundy et al. 1983), and biotinyl-

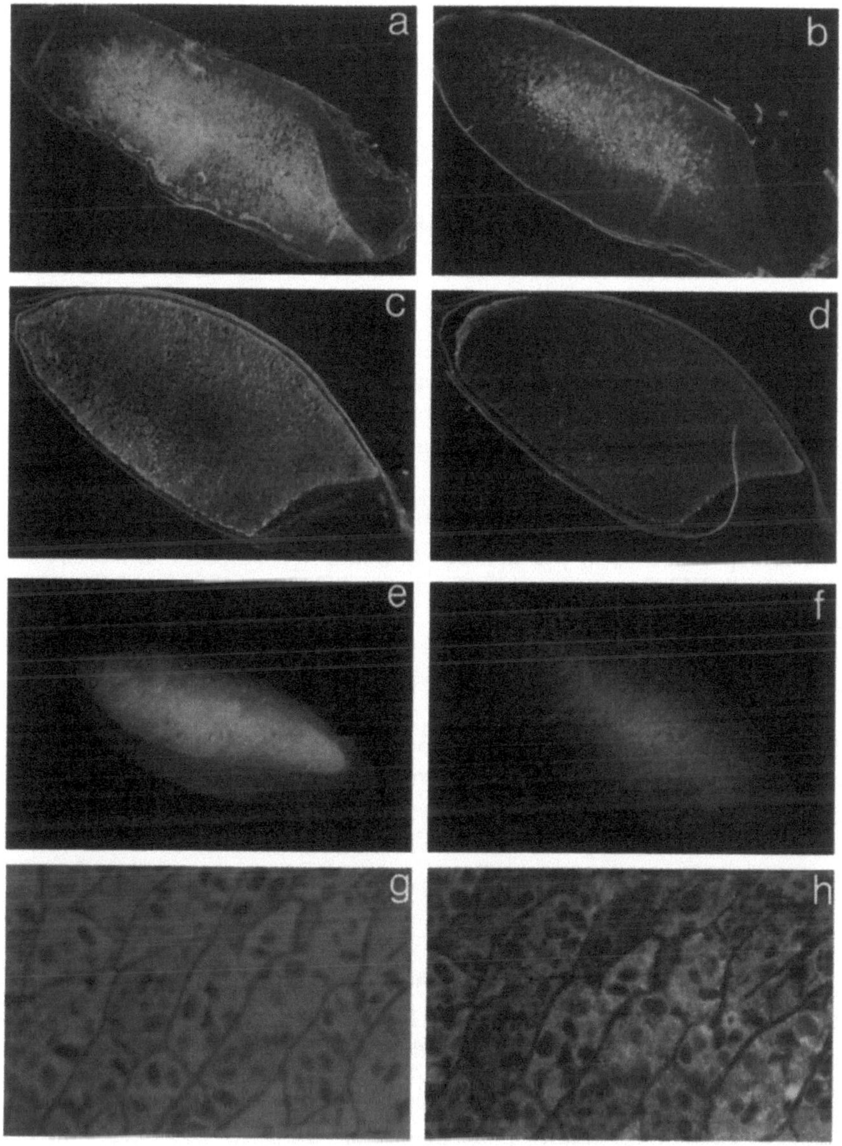

Fig. 7 a–h. Immunofluorescence on barley seeds. **a** and **b** The direct method: Thin section of **a** Hiproly (high content of CI-2), and **b** Hiproly sisterline (low content of CI-2). The sections were treated with FITC conjugated polyclonal antibodies to CI-2. **c** and **d** Amplification systems. A thin section of Bomi was treated with biotinylated monoclonal antibodies to B-hordeins and visualized by FITC conjugated streptavidin (**c**). It is compared to the section in **d** in which FITC conjugated monoclonal antibodies to B-hordeins have been used directly. **e** and **f** Sanded seeds of Hiproly (**e**) and Hiproly sisterline (**f**) treated with FITC conjugated polyclonal antibodies to CI-2. **g** and **h** Two-colour staining. Thin sections of Hiproly treated with a mixture of unlabelled polyclonal antibodies to BASI (barley α-amylase/subtilisin inhibitor), visualized by TRITC conjugated secondary antibody (**g**). A section from the same material was treated with biotinylated monoclonal antibodies to CI-2, visualized by FITC conjugated streptavidin (**h**). **a–f** ×2.5; **g** and **h** ×25

ated monoclonal antibodies, raised in mice and specific for CI-2, are added to the tissue. Addition of TRITC-coupled swine anti-rabbit antibodies and FITC-conjugated streptavidin is then used to visualize the antibody-BASI (Fig. 7 g) and the antibody-CI-2 complexes (Fig. 7 h) respectively. In this case, the staining procedure was performed as described for the indirect methods.

Controls
Individual controls for each staining must be done as described above.

3.6.7 Direct Staining of Sanded Seeds

This method has been worked out for future application which aim at screening thousands of kernels for genetic variants with increased or decreased amounts of specific polypeptides in the endosperm of barley.

The whole procedure is carried out in a 6-ml box, which is shaken mechanically during the reaction.

1. Place the cernit plates upside down in PBS for 30 min.
2. Block non-specific sites in 2% BSA for 20 min.
3. Incubate with FITC-conjugated antibody for 30 min.
4. Wash in PBS 3 × 5 min.

Controls
The same as for the direct test.

This method has been tested with antibodies against the protein chymotrypsin inhibitor 2, at a concentration of (0.5 mg/ml) Fig. 7 e, f.

4 Concluding Remarks

Like other immunochemical techniques, the immunoflourescence method provides a highly sensitive and specific means for the detection of antigens. The advantage of the immunofluorescence method is that it allows a precise localization of antigens in situ.

Immunofluorescence may equally well be adapted to stain thin sections of barley kernels and whole, slightly abraded kernels. However, prior to staining of thin sections, a rather time-consuming sample preparation is required. When the immunofluorescence is adapted for staining of abraded seeds, the sample preparation is considerably simplified. The total analysis time is thus reduced, and many samples can be assayed per day/week. Furthermore, the kernels are still viable if due care is taken during the processing and will germinate.

Acknowledgements. I wish to thank Ing. Pia Vaag for many helpful and stimulating discussions during preparation of this paper. Prof. Diter von Wettstein, Department of Physiology, Carlsberg Laboratory and Dr. Marianne Kjaervig Broe, Dakopatts are thanked for critical reading of the manuscript. Klaus Hansen and Nina Rasmussen are acknowledged for preparing the figures.

References

Aastrup S (1987) Fluorimetric methods in the analysis and control of malting barley and in the analysis of pre-harvest sprouting of all cereals (this volume, pp 56–66).

Amante L, Ancona A, Forni L (1972) The conjugation of immunoglobulins with tetramethylrhodamine isothiocyanate a comparison between the amorphous and the crystalline fluorochrome. J Immunol Meth 1:289–301

Baxter ED, Wainwright T (1979) Hordein and malting quality. J Am Soc Brew Chem 37:8–12

Bayer EA, Zalis MG, Wilchek M (1985) 3-(N-maleimido-propionyl) biocytin: a versatile thiol-specific biotinylating reagent. Anal Biochem 149:529–536

Bergquist NR, Nilsson P (1974) The conjugation of immunoglobulins with tetramethylrhodamine isothiocyanate by utilization of dimethylsulfoxide (DMSO) as solvent. J Immunol Methods 5:189–198

Bollum FJ (1975) Antibody to terminal deoxynucleotidyl transferase. Proc Natl Acad Sci 72:4119–4122

Brandtzaeg P (1973a) Conjugation of immunoglobulin G with different fluorochromes. I. Characterization by anionic exchange chromatography. Scand J Immunol 2:273–290

Brandtzaeg P (1973b) Conjugation of immunoglobulin G with different fluorochromes. II. Specific and non-specific binding properties. Scand J Immunol 2:333–348

Cebra JJ, Goldstein G (1965) Chromatographic purification of tetramethylrhodamine-immune globuline conjugates and their use in the cellular localization of rabbit γ-globulin polypeptide chains. J Immunol 95:230–245

Chaiet L, Wolf FJ (1964) The properties of streptavidin, a biotin-binding protein produced by Streptomycetes. Arch Biochem Biophys 106:1–5

Coons AH, Kaplan MH (1950) Localization of antigen in tissue cells. II. Improvement in a method for the detection of antigen by means of fluorescent antibody. J Exp Med 91:1–13

Coons AH, Creech HJ, Jones RN (1941) Immunological properties of an antibody containing a fluorescent group. Proc Soc Biol Med 47:200–202

Coons AH, Creech HJ, Jones RN, Berliner E (1942) The demonstration of pneumococcal antigen in tissue by the use of fluorescent antibody. J Immunol 45:159–170

Corey HS, McKinney RM (1962) Chromatography of nitrofluorescein, aminofluoresceins, fluorescein isothiocyanate. Anal Biochem 4:57–68

Corey HS, McKinney RM (1965) Paper chromatographic systems for 5- and 6-substituted fluoresceins. Anal Biochem 10:387–394

Craig S, Goodchill DJ (1982) Post-embedding immunolabelling. Some effects of tissue preparation on the antigenicity of plant proteins. Eur J Cell Biol 28:251–256

Dierks-Ventling C, Ventling D (1982) Tissue-specific immunofluorescent localization of zein and globulin in *Zea Mays* (L.) speeds. FEBS Lett 144:167–172

Fahey JL, Terry EW (1978) Ion exchange chromatography and gel filtration. In: Weir DM (ed) Blackwell Scientific, Oxford (Handbook of experimental Immunology, 3rd edn)

Felton LC, McMillion CR (1961) Chromatographically pure fluorescein and tetramethylrhodamine isothiocyanate. Anal Biochem 2:178–187

Forster AC, McInnes JL, Skingle DC, Symons RH (1985) Non-radioactive hybridization probes prepared by the chemical labelling of DNA and RNA with a novel reagent, photobiotin. Nucleic Acids Res 13:745–761

Gibbons GC (1979) On the localisation and transport of α-amylase during germination and early seedling growth of *Hordeum vulgare*. Carlsberg Res Commun 44:353–366

Gibbons GC (1980a) Immunohistochemical determination of the transport pathways of α-amylase in germinating barley seeds. Cereal Res Commun 8:87–96

Gibbons GC (1980b) On the sequential determination of α-amylase transport and cell wall breakdown in germinating seeds of *Hordeum vulgare*. Carlsberg Res Commun 45:177–184

Giloh H, Sedat JW (1982) Fluorescence microscopy: reduced photobleaching of rhodamine and fluorescein protein conjugates by *n*-propyl gallate. Science 217:1252–1255

Goding JW (1976) Conjugation of antibodies with fluorochromes: modifications of the standard methods. J Immunol Meth 13:215–226
Goding JW (1978) Use of staphylococcal protein A as an immunological reagent. J Immunol Methods 20:241–253
Goding JW (ed) (1983) In: Monoclonal antibodies: principles and practice. Academic Press, New York, pp 208–249
Goldstein G, Spalping BH, Hunt WB Jr (1962) Studies of fluorescent antibody staining. II. Inhibition by sub-optimally conjugated antibody globulins. Proc Soc Exp Biol Med 11:416–421
Green NM (1975) Avidin. Adv Protein Chem 29:85–133
Graham TA, Gunning BES (1970) Localization of legumin and vicilin in bean colyledon cells using fluorescent antibodies. Nature 228:81–82
Guesdon J-L, Ternynck T, Avrameas S (1979) The use of avidin-biotin interaction in immunoenzymatic techniques. J Histochem Cytochem 27:1131–1139
Hagman M (1964) The use of disc electrophoresis and serological reactions in the study of pollen and style relationships. In: Linskens HF (ed) Pollen physiology and fertilization. North-Holland, Amsterdam, pp 240–250
Hansen PA (1967) Spectral data of fluorescent tracers. Acta Histochem Suppl VII, pp 167–180
Heide K, Schwick HG (1978) Salt fractionation of immunoglobulins. In: Weir DM (ed) Blackwell Scientific, Oxford (Handbook of experimental immunology, 3rd edn, pp 7.1–7.11)
Hejgaard J, Boisen S (1980) High-lysine proteins in Hiproly barley breeding. Identification, significance and new screening methods. Heriditas 93:311–320
Heltved F, Aastrup S, Jensen O, Gibbons G, Munck L (1982) Preparation of seeds for mass screening. Carlsberg Res Commun 47:291–296
Herbert GA, Pittman B, Cherry WB (1967) Factors affecting the degree of monospecific staining given by fluorescein isothiocyanate labeled globulins. J Immunol 98:1204–1212
Hønberg LS (1984) Probing barley mutants with a monoclonal antibody to a polypeptide involved in photosynthetic oxygen evolution. Carlsberg Res Commun 49:703–719
Jacobsen JV, Knox RB (1973) Cytochemical localization and antigenicity of α-amylase in barley aleurone tissue. Planta 112:213–224
Johnson GD, Aravjo GMdeCN (1981) A simple method of reducing the fading of immunofluorescence during microscopy. J Immunol Meth 43:349–350
Jonassen I (1980a) Characteristics of Hiproly barley. I. Isolation and characterisation of two water-soluble high-lysine proteins. Carlsberg Res Commun 45:47–58
Jonassen I (1980b) Characteristics of Hiproly barley. II. Quantification of two proteins contributing to its high lysine content. Carlsberg Res Commun 45:59–68
Jones RL, Chen R-F (1976) Immunohistochemical localization of α-amylase in barley aleurone cells. J Cell Sci 20:183–198
Karnovsky MJ (1967) The ultrastructural basis of capillary permeability studied with peroxidase as a tracer. J Cell Biol 35:213–236
Kendall C, Ionescu-Matiu I, Dreesman GR (1983) Utilization of the biotin/avidin system to amplify the sensitivity of the enzyme-linked immunosorbant assay (ELISA). J Immunol Meth 56:329–339
Kennett RH, McKearn TJ, Bechtol KB (eds) (1980) Monoclonal antibodies. Hybridomas: a new dimension in biological analysis. Plenum, New York
Knox RB (1970) Freeze-sectioning of plant tissue. Stain Technol 45:265–272
Knox RB, Heslop-Harrison J, Reed CE (1970) Localization of antigens associated with pollen grain wall by immunofluorescence. Nature 225:1066–1068
Kornvall G, Seul US, Finstad J, Williams RC (1970) Phylogenetic insight into evolution of mammalian F_c fragment of γG globulin using staphylococcal protein A. J Immunol 104:140–147
Köhler G, Milstein C (1975) Continuous cultures of fused cells secreting antibody of predefined specifity. Nature 256:495–497

Manen JF, Pusztai A (1982) Immunocytochemical localisation of lectins in the cells of *Phaseolus vulgaris* L. seeds. Planta 155:328–334
Marchylo BA, Kruger JE, Hatcher D (1986) High-performance liquid chromatographic and electrophoretic analysis of hordein during malting for two barley varieties of contrasting malting quality. Cereal Chem 63:219–231
McLean IW, Nakane PK (1974) Periodate-lysine-paraformaldehyde fixative. A new fixative for immunoelectron microscopy. J Histochem Cytochem 22:1077–1083
Mishkind M, Raikhel NV, Palevitz BA, Keegstra K (1982) Immunocytochemical localization of wheat germ agglutinin in wheat. J Cell Biol 92:753–764
Mundy J, Svendsen I, Hejgaard J (1983) Barly α-amylase/subtilisin inhibitor. I. Isolation and characterization. Carlsberg Res Commun 48:81–90
Nairn RC (ed) (1976) Fluorescent protein tracing, 4rd edn. Churchill Livingstone, Edinburg
Okamoto K, Akazawa T (1979) Enzymic mechanism of starch breakdown in germinating rice seeds. Plant Physiol 64:337–340
Rasmussen U (1985) Immunological screening for specific protein content in barley seeds. Carlsberg Res Commun 50:83–93
Reed JE, Chollet R (1985) Immunofluorescent localization of phosphophenolpyruvate carboxylase and ribulose 1,5-biphosphate carboxylase/oxygenase proteins in leaves of C_3, C_4, and C_3-C_4 intermediate *Flaveria* species. Planta 165:439–445
Sexton R, Durbin ML, Lewis LN, Thompson WW (1981) The immunocytochemical locatization of 9.5 cellulase in abscission zones of bean (*Phaseolus vulgaris* cv. Red Kidney). Protoplasma 109:335–347
Shewry PR, Faulks AJ, Parmar S, Miflin BJ (1980) Hordein polypeptide pattern in relation to malting quality and the varietal identification of malting barley grain. J Inst Brew 86:138–141
Steinbach G (1974) Characterization of fluorescein isothiocyanate: Synthesis and testing methods for fluorescein isothiocyanate isomeres. Acta Histochem 49:19–34
Svendsen I, Martin B, Jonassen I (1980) Characteristics of Hiproly barley. III. Amino acid sequences of two lysine rich proteins. Carlsberg Res Commun 45:79–85
The TH, Feltkamp TEW (1970a) Conjugation of fluorescein isothiocyanate to antibodies. I. Experiments on the conditions of conjugation. Immunology 18:865–873
The TH, Feltkamp TEW (1970b) Conjugation of fluorescein isothiocyanate to antibodies. II. A reproducible method. Immunology 18:875–881
Titus JA, Haugland R, Sharrow SO, Segal DM (1982) Texas Red, a hydrophilic, red-emitting fluorophore for use with fluorescein in dual parameter flow microfluorometric and fluorescence microscopic studies. J Immunol Meth 50:193–204
Tokuyasu KT, Singer SJ (1976) Improved procedures for immunoferritin labeling of ultrathin frozen sections. J Cell Biol 71:894–906
Ullrich SE, Rasmussen U, Høyer-Hansen G, Brandt A (1986) Monoclonal antibodies to hordein polypeptides. Carlsberg Res Commun 51:381–399
Vaag P (1988) The enzyme-linked immunosorbent assay (ELISA) in the beverage industries: principles and practice. In: Linskens HF, Jackson JF (eds) Modern methods of plant analysis, vol 8 (in preparation)
Valnes K, Brandtzaeg P (1985) Retardation of immunofluorescence fading during microscopy. J Histochem Cytochem 33:755–761
Warr GW (1982) Purification of antibodies. In: Marchalonis JJ, Warr GW (eds) Antibody as a tool; the applications of immunochemistry. Wiley, Chichester, pp 59–96
Weeke B (1973) Crossed immunoelectrophoresis. In: Axelsen NH, Krøll J, Weeke B (eds) A manual of quantitative immunoelectrophoresis. Methods and applications. Universitetsforlaget Oslo, pp 47–56
Wood BT, Thompson SH, Goldstein G (1965) Fluorescent antibody staining. III. Preparation of fluorescein-isothiocyanate-labeled antibodies. J Immunol 95:225–229
Zee SY, Chan HY, Ma CY (1983) Histochemical studies of scutellum during germination of oat seed by light microscopy and immunofluorescent localization of oat globulin. Acta Bot Sin 25:509–511

Fluorimetric Methods for the Analysis of Malt Modification and Pre-Harvest Sprouting

S. Aastrup and K. G. Jørgensen

1 Introduction

In the study of specific botanical components it is of great help if they can be associated with specific colours. They might have a colour of their own like the tannins in the testa of some sorghum varieties, or be auto-fluorescent like the ferulic acid in the aleurone layer of, for example, barley and wheat. The components of interest might also be rendered visible by staining: Tetrazolium stains living germs, vanillin-HCl stains the proanthocyanidins in the testa of barley. Calcofluor, a fluorescent dye, stains the β-glucan in the endosperm cell walls of barley (Fig. 2B) and oats and fluorescein dibutyrate develops fluorescence if lipase/esterase activity is present in the seed. If no specific stain is directly accessible, more sophisticated methods have to be used. An elegant solution is to raise anti-bodies against the component to be studied followed by conjugation to a fluorochrome, e.g. fluorescein isothiocyanate conjugated with rabbit IgG specific for the high lysine protein CI-2 (Rasmussen this volume, pp 35–55).

Such specific colours are most useful when studying individual kernels, although some are used on flour for quantification of components, e.g. Calcofluor for determination of β-glucan (Jørgensen and Aastrup, this volume, pp 88–108).

In most cases the grains have to be opened in order to locate the components or enzyme activity. This can simply be done by a scalpel. If the methods, however, are to be used for screening or control the sample preparation have to be more automated.

The present chapter will focus on the practical application of two specific stains used on individual half kernels: Calcofluor and fluorescein dibutyrate (FDB) for modification analysis of barley malt and pre-sprouting of cereals, respectively.

Both methods give information on events happened during germination of the individual seeds (Fig. 2C, D). Field sprouting as well as controlled germination, malting, is characterized by initial uptake of water, increasing release of enzymes from scutellum/aleurone into the endosperm, and a resulting progressive breakdown of the endosperm components. FDB demonstrates the presence of enzyme activity in general because the various hydrolytic enzymes are released almost simultaneously, and in addition proceed into the endosperm following the same pattern (Gibbons 1980; Munck et al. 1981). Calcofluor, on the other hand, demonstrates the presence of a substrate, the β-glucan-rich cell walls of barley and oats.

Calcofluor Method. Calcofluor has a high specificity for mixed linkage $(1\rightarrow3)$, $(1\rightarrow4)$-β-D-glucan (β-glucan) (Wood et al. 1983). One of the key scientists in these studies was Dr. R.G. Fulcher, and during his stay at the Carlsberg Research Center in 1979, he and scientists from the Department of Biotechnology studied half barley malt kernels stained with Calcofluor. In UV-light a spectacular fluorescence was seen at the distal end of the kernels. The fluorescence showed the undegraded endosperm cell walls, thus indicating the degree of malt modification. The first attempt to demonstrate that this observation could be used in practice was carried out by Aastrup and Erdal (1980), who showed that it was possible in a reproducible way to follow the average modification during malting. Malt kernels were taken at each day of germination, halved with a scalpel, stained with Calcofluor and the percentage of modified endosperm area measured using an image analyser. The analysis became easier by the introduction of the Malt Modification Analyser – System Carlsberg and a visual evaluation system (Aastrup et al. 1981). The sample preparation was improved by the development of the Carlsberg Seed Fixation System (Heltved et al. 1982). A more convenient and reproducible method was achieved by improvement of the visual evaluation system (Carnielo et al. 1982; Jensen and Aastrup 1985).

FDB Method. Fluorescein dibutyrate is split by lipases and estarases (Guilbault and Kramer 1964). The story of turning this specific staining technique into a practical method for demonstrating pre-harvest sprouting is similar to that of Calcofluor. The principle was shown by Jensen and Heltved (1982) on manually halved seeds of barley, wheat, rye and sorghum germinated for 0 to 5 days. The Seed Fixation System and the Malt Modification Analyser was then part of the final convenient method (Jensen and Heltved 1983; Jensen and Aastrup 1985).

In 1986, both methods were accepted as official method within the European Brewery Convention for evaluation of barley malt modification and pre-germination of barley, respectively.

2 Modification and Homogeneity of Barley Malt

2.1 Field of Application

The method is primarily a check method for commercial malt. Additionally, it can be used as a control method in the malt house and for research. In the latter case, samples of green malt are taken once a day during germination, and prior to analysis dried to a moisture content of 10–15%, e.g. using an IR-lamp.

2.2 Sample Size

Among varieties, but also within varieties, the speed of modification varies considerably due to differences in hardness, β-glucan content, enzyme levels etc. Therefore about 100 malt kernels must be analysed to give reproducible results.

2.3 Reagents and Materials

- 1 g/l aqueous Calcofluor White M2R (Cellufluor) (Polysciences Ltd., 24 Low Farm Place, Moulton Park, Northampton NN3 lHY, England).
- 1 g/l aqueous Fast Green (for counter-staining) (Gurr, England).
- 70% V/V aqueous ethanol.

2.4 Apparatus (Fig. 1)

- Carlsberg Seed Fixation System including a press supplied with a barley matrix and a disc sander (Danbrew Consult Ltd., Rahbeks Alle 21, DK-1801 Copenhagen V, Denmark).
- Modelling clay (Danbrew Consult Ltd.).
- Malt Modification Analyser – System Carlsberg (Danbrew Consult Ltd.).

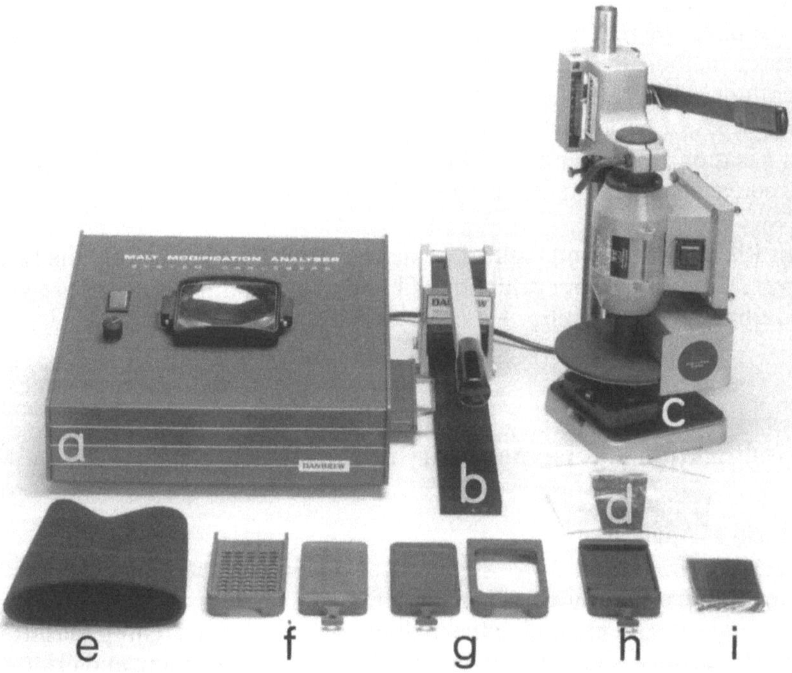

Fig. 1 a–i. The Carlsberg Seed Fixation System and the Malt Modification Analyser-System Carlsberg. **a** Malt Modification Analyser; **b** press; **c** disc sander; **d** chemicals; **e** viewing sheet; **f** barley/malt matrix pair; **g** multimatrix pair; **h** holder for modelling clay block; **i** modelling clay

2.5. Procedure

2.5.1 Fixation and Sanding

Two blocks, each containing 50 halved malt kernels are obtained using the Carlsberg Seed Fixation System.

- Use the over- and undermatrix to place 50 grains in the undermatrix.
- Fix the 50 grains in the modelling clay block by means of the press in such a way that half of each grain is embedded in the clay.
- Sand the grains to half their size with the disc sander.

2.5.2 Staining

- Stain the half grains with the Calcofluor solution for 2 min.
- Rinse with 70% ethanol.
- Remove residual ethanol with filter paper.
- Stain with the Fast Green solution for 2 min.
- Remove residual Fast Green with filter paper.

2.5.3 Examination (Fig. 2)

Place the block in the Malt Modification Analyser and switch on the UV-light.

Non-modified endosperm shows a blue fluorescence (normally adjacent to the tip of the grains), while modified endosperm appears dull blue. The area of the modified endosperm in percentage of the total endosperm area is defined as the modification percentage of a grain (Fig. 2A). The average modification of 100 grains is the modification percentage of a sample.

Judge each grain as being a member of one of the following modification groups describing increasing modification percentages:

$0\% \leqq 0 < 5\%$, $5\% \leqq 1 < 25\%$, $25\% \leqq 2 < 50\%$, $50\% \leqq 3 < 75\%$, $75\% \leqq 4 < 95\%$, and $95\% \leqq 5 \leqq 100\%$.

2.6 Expression of Results (Figs. 2E, F and 3I, II)

The modification (M) of a malt sample is calculated as follows:

$M = (0 \times \text{"0"} + 0.125 \times \text{"1"} + 0.375 \times \text{"2"} + 0.625 \times \text{"3"} + 0.875 \times \text{"4"} + 1 \times \text{"5"})\%$ where "0", "1", "2", "3", "4", "5" represent the number of grains in each modification group (or if less than 100 grains are examined, the percentage of grains in each group). The modification is reported in percent as an integer (rounded up).

The homogeneity (H) of a malt sample is calculated as follows:

$$H = (100 - 2\sqrt{100A - M^2})\%,$$

where $A = (0)^2 \times \text{"0"} + (0.125)^2 \times \text{"1"} + (0.375)^2 \times \text{"2"} + (0.625)^2 \times \text{"3"} + (0.875)^2 \times \text{"4"} + (1)^2 \times \text{"5"}$ and M is the modification percentage. The homogeneity is reported in percent as an integer (rounded up).

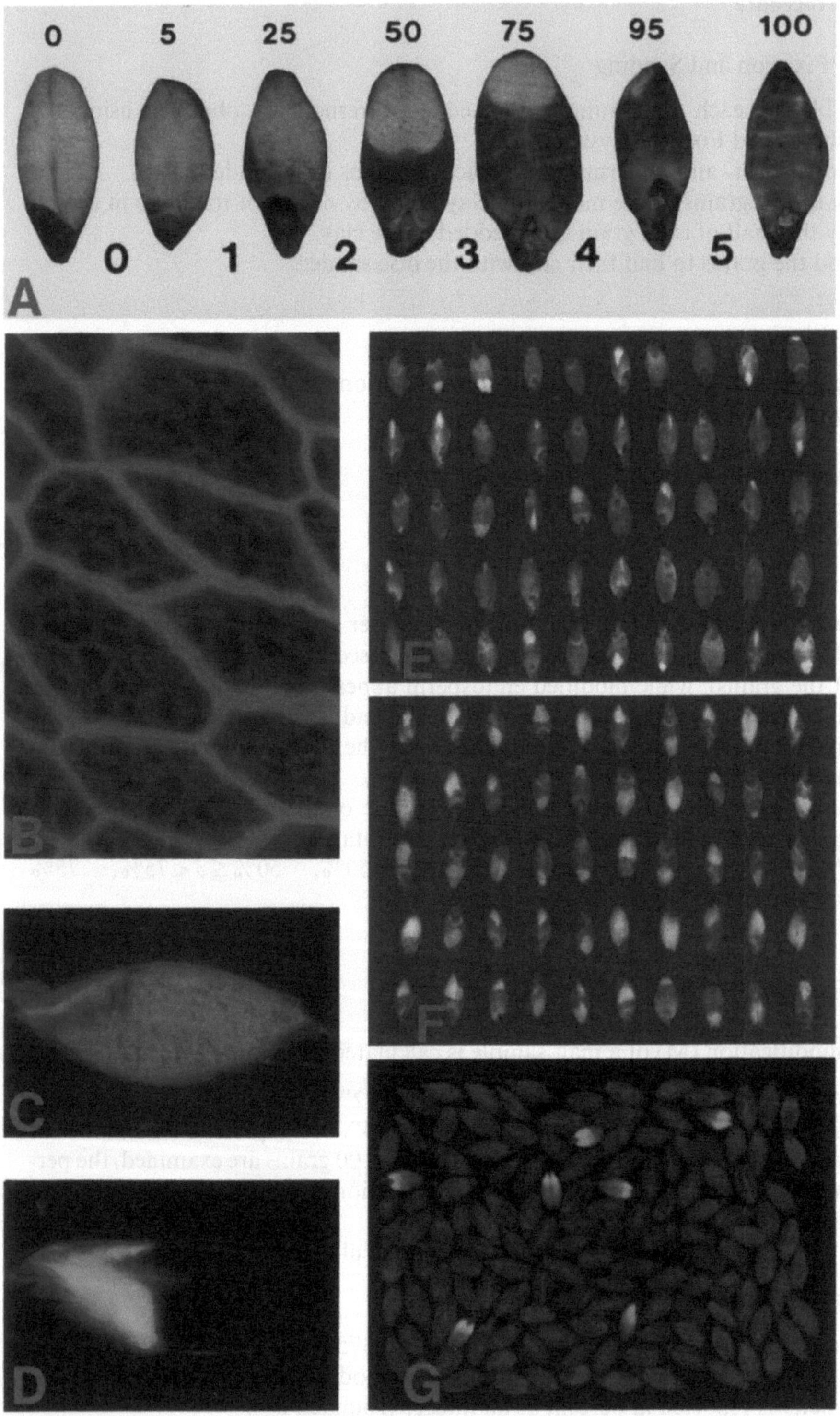
0
5
25
50
75
95
100
0
1
2
3
4
5
A
B
C
D
E
F
G

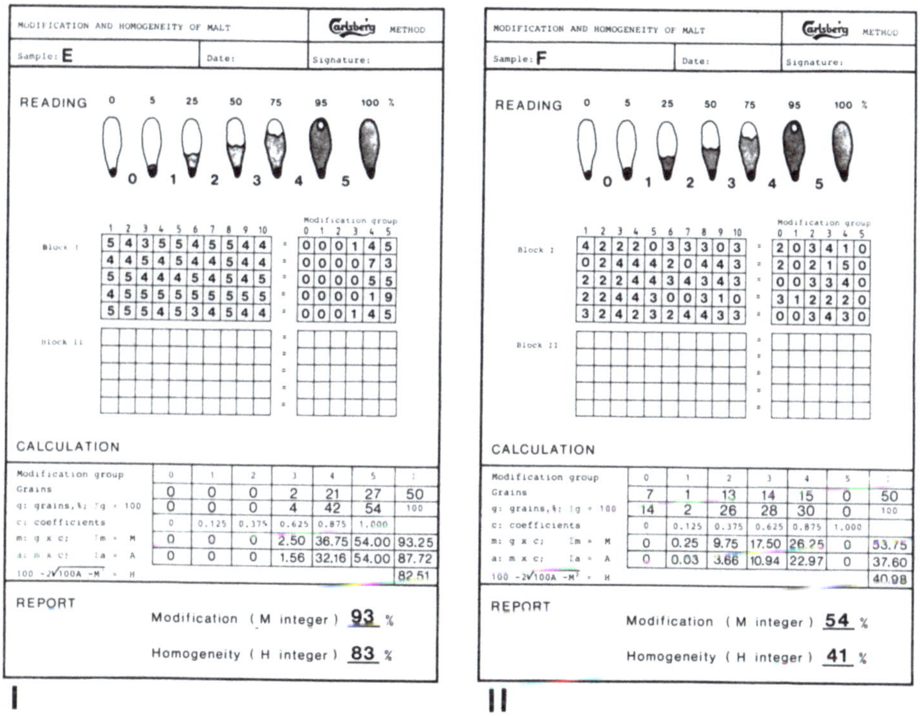

MODIFICATION AND HOMOGENEITY OF MALT — Carlsberg METHOD

Sample: E | Date: | Signature:

READING 0 5 25 50 75 95 100 %

0 1 2 3 4 5

Block I

1	2	3	4	5	6	7	8	9	10		Modification group 0	1	2	3	4	5
5	4	3	5	5	4	5	5	4	4	=	0	0	0	1	4	5
4	4	5	4	5	4	4	5	4	4	=	0	0	0	0	7	3
5	5	4	4	4	5	4	5	4	5	=	0	0	0	0	5	5
4	5	5	5	5	5	5	5	5	5	=	0	0	0	0	1	9
5	5	5	4	5	3	4	5	4	4	=	0	0	0	1	4	5

Block II

CALCULATION

Modification group	0	1	2	3	4	5	Σ
Grains	0	0	0	2	21	27	50
g: grains,%; Σg = 100	0	0	0	4	42	54	100
c: coefficients	0	0.125	0.375	0.625	0.875	1.000	
m: g x c; Σm = M	0	0	0	2.50	36.75	54.00	93.25
a: m x c; Σa = A	0	0	0	1.56	32.16	54.00	87.72
$100 - 2\sqrt{100A - M^2} = H$							82.51

REPORT

Modification (M integer) 93 %

Homogeneity (H integer) 83 %

I

MODIFICATION AND HOMOGENEITY OF MALT — Carlsberg METHOD

Sample: F | Date: | Signature:

READING 0 5 25 50 75 95 100 %

0 1 2 3 4 5

Block I

1	2	3	4	5	6	7	8	9	10		Modification group 0	1	2	3	4	5
4	2	2	2	0	3	3	3	0	3	=	2	0	3	4	1	0
0	2	4	4	4	2	0	4	4	3	=	2	0	2	1	5	0
2	2	2	4	4	3	4	3	4	3	=	0	0	3	3	4	0
2	2	4	4	3	0	0	3	1	0	=	3	1	2	2	2	0
3	2	4	2	3	2	4	4	3	3	=	0	0	3	4	3	0

Block II

CALCULATION

Modification group	0	1	2	3	4	5	Σ
Grains	7	1	13	14	15	0	50
g: grains,%; Σg = 100	14	2	26	28	30	0	100
c: coefficients	0	0.125	0.375	0.625	0.875	1.000	
m: g x c; Σm = M	0	0.25	9.75	17.50	26.25	0	53.75
a: m x c; Σa = A	0	0.03	3.66	10.94	22.97	0	37.60
$100 - 2\sqrt{100A - M^2} = H$							40.98

REPORT

Modification (M integer) 54 %

Homogeneity (H integer) 41 %

II

Fig. 3. Evaluation of the well-modified malt shown in Fig. 2 E (**I**) and the badly modified malt including some unmodified kernels shown in Fig. 2 F (**II**)

2.7 Use of Malt Modification Analysis

The method spotlights the remaining β-glucan-rich cell walls in the endosperm of the individual malt kernels. That is, it is possible to *see* one of the most troublesome components in the brewing industry, namely the β-glucan, and forecast if problems are going to arise or not.

The general use of the method will be for control of final malt, e.g. at the United Breweries in Copenhagen it is a routine analysis carried out on all malts. As a rule of thumb, a good malt will have a modification greater then 80% and a homogeneity greater than 50%. In addition the method can be used to study

Fig. 2. A Halved barley malt kernels stained with Calcofluor and Fast Green, and viewed under UV-light, representing increasing degree of modification. *0* 0–5%; *1* 5–25%; *2* 25–50%; *3* 50–75%; *4* 75–95%; *5* 95–100%. **B** Close-up of barley endosperm cell walls stained with Calcofluor and Fast Green, and viewed under UV-light. **C, D** Two halves of the same barley malt kernel stained with Calcofluor and Fast Green (**C**), and fluorescein dibutyrate (**D**), respectively. Viewed under UV-light. **E** Well-modified malt stained with Calcofluor and Fast Green, and viewed under UV-light. **F** Badly modified malt including some unmodified kernels stained with Calcofluor and Fast Green, and viewed under UV-light. **G** Barley sample containing 7 pre-sprouted grains stained with fluorescein dibutyrate, and viewed under UV-light

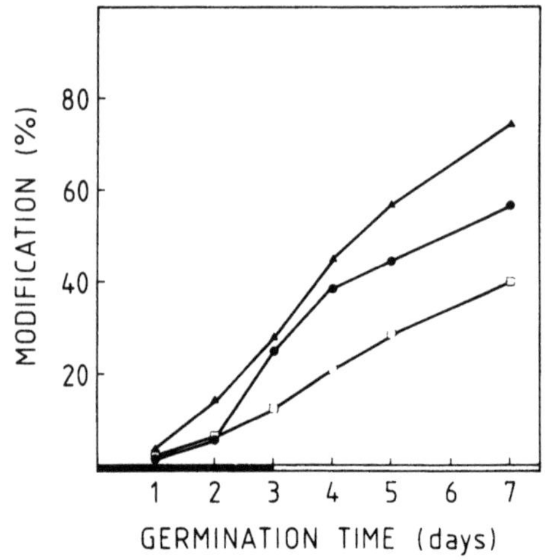

Fig. 4. Relationship between germination time and modification of the high β-glucan barley variety Minerva (□) and its two low β-glucan mutants, M-737 (●) and M-1460 (▲)

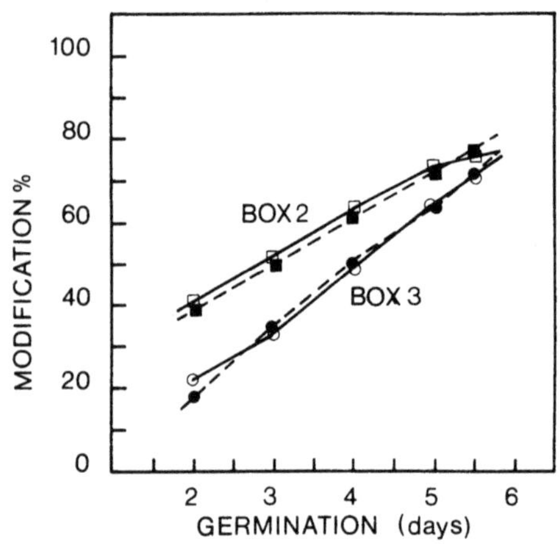

Fig. 5. Relationship between germination time and modification for two different commercial germination boxes (60 tons each). *Solid* and *broken lines* represent replicates of the same variety germinated in the same germination box

the performance of different barleys (Fig. 4, Aastrup 1983) or the influence of different treatments (Fig. 5, Gibbons et al. 1983). In these cases samples of green malt are taken once a day during germination and dried prior to analysis.

The method does also have a great potential in the breeding of malting barleys. To save time of analysis the modification test can be used as a first screening. A typical result is illustrated in Fig. 6 (Aastrup 1987). The barleys showing a modification lower than 70% malted too slowly and gave not acceptable malt analyses. Samples which in addition to the low modification figures show low homogeneity figures might suffer from dormancy (shown as unmodified kernels) and should be tested again after the dormancy has been broken. Barleys showing a modification greater than 80% gave acceptable malt analyses. Samples having a modification greater than 95% are overmodified and should be run again at a shorter programme to give a fair evaluation. Only the samples with a modifica-

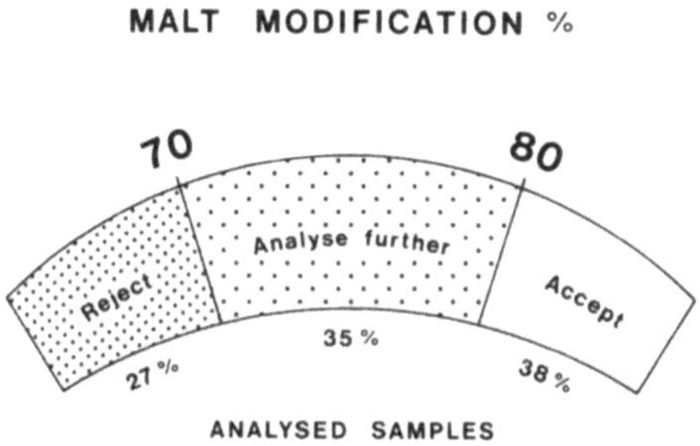

Fig. 6. Trisection of potential malting barleys from a breeding programme using malt modification analysis

tion between 70% and 80% showed no relation between modification and malt analyses, and should consequently be submitted to a complete malt analysis.

3 Pre-Sprouting in Cereal Seeds

3.2 Field of Application

The method is designed to detect enzyme activity which can damage germination ability and baking quality. The method can typically be used as a check method of cereals at elevators before storage.

3.2 Sample Size

A small number of sprouted kernels can be detrimental for baking quality and more than 5% is unacceptable for samples to be used for seed production or malting. To obtain reproducible results 450 kernels have to be analysed.

3.3 Reagents and Materials

5×10^{-3} Molar solution of fluorescein dibutyrate (Serva, W. Germany) in 80% V/V aqueous ethanol.

3.4 Apparatus (Fig. 1)

- Carlsberg Seed Fixation System including a press supplied with a multi-seed matrix and a disc sander. The multimatrix will contain about 150 barley grains, 180 triticale grains, 260 rye grains and 300 wheat grains.
- Modelling clay blocks.
- Malt Modification Analyser equipped with a yellow filter.

3.5 Procedure

3.5.1 Fixation and Sanding

Two or three blocks, each containing 150–300 halved cereal seeds, are obtained using the Carlsberg Seed Fixation System.

- Use the over- and undermatrix to place the grains (counted or estimated by volume or weight).
- Fix the grains in the modelling clay block by means of the press in such a way that half of each grain is embedded in the clay.
- Sand the grains to half their size with the disc sander.

3.5.2 Staining

Stain the half seeds with the fluorescein dibutyrate solution for 2 min. Wait for 15 min for evaluation of the fluorescence [or for quick development, expose the block to hot water (60 °C) for 1 min].

3.5.3 Examination (Fig. 2)

Place the block in the Malt Modification Analyser and switch on the UV light.

Pre-germinated grains show a strong yellow fluorescence in the endosperm adjacent to the embryo, while ungerminated grains do not fluoresce, but might have some white areas (Fig. 2 D, G).

3.6 Expression of Results

The pre-sprouting is reported in percentage as an integer (rounded up).

3.7 Use of Pre-Sprouting Test

The method spotlights in a semi-quantitative way the enzyme activity in the endosperm of the individual seeds, thus pointing out the trouble-makers for the baking industry and the seed producers. The general use of the method will be for quality control of cereals. Grains that are heavily sprouted in the field are not likely to germinate again (Aastrup 1987) making, for example barley samples containing such grains not acceptable for malting. The enzymes produced during germination, in particular α-amylase, have a deleterious effect on the processing quality of milled flour. For example, less than 1% heavily sprouted kernels in a wheat sample have a substantial effect on the falling number (Kruger and Tripples 1980) and thus on baking quality. Table 1 gives a comparison of methods on wheat samples. The FDB method is seen to be well correlated to visual inspection as well as to falling number and α-amylase activity (Jensen et al. 1984).

Table 1. Proportion (%) of sprouted wheat kernels (determined by grain inspectors) in relation to the proportion (%) of sprouted kernels determined by the FDB method, the falling number and the α-amylase activity

Percentage sprouted (assessed visually)	Percentage sprouted (FDB method)	Falling number	α-Amylase activity (mh maltose/min/mg sample)
6.8	8.6	65	525.4
11.8	7.0	105	336.4
3.3	4.9	105	274.7
0.3	0.7	190	80.4
1.4	0.4	210	40.8
0.3	0.4	290	24.3
0.4	0.2	360	17.3
0	0	410	6.2
0	0	530	3.2

4 Further Applications of the Carlsberg Seed Fixation System

The Carlsberg Seed Fixation System and the Malt Modification Analyser were originally developed for malt modification analysis. The sample preparation, however, can be used for any kind of mass screening which includes an opening of the seed. So far the Seed Fixation System, in addition analysis of malt modification and pre-sprouting, has been used for the following purposes:

a) To evaluate the vitrousness of barley grains. This is the simplest test as no staining is involved (Thomsen and Aastrup, unpublished).
b) To distinguish between presence or absence of proanthocyanidins in barley and malt using vanillin-HCl staining (Aastrup 1985; Jende-Strid this volume).
c) To select new proanthocyanidin-free barleys using vanillin-HCl staining (Kristensen and Aastrup 1986; Jende-Strid this volume).
d) To distinguish between dead and living seeds using tetrazolium staining (Aastrup 1987).
e) To select new high lysine barleys using fluorescent antibodies (Rasmussen this volume).

References

Aastrup S (1983) Selection and characterization of low β-glucan mutants from barley. Carlsberg Res Commun 48:307–316
Aastrup S (1985) A test for presence or absence of proanthocyanidins in barley and malt. Carlsberg Res Commun 50:37–42

Aastrup S (1987) Quick, reliable and simple check methods for barley and malt using Carlsberg Seed Fixation System. ASBC J (in press)
Aastrup S, Erdal K (1980) Quantitative determination of endosperm modification and its relationship to the content of 1,3:1,4-β-glucans during malting of barley. Carlsberg Res Commun 45:369–379
Aastrup S, Gibbons GC, Munck L (1981) A rapid method for estimating the degree of modification in barley malt by measurement of cell wall breakdown. Carlsberg Res Commun 46:77–86
Carnielo M, Foucaut A, Moll M (1982) Application of a test for modification in green and kilned malts. Brauwissenschaft 35:168–170
Gibbons GC (1980) On the sequential determination of α-amylase transport and cell wall breakdown in germinating seeds of *Hordeum vulgare*. Carlsberg Res Commun 45:177–184
Gibbons GC, Jensen SAa, Aastrup S, Guildal T, Olsen O, Damstrup J (1983) Practical application of fluorescent seed-analysis systems in a commercial malthouse. ASBC J 41:117–120
Guilbault GG, Kramer DN (1964) Fluorometric determination of lipase, acylase, alfa and gamma-chymotrypsin and inhibitors of these enzymes. Anal Chem 36:409–412
Heltved F, Aastrup S, Jensen O, Gibbons GC, Munck L (1982) Preparation of seeds for mass screening. Carlsberg Res Commun 47:291–296
Jensen SAa, Aastrup S (1985) Determination of malt modification, Cerevisia 10:113–123
Jensen SAa, Heltved F (1982) Visualization of enzyme activity in germinating cereal seeds using a lipase sensitive fluorochrome. Carlsberg Res Commun 47:297–303
Jensen SAa, Heltved F (1983) An improved method for the determination of pregerminated grains in barley. Carlsberg Res Commun 48:1–8
Jensen SAa, Munck L, Kruger JE (1984) A rapid fluorescence method for assessment of pre-harvest sprouting of cereal grains. J Cereal Sci 1:187–201
Kristensen H, Aastrup S (1986) A non-destructive screening method for proanthocyanidin-free barley mutants. Carlsberg Res Commun 51:509–513
Kruger JE, Tripples WH (1980) Relationships between falling number, amylograph viscosity and α-amylase activity in Canadian wheat. Cereal Res Commun 8:97–99
Munck L, Gibbons GC, Aastrup S (1981) Chemical and structural changes during malting. Proc Eur Brew Conv Congr, Copenhagen, pp 11–30
Wood PJ, Fulcher RG, Stone BA (1983) Studies on the specificity of interaction of cereal cell wall components with Congored and Calcofluor. Specific detection and histochemistry of $(1\rightarrow3)$, $(1\rightarrow4)$-β-D-glucan. J Cereal Sci 1:95–110

Hop Analysis

L. C. VERHAGEN

1 Introduction

Hops belong to the botanical family Cannabinaceae, which consists of two classes: *Humulus* and *Cannabis*. Humulus comprises two varieties: ordinary hop *(Humulus lupulus)* and the Japanese hop *(Humulus japonicus)*. There is only one sort of *Cannabis*, the Indian Hemp *(Cannabis sativa)*.

The *Humulus lupulus* grown in Europe and the United States provide the many different bitter substances which are the basis for beer bittering. Hop is a perennial plant, with a long life. It has subterranean parent roots and suckers with a lenght of several meters, with aerial parts, secondary shoots, leaves and cones. For the brewer the female plants are most important, because they have a far greater number of resin-containing lupulin glands than the male plants. The substances of interest for the brewery are found in these lupulin glands. This oily resinous material, which is yellow in color, is at first transparent, but becomes opaque as ripening progresses. It is composed of a mixture of resins which are chiefly responsible for producing the bitter taste of beer and of essential oils which contribute significantly both to the aroma and the flavour of beer. The biosynthesis of bitter substances in hops takes place during the 6–8 weeks between flowering of the hops and harvest (Drawert and Beier 1973). There are several varieties of hops, like Northern Brewer, Hallertau, Saaz, Brewers Gold, Bullian, Jakima, etc., named after the area where they are grown or after the person who introduced them.

2 Chemical Composition of Hops

When harvested, hops normally contain about 80% moisture. This reduces to 10–11% by drying, in order to ensure storability. The chemical composition of air-dried hops is mentioned several times in the literature. An up-to-date version is presented in Table 1, which also gives the relative importance of the various compounds.

The first important chemical report on hops can be attributed to Hayduck (1888), who separated the soft resins into two groups of compounds: those that precipitate as their methanol-insoluble lead salts which he called α-acids and the fraction that is in the filtrate, the β-acids. This latter fraction hardened on standing in contact with air. The general structure of both α- and β-acids was determined by Wieland (1925a, b), Wieland and Martz (1926), and Wöllmer (1925).

Table 1. The chemical composition of hops. (Verzele 1986)

Components	Amount %	Relative importance
α-Acids	2–12	× × ×
β-Acids	1–10	× ×
Essential oil	0.5–1.5	× ×
Polyphenols	2–5	× ×
Oil and fatty acids	Traces up to 25	×
Wax and steroids	–	×
Protein	15	
Cellulose	40–50	
Water	8–12	
Chlorophyll	2	
Salts (ash)	10	

Later it was shown that both α- and β-acids are in fact mixtures of homologues and analogues (Rigby and Bethune 1952 a, b; Verzele 1958). There are three major ones, which are called cohumulone, humulone and adhumulone in case of α-acids, and colupulone, lupulone and adlupulone for the β-acids. Figure 1 gives the structure of both α- and β-acids.

The system of nomenclature that is proposed jointly by the European Brewery Convention (EBC) and the American Society of Brewing Chemists (ASBC) is as follows:

Total resins: that portion of an ether extract of hops that is soluble in cold methanol.

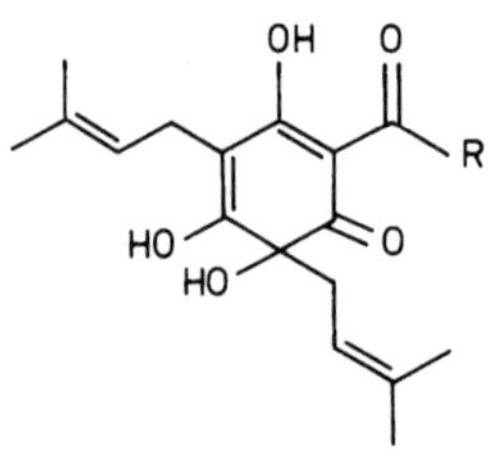

humulones (α-acids)

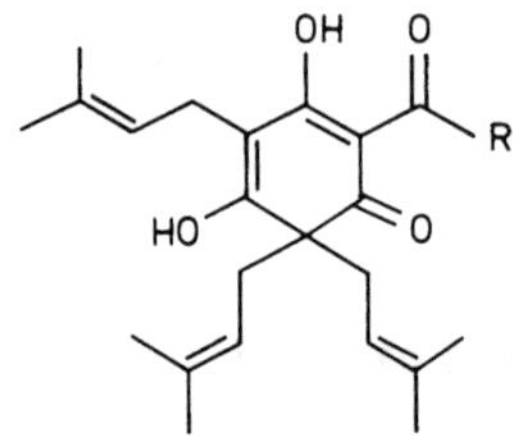

lupulones (β-acids)

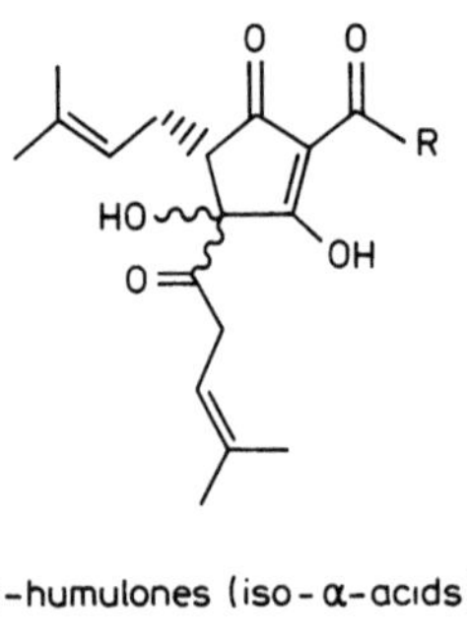

iso-humulones (iso-α-acids)

Fig. 1. Structural formulas of hop bitter acids

$R = -CH_2CH(CH_3)_2$	: –
$-CH(CH_3)_2$	: co
$-CH(CH_3)CH_2CH_3$	: ad
$-CH_2CH_2CH(CH_3)_2$	: pre
$-CH_2CH_3$	: post

Hard resins: that portion of the total resins that is insoluble in hexane.

Soft resins: that portion of the total resins that is soluble in hexane.

Figure 2 schematically represents the different hop fractions (Anderegg 1976):

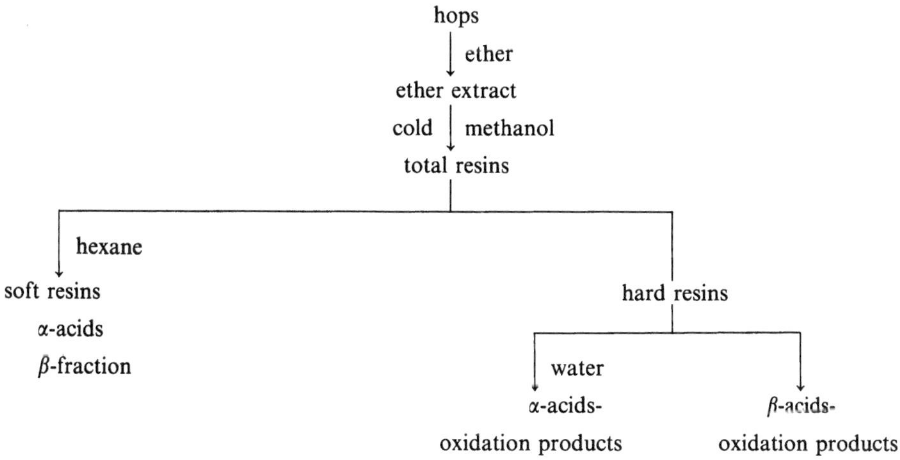

Fig. 2. Hop fractionation scheme

For the brewery the most important compounds of hops are the α-acids. These resinous compounds are chemically converted during the brewing process into the beer bitter substances, which are called the iso-α-acids (Fig. 1). Each α-acid gives a cis/trans stereo-isomeric pair. There are three major α-acids, so beer contains six important iso-α-acids, each contributing to the bitter taste of beer. The total concentration of iso-α-acids in lager beer ranges from 20–30 mg/l.

The importance of the other hop constituents to the brewer is debatable. Some of these will be discussed later in this chapter.

In modern breweries hops are mostly used in the form of extract or pellets. Extracts are prepared by extraction of hops or hop powder with organic solvents. The most commonly used solvents are hexane and methylene chloride. The extracts are concentrated by evaporation of the solvent. This, however, results in extracts that contain small amounts of residual solvents (1–2%). As these solvents are seen as health hazards, the use of other extraction solvents has been advocated. Ethanol and carbon dioxide in particular are interesting alternatives. Carbon dioxide is used either as a liquid or in supercritical form. The polarity of supercritical CO_2 is comparable to that of hexane, so the composition of CO_2 extracts is comparable to that of those prepared with hexane. Ethanol, on the other hand, is very polar. This implies that not only the soft resins are extracted from the hops but also the more polar compounds. The manufacturing of products from hops and their use in brewing has been described by Hildebrand (1979).

3 Analysis

For the brewing industries, the α-acids are of primary importance as these compounds are converted during the brewing process into the so-called iso-α-acids, which give beer its characteristic bitter taste. Therefore analyses of hops are in the first place analyses of the resin content. The analysis of α- and β-acids in hops and hop products is a problem which has confronted workers in the field of hop chemistry and brewing practice. The methods described or referred to in this chapter are only a selection of the numerous contributions in this field. The development of hop analysis methods has always been made difficult, as absolute calibration is impossible because the different α- and β-acids are not available in a pure form. It is possible, however, to isolate and purify the compounds. Of the α-acids the most easily available homologue is the humulone. It can be isolated from the α-acid mixture by repeated crystallization of the 1:1 complex with O-phenylene diamine. After at least ten repeated crystallizations, a small portion of the 1:1 complex remains, which is decomposed in a two-phase system of dilute hydrochloric acid and freshly distilled ether, giving a fairly pure humulone in the ether layer. The possibilities for synthezing α-acids are very limited. Synthesis of hop bitter acids was carried out mainly by Riedl (1951, 1954). The yields in these syntheses were fairly low. Recently a method was published to increase the yield of such synthesis (Pfenninger et al. 1975).

The chemistry of hops and its relevance to brewing has been reviewed extensively by Stevens (1967) and Verzele (1979, 1986).

3.1 Resin Analysis of Hops

Analysis methods of hops can be divided into specific and unspecific methods. Early unspecific methods for α- and β-acid analysis are spectrophotometry (Alderton et al. 1954), polarometry and nuclear magnetic resonance (Alderweireldt and Verzele 1964). In order to determine individual hop acids, separation techniques were developed: countercurrent distribution (Howard and Tatchell 1953) and indirect gas-chromatography (Howard and Tatchell 1957). Some of these methods have been reviewed by Hudson (1963) in the previous edition of this series of books.

For a long time, countercurrent distribution (CCD) served as the reference method (Verzele et al. 1967a). It enables good separation of bitter substances, but is very time-consuming and requires large amounts of expensive and analytically pure solvents. Another reference method that has been used is the "paper-strip" method (Verzele et al. 1967b), that enables the separation of α- and iso-α-acids based on their pK value. The method was rather complicated and therefore not suitable for routine purposes.

Thin layer chromatography is also used to separate and quantify bitter substances (Aitken et al. 1968; Franiau and Mussche 1974; Schur and Pfenninger 1975). This method gives a general view of the composition of the sample. Quantitative results, however, are more difficult to obtain.

Separation of bitter acids by means of column chromatography has been widely used and developed. Of great significance in this field was the use of ion-exchange resins (Kokubo et al. 1968). In central Europe much emphasis is lead on the measurement of the total resin, soft and hard resin content of hops.

Recommended methods of analysis of hops and hop products are described by several groups: The European Brewery Convention, EBC (1975), and the Mitteleuropäische Brautechnische Analyse Kommitee MEBAK (1978) and the American Society of Brewing Chemists, ASBC (1976).

There are two methods in this field which differ in their sample pretreatment. One method extracts the bitter substances by distributing them between an acidic-aqueous-methanolic phase and diethylether (Pfenninger et al. 1973; Ganzlin 1975 a–c; Pfenninger and Schur 1968; Schur and Pfenninger 1979). The bitter substances that are extracted by ether are subsequently fractionated according to their varying solubility in cold methanol and hexane (Wöllmer 1930) as total soft and hard resin. In the soft resin fraction, the import α-acids are separated from the β-fraction by the formation of lead salts of the α-acids (conductometric value). The procedures for hops, hop powder and extracts described by the EBC (1975) are stated below (reproduced with permission).

Preparation of Stock Solution. About 10 g of finely ground hops or hop powder is extracted for 30 min with a mixture of 20 ml of methanol and 100 ml of ether. Following the addition of 40 ml of 0.1 N hydrochloric acid, the shaking is continued for another 10 min. After phase separation, 50.0 ml of the ether layer is pipetted and mixed in a flask with 25 ml of methylene chloride. The solvents are removed by rotary evaporation at 70 °C.

The residue is dissolved in 5 ml of methanol added through the gas inlet tube and transferred quantitatively into a 50 ml graduated flask using methanol to rinse the flask and funnel.

The solution is filled up to the mark at 20 °C with methanol. Subsequently the flask is allowed to stand at 0 °C for 1 h to precipitate the hop waxes. The cold methanolic extract solution is quickly filtrated through a filter funnel, fitted with a dry fluted filter paper, placed in the neck of a 50 ml graduated flask. The filtrate is used as the stock solution for further analysis.

Determination of Total Resin. Stock solution (10.0 ml) is pipetted into a clean, dry (kept in a desiccator for at least 30 minutes) and tared 100 ml flat-bottomed flask (neck not greased). The methanol is distilled by using a rotary evaporator at about 70 °C under slight vacuum. To dry the residue in the flask it is removed from the rotary evaporator and attached to a vacuum adaptor. The residue in the flask is dried at 70 °C for 6 min under a high vacuum (less than 15 Torr). The vacuum is released and, using the water pump, air is briefly drawn through the flask. The flask is dried carefully with a soft cloth, then kept for 30 min in a desiccator and weighed.

Calculation:

$$\text{Total resin \% w/w} = \frac{\text{weight of residue (g)} \times 1000}{\text{initial weight (g)}}.$$

Determination of Soft Resin. Methanolic stock solution (10.0 ml), 5.0 ml of methanol and 40.0 ml of n-hexane are mixed in a glass centrifuge tube. 10.0 ml of 0.1 N hydrochloric acid is added. The tube is stoppered and vigorously shaken for about 30 s by hand or 5 min using a shaker, after which it is centrifuged for 2 min at 2000 rpm to separate the phases. 30.0 ml is pipetted into a clean dry 100-ml flask and evaporated in the same way as described for total resin.

Calculation:

$$\text{soft resin \% w/w} = \frac{\text{weight of residue (g)} \times 1333}{\text{initial weight (g)}}.$$

Determination of Conductometric Value. Stock solution (10.0 ml) is pipetted into a wide-necked Erlenmeyer flask or tall beaker followed by 40 ml of dimethylsulphoxide ethanol mixture (6:94). A stirring bar is added and the flask or beaker stirred using a magnetic stirrer. The measuring cell of the automatic titrator is immersed to the correct depth in the liquid. It may be necessary to add more dimethylsulphoxide ethanol mixture. The burette containing lead acetate solution is placed so that the fine tip is immersed to 3 to 4 cm into the test solution. The automatic titrator is switched on and the plot recorded. The intersection of the straight lines for salt precipitation and addition of excess lead acetate solution is the equivalence point. To establish the second straight line it is necessary to continue the titration until the curve shows a second inflection.

Manual Titration. The lead acetate solution is added in 0.25 ml portions and the conductivity measured after each addition is plotted against ml of lead acetate solution used. The titration is continued beyond the equivalence point until a straight line has been obtained. From the graph the volume of lead acetate solution used for precipitation is employed to calculate the conductometric value.

Calculation:

$$\text{conductometric value \% w/w} = \frac{\text{volume of lead acetate (ml)} \times T \times 18.85}{\text{initial weight}}$$

T = titre of lead acetate solution.

Lead Acetate 2%. 20 g $Pb(C_2H_3O_2)_2 \cdot 3H_2O$ is dissolved in 1 l methanol containing 0.5 ml acetic acid or in ethylene glycol monoethyl ether (e.g. Titrisol Merck Cat. No 9897, diluted with one of the solvents mentioned).

Before use, the lead acetate concentration is measured by conductometric titration against 2.0 ml of 0.1 N sulphuric acid in 40 ml of dimethyl sulphoxide ethanol mixture

$$\text{Titre of lead solution (T)} = \frac{1.897}{\text{ml lead acetate solution used}}.$$

The β-fraction and the hard resins are determined as:
β-fraction % w/w = soft resin − conductometric value
hard resin % w/w = total resins − soft resin.

For hop extracts the procedure is very similar:

Preparation of Stock Solution. A sample of hop extract corresponding to about 0.6%–0.7 g total resin is thinly streaked using a spatula on the inside of a stoppered, tared glass centrifuge tube and weighed accurately to 1 mg. 20.0 ml of 0.1 N hydrochloric acid, and 50.0 ml of ether are pipetted into the tube. The centrifuge tube is tightly stoppered and either shaken for 5 min using an automatic shaker or shaken manually until all of the extract on the glass wall has dissolved. The centrifuge tube is carefully opened and 10.0 ml of methanol is added. The centrifuge tube is carefully stoppered and shaken for at least a further 15 min at optimal speed on the shaker, after which it is centrifuged for 2 min at 2000 rpm. 40.0 ml of the clear ether phase is pipetted into a flat-bottomed flask and 20 ml of methylene chloride is added. The solvents are removed by rotary evaporation at 70 °C. Then continue as described for hops.

Calculation:

$$\text{total resin \% w/w} = \frac{\text{weight of residue (g)} \times 625}{\text{initial weight of extract (g)}}$$

$$\text{soft resin \% w/w} = \frac{\text{weight of residue (g)} \times 833}{\text{initial weight of extract taken (g)}}$$

$$\text{conductometric value \% w/w} = \frac{\text{volume of lead acetate (ml)} \times T \times 11.78}{\text{initial weight (g)}}$$

3.2 The Conductometric Lead Value of Hops

The bittering capacity of freshly harvested hops is almost entirely due to the α-acids.

However, when hops deteriorate during storage, the α-acid content declines, mainly due to oxidation. Some oxidation products of both α- and β-acids contribute to the beer bitterness. Many oxidation products also react with lead acetate. Therefore, especially for old hops the "conductometric lead value" probably gives a more reliable indication of bittering capacity. It is determined as follows:

About 10 g of hops or hop powder is weighed and broken up in a square glass jar containing 100 ml of toluene. The jar is flushed with nitrogen closed and macerated for ten periods of 30 s. each. The content is allowed to settle for 5 min and the supernatant decanted through a funnel with glass wool. 15 ml is pipetted into a beaker and 40 ml of methanol is added. This mixture is titrated conductometrically with lead acetate in methanol.

Calculation:

$$\text{L.V.C.} = 12.60 \times \frac{V}{W} \times T$$

V = ml of lead acetate
W = weight of sample
T = titre of lead acetate solution.

3.3 Verzele Conductometric Analysis of α-Acids

The EBC also describes another conductometric method for α-acid analysis in hop (Verzele and Van Dyck 1971). In this method hop extract is shaken with toluene, together with a buffer solution at pH 7. This gives a selective extraction of α-acids into the toluene, while other bitter substances pass into the buffer. The toluene phase is diluted with methanol and dimethylsulfoxide (DMSO) and titrated conductometrically with lead acetate. Instead of fractionation of the hop resins before titration based on solubility of the α-acids, selectivity is reached by using a pH 7 buffer to extract interferences into the aqueous phase. This method has been criticized by Forster et al. (1977), who stated that extraction of the toluene phase with buffer pH 7 resulted in losses of α-acids while non-α-acids (oxidation products) were also co-extracted and titrated. The method was subsequently reexamined (Verzele et al. 1980). By HPLC it was possible to prove that at pH 7 α-acids were indeed extracted from the toluene phase. Originally the use of pH 7 was chosen to remove oxidation products that are also lead-titratable. It is not known whether these oxidation products contribute to beer bitterness in a positive or negative sense. The improved procedure for α-acid analysis of hops according to Verzele is as follows (Verzele et al. 1980):

10 g of hops, hop pellets, milled hops, hop powder, etc., are placed in a plastic container of 250 ml together with five 1-inch glass beads and 100 ml of toluene containing 1% butylhydroxytoluene (BHT). The container is shaken vigorously for 30 min. Sufficient liquid is poured into a closed centrifugation tube and centrifuged for a short time. 25 ml of the clear toluene solution are pipetted into a 80–100 ml centrifuge tube together with 10 ml pH 6.0 buffer. The toluene layer is shaken with the buffer for a few minutes. 10 ml of the toluene layer are pipetted in a beaker containing DMSO (10 ml), methanol (10 ml) and toluene (10 ml). This mixture is conductometrically titrated with lead acetate. The α-acids are calculated according to the formula:

$$\%\,\alpha\ \mathrm{w/w} = \frac{\mathrm{ml\ Pb(Ac)_2} \cdot 50 \cdot \mathrm{titre\ of\ Pb(Ac)_2} \cdot 357.8}{\mathrm{weight\ of\ sample\ (g)} \cdot 20 \cdot 379.3}.$$

The conductometric methods described above and the many modifications made by several groups have a disadvantage in that they are only moderately specific. Determination of individual α-acids is not possible and on ageing of hops the results become debatable, as oxidation products of α and/or β-acids also respond to the conductometric method. Reliable figures on the α-acid content of hops with the above methods can therefore be obtained only on fresh hops.

In order to determine the α-acids more specifically, or to determine them individually, it is necessary to separate the hop bitter acids by chromatographic methods.

Separation of the different hop bitter acids on ion exchange resins is often used and several modifications in the original work by Kokubo et al. (1968) and Hansen et al. (1971) have been developed. Otter et al. (1972) used a solvent gradient to elute the hop bitter acids. The separated compounds were collected and quantified by measuring the absorbance at the UV maximum of each compound. This method was modified by Anderegg et al. (1977), who also determined the extinction coefficient of α-acids with pure humulone, and Ganzlin (1978), who improved the separation in order to also determine the β-acids more specifically and modified the method to allow determination of specific hop compounds, viz. α-, β- or iso-α-acids.

Both Analytica EBC III and MEBAK describe the determination of specific hop compounds and their derivatives (International Method). One method describes the separation of individual compounds such as α-, iso-α-, humulinic and β-acids on an ion-exchange resin column (QAE Sephadex A 25), by elution with an acetic acid-methanol gradient. The eluted fractions containing the individual components are collected and their concentration measured by spectrophotometry at the absorption maximum. Sample preparation for hops and hop powder is essentially the same as described before, that is distribution of the bitter acids between an acidic-aqueous-methanolic phase and di-ethyl ether. An aliquot of the ether phase is diluted with methanol and applied to the column. Hop extracts are dissolved in methanol.

Specific α-acid analysis can be performed by isocratic elution of the column with 10% acetic acid-aqueous methanol.

Although the above methods improved the selectivity of the analysis they are still rather complicated and laborious. Application of the more specific HPLC analysis of hops will be discussed in Sect. 3.5.

3.4 Beer Bitterness Analysis

The measurement of the bitterness of beer is classically done by spectrophotometry of an iso-octane extract of beer. The procedure described by the EBC (1975) is as follows:

Beer is degassed without loss of foam. 10 ml of the sample is pipetted into a 35-ml centrifuge tube, and 0.5 ml of 6 N hydrochloric acid is added, followed by 20 ml iso-octane and 2 or 3 glass balls. The tube is screwcapped and shaken for 15 min at 20 °C $\pm$ 1 °C using a rotary shaker set at 130 $\pm$ 5 rpm. The tube is then centrifuged for 3 min at 3000 rpm. The extinction of the iso-octane layer in a 1 cm cell is measured at 275 nm, using pure iso-octane in the reference cell.

Calculation:

bitterness units (BUs) $= 50 \times E_{275}$.

The results are influenced by the presence of substances in the beer that are extracted into iso-octane and absorb at 275 nm. Therefore the selectivity of the analysis method is not very high. This can be increased by using separation techniques such as HPLC.

3.5 High Performance Liquid Chromatography (HPLC)

The main drawback of the above methods for hop analysis is either a lack of selectivity, insufficient accuracy or laboriousness. Also, these methods are not equally applicable to all forms of hops and hop-derived products. The development of HPLC as an analytical tool has provided an excellent method for separation and quantification of hop resin constituents present in hops, hop products and beer. There are numerous papers on the analysis of bitter acids by HPLC.

Different types of separation can be employed in HPLC, such as ion-exchange chromatography (IEC), liquid-solid or adsorption chromatography (LSC), liquid-liquid partition chromatography (LLC), and gel-permeation chromatography (GPC).

Except the latter, all separation mechanisms mentioned above have been employed for the separation of hop bitter compounds by HPLC. The methods described in the previous part of this chapter were used to determine only a certain class of bitter acids in hops or hop products. HPLC has equally provided methods for the separation and quantification of quite a range of bitter acids in a relatively short time. Several of the papers therefore are dealing with the separation of both α-, β-, and iso-acids. One of the first papers dealing with HPLC of bitter acids (Molyneux and Wong 1973) described the separation of different bitter acids on a 1 m long column packed with silica gel (Corasil II) with a solvent system consisting of various proportions of iso-octane-ethylacetate. This system allowed the separation of α- and β-acids in about 6 min.

It was not possible, however, to separate the individual homologues of either α- or β-acids. Adsorption chromatography on silica columns was also used by Kleber and Hums (1974) and Siebert (1976). The separations on silica gel are essentially a translation of the thin layer chromatographic technique that has been developed for hop compounds. Problems in adsorption chromatography are described by Siebert (1976). It is often difficult to cause retention for both β-acids and α-acids. Another problem is the difference in solvent strength required to elute both α- and iso-α-acids. In order to elute the different compounds the use of gradient elution was suggested. Gradient elution adsorption chromatography, however, has a major disadvantage in that it is often quite irreproducible. A serious disadvantage of adsorption chromatography is the profound effect of the water content of the mobile phase on the retention of the compounds. All these effects cause the need for a very long re-equilibration period between gradient runs. Another problem in adsorption chromatography is the considerable peak tailing of the hop acids. Several authors have described the use of buffered silica gel columns to reduce peak tailing (Slotema et al. 1977; Claus et al. 1978; Schulze et al. 1979; Schwarzenbach 1979). Buffered silica gel columns can be prepared by treating silica gel with a solution of buffer salts and acids. The treated silica gel is subsequently dried and the dry material is packed in a column. Alternatively, the silica column can be flushed with an aqueous buffer solution followed by an immiscible solvent. Separation between α-acids, iso-α-acids, and β-acids was described by Gill (1979), who used a solvent containing di-n-butylammonium acetate which also reduced peak tailing.

Ion exchange chromatography has been applied quite often, especially in the starting period of HPLC. As described earlier in this chapter, low pressure ion-exchange chromatography has found wide application in the separation of hop bitter compounds, primarily as a result of the work of Kokubo et al. (1968). One of the first papers on this subject described the separation of hop acids on a polystyrene-divinyl benzene resin (Vanheertum and Verzele 1973). They were able to separate the major groups of hop compounds, but a very long column was required (4 m). The elution order of the different compounds was found to be quite different from that reported on low pressure IEC. Siebert (1976) reported on the use of a small macro reticular bead anion exchange resin, Aminex A28. Separation between the hop compounds was achieved with a potassium acetate-acetic acid buffer in 80% methanol-water in about 2 h time. The peak shape of the different compounds was rather poor however. Ion-exchange chromatography was also described by Slotema et al. (1977) using a commercial 10 μ anion exchange material (Partisil SAX). Elution with 90% methanol/buffer pH 4.8 resulted in separation of the different hop compounds, such as colupulone, humulone, cohulupone, trans-iso-humulone and trans- and cis-humulinic acid. It was reported that the elution order on other anion exchange resins (Nucleosil and RSL) was quite different.

Otter and Taylor (1977) described the development of a totally porous weak ion-exchange silica (phenyl diethanolamine silica) which had useful properties for the resolution of hop components by HPLC. The mobile phase used consisted of 0.001 *M* or 0.1 *M* citric acid in varying methanol-water mixtures. It was described that reversed-phase partition is the major effect necessary for the separation of

α-, β- and humulinic acids, and ion pairing is important for the resolution of the iso-α-acids.

The ion-exchange column was used to analyze the ratio of the humulone homologues in different hop varieties and the conversion of α-acid into iso-α-acid during the hop boil (Otter and Taylor 1978).

Although the above methods were a distinct improvement compared to classical analysis, they did not show the performance that might be expected from modern liquid chromatography. By far the best results are achieved with reversed phase chromatography, using chemical bonded octadecyl modified silica gel as stationary phase. These phases are quite stable and operationally convenient. Several investigators used reversed phase systems for separating hop bitter acids (Siebert 1976; Slotema et al. 1977).

Superior systems for HPLC separation and quantification of hop bitter acids were described by Verzele and De Potter (1978). They studied an amino phase, a cyano phase, and 12 different ion-exchangers based on chloropropyl silica gel and an apolar octadecyl phase chemically bonded to microparticulate silica gel, for the chromatography of the hop compounds. Both the amino and cyano phases were found to give poor results. Some separations could be performed on the ion-exchange materials, but poor column efficiency and strong tailing of peaks caused a considerable loss of resolution. Octadecyl silica gel was found to be the most useful stationary phase for the analysis of hop bitter acids. Elution of a 5- µm SiO_2–C_{18} phase with buffered mobile phases with methanol and water led to complete separation of all the individual hop acids. Also the five-membered ring iso-α-acids could be separated well using a 1 *M* citrate buffer pH 3.0 in methanol-water 60/40. It was noted that slight differences in the composition of the solvent caused major shifts for some compounds, with resulting loss of resolution.

A quantitative analysis was described using chalkone as an internal standard on 10 µm SiO_2-C_{18}. Detection of the compounds was performed at 334 nm with a variable wavelength detector. At this wavelength the three major α-acids show maximum light absorption. As chalkone has a maximum absorption at 313 nm it was necessary to pay attention to the correct calibration of the detector monochromator. The choice of 334 nm increased the selectivity of the method because iso-α-acids and most of the oxidation products of the bitter acids do not absorb at 334 nm. Calibration was performed with pure humulone. The other homologues, cohumulone and adhumulone, were determined with the same calibration curve since their extinction coefficients in the solvent were found to be the same. The method enabled the analysis of α- and β-acids from the same chromatogram. Although the resolution of the α- and β-homologues was very good, a serious drawback was the long analysis time (90 min). A faster separation was achieved with methanol-water-acetic acid (85 : 15 : 1), but humulone and adhumulone were not separated under these conditions and chalkone was less suitable as an internal standard. For this phase system β-phenylchalkone was used as an internal standard. Detection at 324 nm was used for this solvent system. The method showed a standard deviation of 2–3%, which was rather high, probably due to oxidation of the bitter acids during the chromatographic run. Addition of butyl hydroxy toluene (BHT) to the sample and to the eluting solvent improved the results and a relative standard deviation of 1% was reached. (Verzele and De Potter 1978). An extensive discussion of the HPLC system and the sample preparation was given by Verzele et al. (1980). Slight changes in the mobile phase enabled the use of the more readily available chalkone as internal standard. The mobile phase consisting of methanol-water-phosphoric acid enabled analysis of α- and β-acids in 10 min.

The extinction coefficients of the bitter acids in this solvent system appeared to differ from those published previously (Claus et al. 1978) and were again determined. The three major α-acids were found to have an extinction coefficient of 208 for humulone and adhumulone and 198 for cohumulone at 314 nm, which is near or at a maximum wavelength for both the α-acids and β-acids. Calibration was performed by constructing a calibration curve with increasing weights of pure humulone. This humulone was prepared and purified as described above. In order to obtain quantitative figures for cohumulone, a correction factor for the area was introduced which was the ratio of the extinction coefficient values of humulone over cohumulone $208/198 = 1.050$. For colupulone the value was 163 and for lupulone-adlupulone it was 174. Thus, these β-acids were calculated using corrections of 1.28 and 1.20 respectively. It was stated that wavelength calibration was essential and a method was described using naphtalene in iso-octane. The HPLC column was packed with an octa-decylated silica that was carefully demineralized by boiling the packing material three times with methanol/2 N HCl 60/40 and extensive washing.

This was necessary as α-acids tend to chelate with metal ions. (Verzele and Dewaele 1981). The use of phosphoric acid improved the results as it is a complexing agent. The method for the analysis was as follows:

Hops, Hop Powder, Hop Pellets: Ten g is shaken for ½ h with five 1-inch glass beads in 100 ml toluene containing 1% of BHT. A portion of the mixture is centrifuged in a stoppered tube for a short time (1–2 min). 2 ml of the clear solution is evaporated in a rotavapor and the residue dissolved in 50 ml (or 100 ml if the α-acid level is expected to be above 7–8%) methanol containing 0.25% H_3PO_4 85%, 1% BHT, and a precisely known amount of the internal standard chalkone (15 ml/l).

Hop Extracts. 50–100 mg of the extract is weighed on a polythene foil and dissolved in 100 ml of methanol containing 0.25% H_3PO_4 85%, 1% BHT and 15 mg chalkone per litre.

10 µl of the sample is injected into the HPLC column (25 × 0.46 cm 10 µ RSil-C_{18}). Elution with methanol-water-phosphoric acid 85:17:0.25 containing 1% BHT. Detection at 314 nm.

The authors used the described method for a detailed reexamination of the classical conductometric α-acids analysis methods and compared HPLC results with those obtained by conductometry.

Reversed phase chromatography was also used by Schulze et al. (1981) for separation of humulinic acids and the oxidation products of iso-humulone with a mobile phase system of methanol-water-tetrahydrofuran (THF).

Lance et al. (1981) described a reversed phase system with a mobile phase containing citric acid acting as a complexing agent to overcome irreversible adsorption by metal traces in the packing material. They used octanal-2,4-dinitrophenylhydrazone as internal standard. This compound showed maximum absorption at

362 nm, and eluted after all the major hop compounds. The authors noted a certain instability of the hydrazone, probably due to cis-trans isomerization. Despite this problem relative standard deviations of 1.0% and 2.1% for α- and β-acid respectively were reported as well as general agreement with polarimetric α-acid analysis results. Samples were prepared for both hops and hop extract by dissolving in methanol and filtration. Routine HPLC methods were also reported by Knudson and Siebert (1983), who used simplex optimization as a method for efficient determination of optimal mobile phases for separation of hop-bittering compounds in hops, hop extracts, wort and beer.

A mixture of methanol, acetonitrile, tetrahydrofuran and water containing tetramethyl ammonium hydroxide and phosphoric acid was found to be the optimal mobile phase. Two different eluents were necessary for optimal separation of iso-α and α-acids from beer and wort samples and of α- and β-acids from hops and hop extracts. Benzophenone and 4-methyl benzophenone were useful as internal standard for detection at 280 or 300 nm. Fast HPLC analysis of hop bitter compounds was described by Verzele et al. (1983). Analysis time was reduced to 5 min by using a 10 cm column packed with 5 μm spherical octadecyl silica gel. The internal standard formerly used, chalkone, was not useful in this system as it elutes before the major α acids. Use was made of an internal standard that elutes after the last hop bitter compound, p-nitro anilide of myristic acid. This compound shows maximum absorption at about 314 nm. By using an internal standard which elutes after the hop bitter acids, interference with oxidation products, early eluting in the chromatogram, is avoided. Although the anilide was found to be stable for weeks when dissolved in methanol under neutral conditions, it was noted that under acidic conditions appreciable amounts of methylmyristate may form (Verzele et al. 1984). The same article describes a routine analysis method for iso-α-acids in beer. Iso-octane extracts of acidified beer are directly injected into the HPLC column. For this analysis 270 nm was used for detection and 2,6-ditertiary butylphenol served as internal standard. Figure 3 shows a chromatogram of a beer sample prepared as follows:

14 g beer is weighed in a screwcap tube containing about 5–10 small glass beads. 1 ml of a solution of 2,6-ditertiary butylphenol is added, followed by 1 ml of 3N HCl and 1 ml of iso-octane. The tube is carefully filled up with water, capped, and rotated mechanically for 15 min. Then the tube is allowed to stand for 5 min and slowly opened. Of the upper iso-octane layer 10 μl is injected.

Simultaneous determination of α-, β-, and iso-α-acids in hops and hop products was described by Silvester (1984) using gradient elution. Gradient elution separation of the different hop compounds was also used by Ono et al. (1984, 1985) using mobile phases containing tetraethylammonium hydroxide. They used the method for hop evaluation showing differences in pellets from several varieties (Saazer, Yakima, Cluster and Target) and for studying the utilization of hop α-acids during the brewing process.

Narziss and Scheller (1984, 1985) used reversed phase HPLC for a detailed study of the composition of hops and hop products and their changes during the brewing process. Buckee (1985) used reversed phase HPLC for analysis of α-, β-, and iso-α-acids in hops, hop products and beer with isocratic elution on a polythene cartridge column (radial compression) with 5- μ Nova Pak C_{18} as the sta-

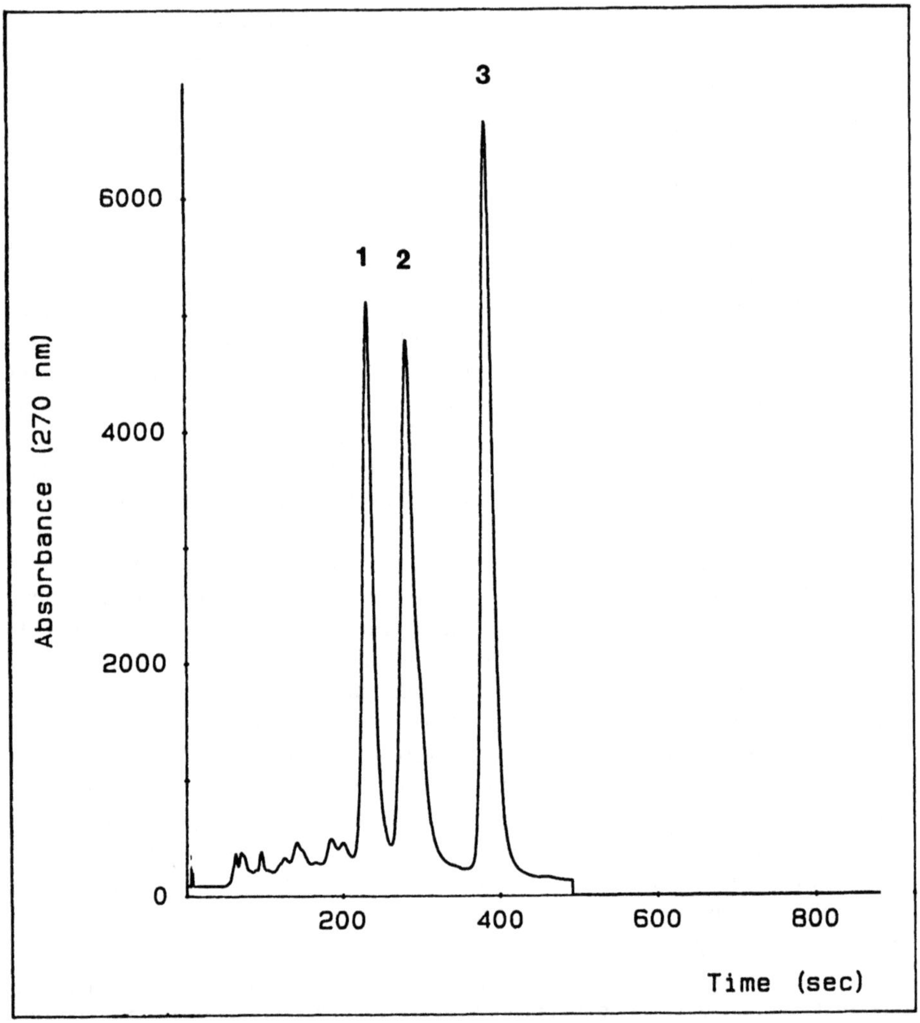

Fig. 3. Chromatogram of an iso-octane extract of beer. Column: 10 × 0.46 cm 5 μ ROSil–C_{18}-D (demineralized). Mobile phase: CH_3CN–CH_3OH–H_2O 45–25–30, 1% H_3PO_4 (85%), 0.5% of cetyltrimethylammoniumhydroxide. Flow rate: 2 ml/min, detection: 270 nm, injection 10 μl, *1* cis/trans iso-cohumulone, *2* cis/trans iso-humulone/adhumulone, *3* internal standard

tionary phase and with the p-nitro-anilide of myristic acid as internal standard. They noticed that after 1000 analyses the column showed no significant loss in resolution.

The HPLC methods described in the literature for analysis of hops and hop extracts have shown that for routine application reversed phase separation with methanol-water-phosphoric acid as mobile phase is the most appropriate. For commercial transactions an accurate analysis method for α-acid is very important. The literature has shown that quantitation of α- and β-acids is hampered by

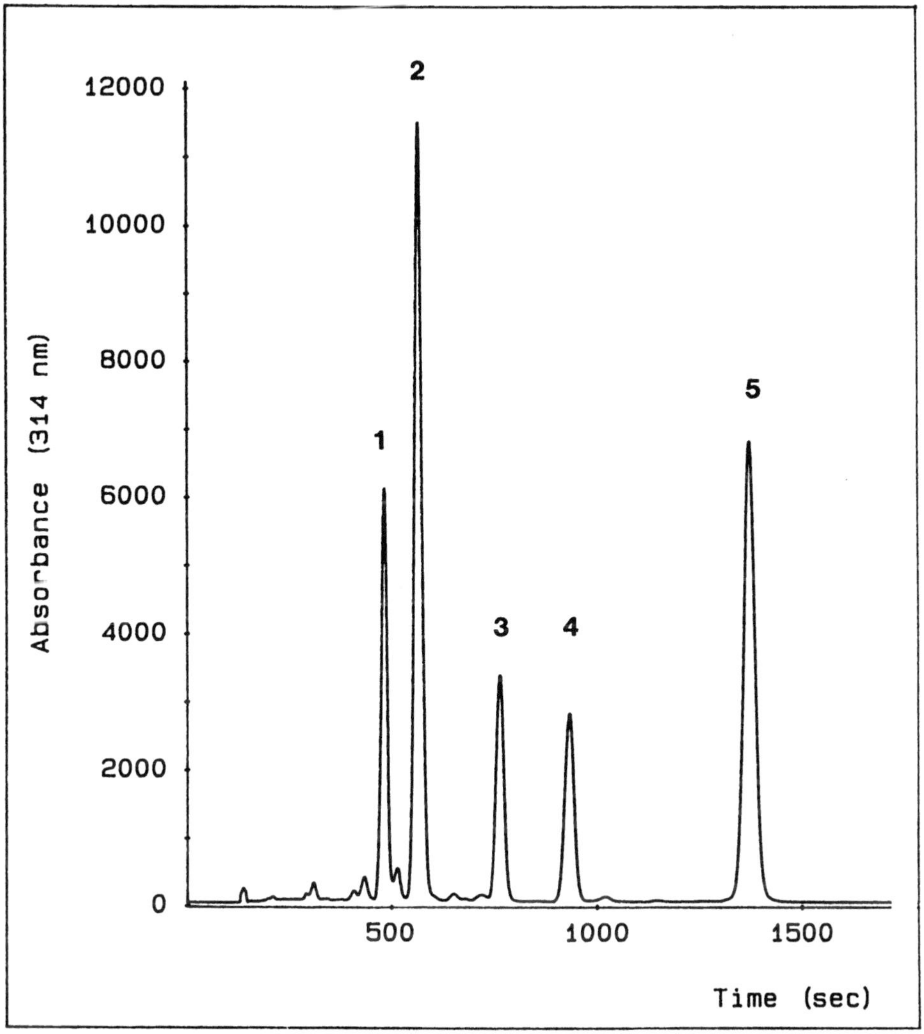

Fig. 4. Chromatogram of hop extract. *1* Cohumulone; *2* humulone + adhumulone; *3* colupulone; *4* lupulone + adlupulone; *5* p-nitro-anilide myristic acid. Column: 25 × 0.4 cm nucleosil 5-C_{18}. Mobile phase: CH_3OH–H_2O–H_3PO_4 (85%) 85–17–0.25. Flow rate: 0.8 ml/min, detection: 314 nm, injection 10 μl

several problems. Quantification is difficult due to the lack of reference compounds. These can only be prepared by isolation and purification from hops. Also the type of stationary phase seems to be important (demineralization and selectivity). Certain types of reversed phase columns do not separate the small peaks that elute just before and after the cohumulone peak. With modern methods it has been possible to identify these compounds as the desoxy-α-acids (van de Velde 1986). It is debatable whether these compounds are transferred into iso-α-acid during wort boiling. Therefore these desoxy-compounds should not be taken as α-acids. The choice of the internal standard is of influence. Ideally, it should have maximum absorbance at the same wavelength as the α- and β-acid, that is at 314 or 340 nm. The most suitable compounds are p-nitro-anilide of myristic acid and the 2,4-dinitrophenyl hydrazone of octanal or decanal.

Figure 4 shows the chromatogram obtained on good quality column with methanol-water-85% H_3PO_4, 85:17:0.25 and the p-nitro-anilide of myristic acid as internal standard. Another approach to quantification is the use of an external standard. This has the disadvantage that sample preparation needs more attention, but instrument and column influences and effects are compensated.

The use of such an external standard has been advocated by the AHA (Arbeitsgruppe für Hopfen-Analyse), who has initiated several ring analyses of hop pellets and hop extracts both with the p-nitro-anilide of myristic acid as an internal standard and with a calibration hop extract as an external standard (Anderegg and Pfenninger 1985). The calibration extract, supplied by the Versuchsstation Schweiz. Brauereien, is claimed to be very stable when stored cold. The results showed that the use of the external standard gave a better standard deviation and interlaboratory agreement.

These results led to the method having been adopted by the EBC as a reference method for α-acid analysis of hops (EBC 1986):

A known amount of hops or hop powder (about 10 g) is extracted with a mixture of 20.0 ml methanol and 100 ml diethylether for 30 min. Then 40.0 ml of 0.1 N Hydrochloric acid is added and extraction is continued for another 10 min.

After phase separation, 5.0 ml of the upper ether phase is pipetted into a 50 ml graduated flask and filled up to the mark with methanol. For extract about 0.5 g of properly homogenized extract is transfered into a 100 ml graduated flask and 50 ml of methanol is added. The extract is dissolved by placing the flask in an ultrasonic bath and subsequently filled up to the mark with methanol.
The solution is centrifuged for 2 min in a closed tube.

Ten μl of the sample is injected into the HPLC column and separated under the conditions given for Fig. 4, but without the internal standard. By injection of an identically prepared sample fo the calibration hop extract at the beginning and the end of the day, specific peak areas, A_s, are determined. The samples are calculated with these specific factors.

As was described earlier, the small peaks just before and after the cohumulone should be separated from it. Slight changes in the water content of the mobile phase may be necessary to keep the separation optimal.

4 Essential Oil of Hops

The essential oils of hops are responsible for the so-called hop flavour of beer. Hop cones contain 0.4–2% by weight of dry matter of essential oils which are mainly localized in the lupulin glands. Hop oil is a very debatable group of substances in hops. The main hop oil constituents can be divided into non-polar hydrocarbons, including the major mono- and sesquiterpenes and the polar oxygen-containing compounds. Of the terpenes the most important compounds are myrcene, humulene and caryophylene, while of oxygen-containing products linalool, humulene epoxide and caryophylene epoxide are seen as important.

Hop oil is relatively volatile. The terpenes, being non-polar, are only slightly soluble in water and therefore steam distilled out of the wort, during wort boiling. The typical aroma of a given hop is mainly due to specific low-boiling constituents of the volatile oil. In beer only traces of the hop essential oils are found, so it is evident that hop aroma substances which are introduced in the wort are removed or chemically changed during the process.

Studies following the boiling of wort with hop cones showed that the flavour effects are due to oxidation products rather than to unchanged hop oil. The chemistry of hop aroma in beer has been described by Peacock and Deinzer (1981).

The composition of the essential oils of hops depends on the variety. Individual varietal differences are related in the first place to the terpene content (Pfenninger et al. 1979).

Determination of the percentage of essential oil in hops is of importance for the brewer.

Steam distillation has been applied by Howard (1970) and Krüger and Baron (1975). Steam distillation-extraction followed by gas chromatography has been used by Godefroot et al. (1981). An official method has been given by the MEBAK (1982):

Fifty g of hops are milled with 1 l of water. The mixture is steam-distilled in a specially designed apparatus containing diethyl ether/pentane 1/1.

The organic phase is separated, dried and concentrated via a Vigreux column.

The extract is preseparated into a hydrocarbon fraction and an oxygen containing fraction by column chromatography on Silica gel (Kieselgel 60) by elution with pentane and ether respectively.

The separated fractions are concentrated and injected into a gas chromatographic column (3 m × ¼ inch) packed with 10% FFAP and eluted with a temperature programme from 80–250 °C.

Calculation is done by normalization to 100%. The peak areas are measured and summed, and the relative amounts of the hydrocarbon fraction and of the oxygen-containing fraction is determined.

A large number of papers has been published dealing with the nature of hop essential oils and their contribution to beer aroma.

Capillary gaschromatography combined with mass-spectrometry has shown the complexity of hop essential oils.

Recent contributions to the analysis of hop essential oils have been made by Tressl et al. (1983), Seeleitner et al. (1985), Narziss et al. (1985a, b), and Lam et al. (1986).

The latter described a solid phase rapid solvent extraction method for hop essential oil based on a selective retention of the bitter components by adsorption on alumina which was premixed with an aqueous potassium hydroxide solution followed by elution of the essential oil. The improvement of analytical techniques will probably result in better information on the importance of hop essential oil constituents for beer flavour and answer the question whether hop essential oils must be seen as a hop quality factor.

5 Polyphenols

The polyphenols of hops are of interest because of the role these substances play in the formation of hazes and in the colour and taste of beer. They are also thought to serve as anti-oxidants, thus preventing staling. All plants contain phenolic acids such as coumaric acid, gallic acid, caffeic acid, etc. Polyphenols in hops include a complex mixture. Flavanols and flavanediols are transferred from hops to wort and although they are only sparingly soluble, some of them remain in the final beer. They can slowly convert into dimers and polymers, which have tanning properties. Hops also contain the typical chalkone-structure polyphenol xanthohumol. It is readily oxidized during storage of hops. Analytical methods for polyphenols have been described by Dadic (1971).

Analysis of polyphenols by standard methods suffer in specificity, speed and precision. High performance liquid chromatography has provided the tool to measure a variety of simple phenolic compounds including flavonoids in plants.

McMurrough (1981) used reversed phase columns to separate several flavonoids in flower cones of hops with gradient elution. Jerumanis (1985) described quantitative analysis of flavanols in hops with an acetic acid gradient.

The importance of the analysis of polyphenols in hops to the brewer is limited. The modern way of hopping by using hop extracts has eliminated several compounds that were originally present in the whole hop cones, specially when the extract is prepared by non-polar solvents such as hexane or carbon dioxyde. Extraction with the polar ethanol results in extracts containing more of the polar phenolics.

6 Prospect

Improvements in column efficiency and detection instruments in HPLC, will expand our knowledge of hops and of its impact on beer bitterness in the near future. A very promising development is that of the diode array detector, which can give information on the many still unknown minor compounds in both hops and hopped beer. Also the introduction of liquid chromatography-mass spectrometry will greatly increase our knowledge of hops.

References

Aitken RA, Bruce A, Harris JO, Seaton JC (1968) Quantitative analysis of beer bittering substances and hop resins by thin-layer chromatography. J Inst Brew 74:436–440

Alderton G, Bailey GF, Lewis JC, Stitt F (1954) Spectrophotometric determination of humulone complex and lupulone in hops. Anal Chem 26:983–992

Alderweireldt F, Verzele M (1964) Nuclear Magnetic Resonance spectroscopy and the analysis of the alpha acids in hops. Wallerstein Lab Commun 27:19–28

Anderegg P (1976) Chemie der Hopfen-Inhaltsstoffe. Schweiz Brau Rundsch 87:4–8

Anderegg P, Pfenninger H (1985) Bestimmung von α- und β-Säuren in Hopfenprodukten mit HPLC. Ergebnisse von Ringversuchen. Schweiz Brau Rundsch 96:201–208

Anderegg P, Schur F, Pfenninger H (1977) Säulenchromatographische Bestimmung von Hopfenbitterstoffen. Schweiz Brau Rundsch 88:301–305

ASBC, American Society of Brewing Chemists (1976) Methods of analysis, 7th edn. St. Paul, Minn

Buckee GK (1985) Estimation of α-, β-, and iso-α-acids in hops, hop products and beer using high performance liquid chromtography. J Inst Brew 91:143–147

Claus H, van Dijck J, Verzele M (1978) Photometric constants of hop bitter acids. J Inst Brew 84:218–220

Dadic M (1971) Analytical methods for polyphenols in brewing. Proc Am Soc Brew Chem 29:149

Drawert F, Beier J (1973) Über die Biosynthese der Hopfenbitterstoffe. Brauwissenschaft 26:357–366

EBC, European Brewery Convention (1975) Analytica – EBC, 3rd edn. E45–E54 (Suppl 1980), E60 Schweiz Brau Rundsch

EBC, European Brewery Convention (1986) Analytica IV. Schweiz Brau Rundsch (in preparation)

Forster A, Schur F, Pfenninger H (1977) Überprüfung der konduktometrischen α-Säurenbestimmung nach Verzele in Hopfenextrakt. Schweiz Brau Rundsch 88:239–245

Franiau R, Mussche R (1974) Quantitative determination of hop bitter substances and their derivates in hop extracts by thin-layer chromatography. J Inst Brew 80:59–67

Ganzlin G (1975 a) Zur Hopfenextraktanalyse: Vereinfachung der Extraktionsmethode. Brauwissenschaft 28:205–213

Ganzlin G (1974 b) Zur Analyse von Hopfen und Pellets und Hopfenpräparaten, Vereinfachung der Extraktionsmethode. Brauwissenschaft 28:231–239

Ganzlin G (1975 c) Verbesserte Methode zur Hopfen-Weichharzbestimmung. Brauwissenschaft 28:344–346

Ganzlin G (1978) Verbesserte Methode zur spezifischen Hopfenbitterstoff-Analyse. Brauwissenschaft 31:128–139

Gill R (1979) Hop resin analysis by high-performance liquid chromatography on microparticulate silica. J Inst Brew 85:15–20

Godefroot M, Sandra P, Verzele M (1981) New method for quantitative essential oil analysis. J Chromatogr 203:325–335

Hansen GL, Hetzel D, Miller R (1971) Analysis of hop components by ion-exchange chromatography. 1. hops and hop extracts. Proc Am Soc Brew Chem 246–254

Hayduck M (1888) Über die bitteren und harzigen Bestandteile des Hopfens. Wochenschr Brau 47:937–947

Hildebrand RP (1979) Manufactured products from hops and their use in brewing. In: Pollock JRA (ed) Brewing science, vol 1. Academic, London, p 325

Howard G (1970) Institute of brewing analysis committee: the determination of hop oil. J Inst Brew 76:381–386

Howard G, Tatchell A (1953) Further observations on the occurrence and isolation of cohumulone. J Inst Brew 59:491–498

Howard G, Tatchell A (1957) Evaluation of hops. Influence of variety on the composition of the α-β acids. J Inst Brew 63:138–153

Hudson JR (1963) Humulones, Lupulones, and other constituents of hops. In: Linskens HF, Tracey MV (eds) Moderne Methoden der Pflanzenanalyse 6. Bd. Springer, Berlin Göttingen Heidelberg, p 135–154

Jerumanis J (1985) Quantitative analysis of flavanoids in barley, hops and beer by high-performance liquid chromatography. J Inst Brew 91:250–252

Kleber W, Hums N (1974) Die Auftrennung von Bitterstoffen aus Hopfen, Hopfenextrakten, Würzen und Bieren mit Hilfe der Niederdruckflüssigkeitschromatographie. Brauwelt 114:414–418

Knudson EJ, Siebert KJ (1983) High performance liquid chromatographic analysis of hop bittering components in hops, hop extracts, wort and beer. J Am Soc Brew Chem 41:51–57

Kokubo E, Kuroiwa Y, Oda K (1968) Ion-exchange chromatographic assay for hop resins and hop extracts. Proc Am Soc Brew Chem 94–100

Krüger E, Baron G (1975) Die Bestimmung des Hopfenölgehaltes in Hopfen und Hopfenpräparaten. Monatsschr Brau 28:118–120
Lam KC, Nickerson GB, Deinzer ML (1986) A rapid solvent extraction method for hop essential oils. J Agric Food Chem 34:63–66
Lance DG, Kavanagh TE, Clarke BJ (1981) The determination of α- and β-acids in hops and hop products using HPLC. J Inst Brew 87:225–228
McMurrough I (1981) High-performance liquid chromatography of flavonoids in barley and hops. J Chromatogr 218:683–693
MEBAK (1978) Methodensammlung der mitteleuropäischen brautechnischen Analysen-Kommission. Drawert F (ed) Freising-Weihenstephan, S 201–228
MEBAK (1982) Methodensammlung der mitteleuropäischen brautechnischen Analysen-Kommission. Drawert F (ed) Freising-Weihenstephan, S 465
Molyneux RJ, Wong Y (1973) High-pressure liquid chromatography in the separation and detection of bitter compounds. J Agric Food Chem 21:531–535
Narziss L, Scheller L (1984) Über die Bitterstoffzusammensetzung von Hopfen und Hopfenprodukten und ihre Veränderung beim Brauprozeß. Monatsschr Brau 12:496–504
Narziss L, Scheller L (1985) Über die Bitterstoffzusammensetzung von Hopfen und Hopfenprodukten und ihre Veränderung beim Brauprozeß. Monatsschr Brau 1:4–12, 2:86–96, 6:248–261
Narziss L, Miedaner H, Gresser A (1985a) Routinemäßige Bestimmung von Hopfenölen in Hopfen und Hopfenprodukten sowie in Würze und Bier. Monatsschr Brau 8:324–330
Narziss L, Miedaner H, Gresser A (1985b) Vergleich des Hopfenölspektrums verschiedener Hopfensorten. Brauwelt 36:1794–1798
Ono M, Kakuko Y, Yamamoto Y, Nagami K, Kumada J (1984) Quantitative analysis of hopbittering components and its application to hop evaluation. J Am Soc Brew Chem 42:167–172
Ono M, Kakudo Y, Yamamoto Y, Nagami K, Kumada Y (1985) Simultaneous analysis of hop bittering compounds by high performance liquid chromatography and its application to practical brewing. J Am Soc Brew Chem 43:136–144
Otter GE, Taylor L (1977) The development of a bonded weak anion-exchange packing suitable for the high-performance liquid chromatography of hop resins. J Chromatogr 133:97–106
Otter GE, Taylor L (1978) The high performance liquid chromatography of hop resins. J Inst Brew 84:160–164
Otter GE, Silvester OJ, Taylor L (1972) Automatic analysis of hop resins by ion-exchange chromatography. J Inst Brew 78:57–64
Peacock VE, Deinzer ML (1981) Chemistry of hop aroma in beer. J Am Soc Brew Chem 39:136–141
Pfenninger H, Schur F (1968) Bewertung und Standardisierung von Hopfenextrakt. Schweiz Brau Rundsch 79:241–245
Pfenninger H, Schur F, Senn R (1973) Problematik der Analyse von Hopfen und konventionellen Hopfenprodukten. Schweiz Brau Rundsch 84:183–190
Pfenninger HB, Schur F, Vaterlaus BP, Sigg T, Wild J (1975) Synthetische Hopfenbitterstoffe zur Bierherstellung. Proc Eur Brew Conv 15:159–168
Pfenninger H, Schur F, Anderegg P (1979) Composition and analysis of hops and derived products. In: Pollock JRA (ed) Brewing science, Academic, London, p 451
Riedl W (1951) Über die Synthese der Alpha-Hopfen-Bittersäuren. Brauwissenschaft 6:81–85
Riedl W (1954) Mitteilung über Hopfen. Synthese einiger Lupulon-Analoga mit abgewandeltem Acyl-Rest. Annalen 585:38–42
Rigby FL, Bethune JL (1952a) Cohumulone, a new hop constituent. J Am Chem Soc 74:6118–6119
Rigby FL, Bethune JL (1952b) Countercurrent distribution of hop constituents. Proc Am Soc Brew Chem 98–105
Schulze WG, Ting PL, Goldstein H (1979) Separation of hop compounds by reserved phase HPLC. In: Charalambous G (ed) Liquid chromatographic analysis of food and beverages, vol 2. New York, p 441

Schulze WG, Ting PL, Henckel LA, Goldstein H (1981) Separation of humulinic acids by reversed phase high performance liquid chromatography. J Am Soc Brew Chem 39:12–15
Schur F, Pfenninger H (1975) Analysis, evaluation and application of a new hulupone product. Proc Am Soc Brew Chem 33:4–7
Schur F, Pfenninger H (1979) Offizielle Analysenmethoden für Doldenhopfen, Pulverprodukte und Extrakte. Brauwissenschaft 32:155–159
Schwarzenbach R (1979) HPLC of hop acids on buffered silica gel systems. J Am Soc Brew Chem 37:180–184
Seeleitner G, Seif P, Püspök I (1985) Ein Fortschritt in der Hopfenölanalytik. Monatsschr Brau 9:367–371
Siebert KJ (1976) High performance liquid chromatography of hop compounds in brewing. J Am Soc Brew Chem 34:79–90
Silvester DJ (1984) Simultaneous determination of α-, β-, and iso-α acids in hops and hop products. J Inst Brew 90:319–322
Slotema FP, Verhagen LC, Verzele M (1977) Zur Bestimmung der Hopfenbittersäuren in der Brauerei mittels Hochleistungs-Flüssig-Chromatographie (HLFC). Brauwissenschaft 30:145–149
Stevens R (1967) The chemistry of hop constituents. Chem Rev 67:19–71
Tressl R, Engel KH, Kossa M (1983) Charakterisierung neuer Aromastoffe in Hopfen, Hopfenextrakten und Bier. Monatsschr Brau 11:432–438
Van de Velde (1986) Hoge druk vloeistof chromatografie van hopzuren. Thesis, Gent University, Belgium
Vanheertum R, Verzele M (1973) Quantitative analysis of hops and hop extracts by high pressure ion exchange chromatography. J Inst Brew 79:324–331
Verzele M (1958) Posthumulone, a new „alpha acid". Bull Soc Chim Belg 67:278–279
Verzele M (1979) The chemistry of hops. In: Pollock JRA (ed) Brewing science. Academic, London, p 279
Verzele M (1986) 100 years of hop chemistry and its relevance to brewing. J Inst Brew 92:32–48
Verzele M, De Potter M (1978) High-performance liquid chromatography of hop bitter substances. J Chromatogr 166:320–326
Verzele M, Dewaele C (1981) Stationary phase characterization in high-performance liquid chromatography. A test for trace metal activity in octadecyl bonded silica gel. J Chromatogr 217:399–404
Verzele M, Van Dyck J (1971) Analytical evaluation of alpha acids in hop extract and in hops. J Inst Brew 77:529–538
Verzele M, Claus H, Van Dyck J (1967 a) Analysis of hop extracts by simplified counter current distribution methods. J Inst Brew 73:39–43
Verzele M, Claus H, van Dyck J (1967 b) Paper strip analysis of substances derived from hops. J Inst Brew 73:298–306
Verzele M, van Dyck J, Claus H (1980) On the analysis of hop bitter acid. J Inst Brew 86:9–14
Verzele M, Dewaele C, Van Kerrebroeck M, Strating J, Verhagen L (1983) Fast high performance liquid chromatography analysis of hop bitter compounds. J Am Soc Brew Chem 41:36–40
Verzele M, Dewaele C, Van Kerrebroeck M, Strating J, Verhagen L (1984) Fast high-performance liquid chromatography analysis of hop bitter compounds: correction and addendum. J Am Soc Brew Chem 42:94–95
Wieland H (1925 a) Über die chemische Natur der Hopfenharz-Säuren (I). Chem Ber 58:102–112
Wieland H (1925 b) Über die chemische Natur der Hopfenharz-Säuren (II) Chem Ber 58:2012–2017
Wieland H, Martz E (1926) Über die chemische Natur der Hopfenharz-Säuren (III). Chem Ber 59.2352–2356
Wöllmer W (1925) Über die Bitterstoffe des Hopfens. Chem Ber 58:672–678
Wöllmer W (1930) Die Analyse des Hopfens. Wochenschr Brauerei 47:521–525

Determination of β-Glucan in Barley, Malt, Wort and Beer

K. G. Jørgensen and S. Aastrup

1 Introduction

The polysaccharide (1–3)(1–4)-β-D-glucan, hereafter referred to as β-glucan, is present in most cereals. Highest levels are found in grains of barley and oats, where β-glucan is the major component of the endosperm cell walls, enclosing the storage components starch and protein. Compared with the total carbohydrate content of the grain, the β-glucan constitutes only a minor fraction. However, as a high molecular weight water-soluble polymer, the β-glucan has properties of both technological and nutritional significance. β-glucan is undigested by monogastric animals and has been reported to be responsible for the low utilization of barley, when this cereal is used as a poultry feed. On the other hand, as a soluble dietary fiber, the β-glucan has, from a human nutritive point of view, some beneficial effects (for a review see Wood 1984).

Most attention towards the β-glucan has been for the barley β-glucan within the brewing industry. Several problems have been associated with increased levels of β-glucan, such as: retardation of malting, increased viscosity of wort and beer which increases the process time needed for wort separation and beer filtration, or formation of precipitates in the final beer (for a review see Bamforth 1982). In addition the β-glucan, as a cell wall component, is a barrier for the extraction of the malt during the mashing process, and hence increased levels lead to a decrease in the brewhouse yield. β-glucan represents therefore a very important marker for the evaluation of malt quality (see also Aastrup and Jørgensen this Volume).

In order to understand and control the β-glucan in all levels of the brewing process, the β-glucan must be quantified.

In this chapter we will consider the numerous methods which have been developed to measure β-glucan, mainly within the brewing industry. In addition we will present our own contributions to an automatic β-glucan analyzer system. A detailed survey of all methods will not be given, as methods have been improved together with the knowledge about the β-glucan. Some steps involved in older methods may, however, be useful information for other contexts, e.g. extraction and purification of β-glucan for standardization etc. The chapter is introduced by a short review of the nature of β-glucan in order to understand the different approaches which have been taken for quantification.

2 The Nature of β-Glucan

The barley endosperm cell walls consist of 75% β-glucan, the rest being protein and pentosans (Fincher 1975). The basic chemical structure of β-glucan is well established. It is a glucose polymer in which about 30% of the linkages are (1→3)-β-glucosidic, the remainder being (1→4)-β-glycosidic (Clarke and Stone 1963; Fleming and Kawakami 1977). The greatest part of the polymer is built up of cellotriosyl and cellotetraosyl units randomly distributed throughout the chain and joined by single (1→3)-β-linkages (Fleming and Kawakami 1977; Staudte et al. 1983; Woodward and Fincher 1983). Minor parts may, however, consist of longer chains of (1→4)-β-linkages, but whether the polysaccharide also contains repetitive (1→3)-β-linked glucose units has not been established conclusively.

When situated in the cell walls, the β-glucan is believed to be covalently bound to protein, forming very large molecules of approximately 10^7 dalton (Forrest and Wainwright 1977a). Somewhat depending upon the barley variety and the extraction conditions, 10–50% of the barley β-glucan is water-soluble. The water-soluble fraction has often been referred to as the barley "gums", whereas the insoluble part has been termed "hemicelluloses" (Preece and MacKenzie 1952; Schuster et al. 1967; Morrall and Briggs 1978).

During germination of barley, hydrolytic enzymes are produced including (1→3)(1→4)-β-glucanases (β-glucanase) which are responsible for the depolymerization of the β-glucan. The great interest towards the β-glucan in brewing is closely connected with the action or the lack of action of these enzymes. Malt β-glucanases are very heat-labile and are inactivated during the early stages of mashing (Erdal and Gjertsen 1971). A substantial quantity of high molecular weight β-glucan can therefore be found in wort and in the final beer, especially if poorly modified malt and/or barley is used. An additional enzyme has been reported (referred to as solubilase), which is more heat-stabile than the β-glucanases and which cleaves the protein bound to β-glucan resulting in the solubilization of β-glucan (Bamforth et al. 1979; Bamforth and Martin 1981). Definitive proof for the existense of this enzyme has not yet been provided.

Only the high molecular weight β-glucan fraction in malt, wort and beer is of analytical interest, as this is the troublesome fraction (Forrest and Wainwright 1977b). In wort, β-glucan with a molecular weight of 10^4–10^7 has been found (Bamforth and Martin 1983; Forrest and Wainwright 1977b; Foldager and Jørgensen 1984) and precipitates isolated from beer have been reported to be 7×10^4 in molecular weight (Gjertsen 1966). Any analytical method used to measure the β-glucan content in malt, wort or beer should therefore specifically measure the high molecular weight fraction.

3 Quantification of Barley β-Glucan

Individual barley and malt grains vary significantly in their content of β-glucan. A determination of the β-glucan content of a sample represents thus a mean value of grains. A representative sampling must therefore be established (e.g. Wainwright and Buckee 1977), and normally a flour representing 200–1000 kernels (10–50 g) is made. Grinding of the kernels is also important, since improper grinding may prevent some of the β-glucan present from being accessible to extraction, enzymatic attack etc., and the content may therefore be underestimated. A proper grinding is normally performed so that all flour particles may pass a 0.5–0.8 mm sieve. Before a flour sample is weighed for analysis, careful mixing must be done. The amount of flour which should be used for the analysis will depend on the analytical protocol.

3.1 Methods Based on Extraction

The majority of methods published for β-glucan analysis involve a solubilization of the β-glucan polymer either in the native form or degraded. Methods based on extraction without a forced degradation of the β-glucan will be considered in the following.

3.1.1 Viscosity Methods

Determination of β-glucan by viscosity of barley extracts is a simple method and has been used as a screening technique in barley breeding programmes (Greenberg and Whitmore 1974; Bendelow 1975; Morgan and Gothard 1977; Aastrup (1979 a) and also as a method for evaluation of malting quality (Smith et al. 1980). The results of such methods might differ according to the molecular weight of the extracted β-glucan and co-extraction of other components which also increase the viscosity (e.g. pentosans, starch). Possible action of the β-glucanases present in the barley during extraction should be avoided. A common procedure minimizing interference has been the use of an acid extraction at low temperature. By this procedure a good correlation to soluble β-glucan is achieved (Fig. 1), but there is no

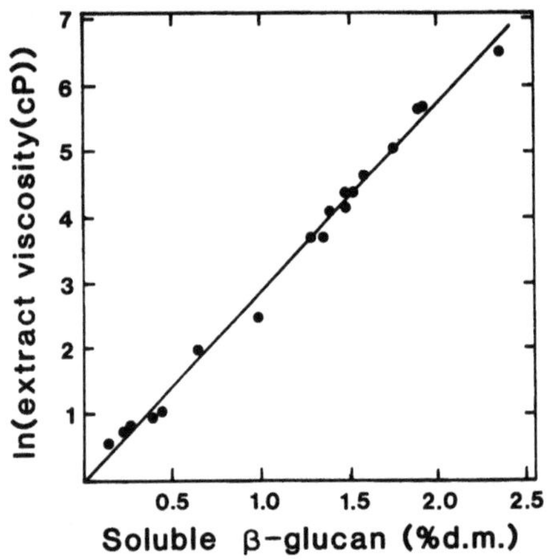

Fig. 1. Relationship between the viscosity and the β-glucan content of acid flour extracts from 18 barley genotypes. (Aastrup 1979 a)

direct relationship to the total β-glucan content of the barley (Aastrup 1979a). The method has, however, been used successfully to find a low β-glucan mutant in a high β-glucan barley line (Aastrup 1983).

Acid Extraction

a) 0.2 g barley flour is poured into a plastic tube and mixed with 2 ml of a HCl–KCl solution pH 1.5 (82.8 ml 1 *M* HCl + 7.46 g KCl per litre).
b) β-glucan is extracted at 20 °C for 1 h with continuous shaking.
c) The extract is centrifuged (1000 *g* 20 min) and the viscosity of the supernatant measured.

3.1.2 Difference Methods

These methods are based upon the assumption that the only glucose polymers present in the extracts are derived from β-glucan or starch. By measuring the total carbohydrate content (as glucose) and subtracting the glucose derived from specific enzymatic degradation of starch with amyloglucosidase, an estimate of β-glucan content is achieved. It is evident that methods based on this principle are accurate only if the extraction method employed does not extract too much of the starch. Extraction is thus a compromise between extraction of the β-glucan and little co-extraction of the starch fraction. Although much of the β-glucan can be extracted, significant levels will not be solubilized. An investigation by histochemical techniques of the most commonly used alkali method (Wood et al. 1977), has shown that especially the β-glucan present in the aleurone and subaleurone layer still remain after repeated extractions (Wood et al. 1983). Further, the starch is degraded by commercially available enzymes, and as pointed out by Wood et al. (1977), great precaution should be taken to assure that these preparations do not contain impurities of β-glucan degrading enzymes. This is normally the case with many of the enzymes of fungal origin.

Fleming et al. (1974) have used repeated 40 °C water extraction. Barley flour was boiled in 80% ethanol prior to extraction to inactivate enzymes. A better method seems to be the extraction procedure proposed by Wood et al. (1977), using carbonate buffer avoiding a pre-inactivating step. A method for the extraction of oats and barley β-glucan is described below.

Carbonate Extraction (Wood et al. 1977)

a) 1 g of flour is suspended in 10 ml 20% (w/v) sodium carbonate-bicarbonate buffer pH 10.
b) Extraction is performed with continuous stirring at 45 °C for 0.5 h.
c) The sample is centrifuged (24,000 *g* 5 min) and the supernatant collected. The extraction is repeated three times.

Starch is measured by a specific enzymatic glucose assay after degradation with amyloglucosidase. Total carbohydrate present in the extract is measured using a H_2SO_4/cysteine reaction listed below, which for barley and oats is found to be specific for glucose.

Glucose Determination by Cysteine/H_2SO_4 (Wood et al. 1977)

a) 2 ml ice-cold sulfluric acid 86% (v/v) containing L-cysteine hydrochloride (700 μg/ml) is added to 400 μl sample (containing 0–25 μg glucose) and mixed thoroughly.
b) The sample is kept at 100 °C for 3 min and thereafter at 20 °C for 40 min.
c) Absorbance is read at 528 nm.

3.1.3 Selective Precipitation Methods

In order to avoid removal of the starch by enzymes, selective precipitation techniques have been described. Since the pioneer work of Preece and McKenzie (1952) studying the "gums" present in cereals, selective precipitation of the β-glucan with ammonium sulphate (AMS) has been widely used.

Wood and Weiz (1984) have used the specificity of the optical brightener Calcofluor to complex and precipitate oat β-glucan from carbonate extracts (the specificity of this dye is further discussed in Sect. 5).

Precipitation of Soluble β-Glucan with Calcofluor (Wood and Weiz 1984)

a) 2.5 ml carbonate extract (see above) is mixed with 0.25 ml of a 20 mg/l aqueous Calcofluor solution.
b) The solution is stirred vigorously for 0.5 h and the precipitate recovered by centrifugation (12,000 *g* 20 min).
c) The recovered precipitate is hydrolyzed by acid and the glucose produced is determined by HPLC.

Madacsi et al. (1983) have used a specific precipitation employing a $CuSO_4$ reagent also from carbonate extracts. The method was used to measure β-glucan in sorghum. Extraction is performed at high temperature (80 °C) with 0.1 *M* $NaBH_4$ added to prevent alkali degradation of the β-glucan.

Precipitation of Soluble β-Glucan with $CuSO_4$ (Madacsi et al. 1983)

a) 10 ml carbonate extract is mixed with 2.0 ml 2.5 *M* NaOH and 2.0 ml $CuSO_4$ reagent (0.3 g $CuSO_4 \cdot 5H_2O$ + 25 g Na_2SO_4 + 3 g Na citrate in 200 ml of water).
b) The sample is heated for 5 min in boiling water, allowed to cool to room temperature and centrifuged (5000 *g* 25 min).

3.2 Non-Extraction Methods

3.2.1 Near Infrared Reflectance

Near Infrared Reflectance (NIR) is a well-known non-destructive technique commonly used in the food and feed industry (McDonald 1986). The method is fast (normally less than 1 min per sample) and often more than one component can be determined at the same time. The speed of the analysis can therefore in many cases, such as for plant breeding, compensate for a lower accuracy. NIR analysis is based on the fact that the information stored in the spectra is specific for the component to be investigated, so that a valid computer model can be set up. Normally such models are only valid for one type of sample, e.g. one calibration for barley proteins and another for wheat proteins. Widely used are models based on Multiple Linear Regression analysis, but modelling techniques are still improving (Martens 1985), and as computer power still increases, better calibrations will undoubtedly be achieved in the future.

Only small differences are found between recorded spectra of different sugars and polysaccharides. Henry (1985) has reported the use of NIR for detection of total β-glucan content of barley. An instrument calibration based on three filters (1656 nm, 1676 nm, and 1752 nm) and a calibration set of 50 barley samples with β-glucan measured by an enzymatic method was made. It was possible to predict the β-glucan contents of 25 other samples with a root mean square of 0.403%.

3.2.2 Calcofluor Staining in Flour Suspensions

The fluorochrome Calcofluor specifically stains the β-glucan. In estimation of malt modification, described in this book by Aastrup and Jørgensen (1986), this is done by staining intact cell walls of halved grains and visual inspection in UV light. A technique has been described by Jensen and Aastrup (1981), where a fine flour is stained by Calcofluor suspended in glycerol and the fluorescence intensity measured in a spectrofluorimeter. A good correlation was found between the fluorescence intensity obtained from the flour suspensions and the total β-glucan content determined by an enzymatic method. Not all β-glucan of the cell walls might, however, be equally accessible to staining and are therefore not stained uniformly. Despite the fact that the relationship is not ideal, the method can easily judge between low and high β-glucan barleys. A major drawback of the method is that it must be calibrated to another independent method (e.g. an enzymatic method).

Calcofluor Staining of β-Glucan in Flour Suspensions (Jensen and Aastrup 1981)

a) A 100-mg subsample of a coarse flour is further milled in a ball mill for 15 min to obtain a fine homogeneous flour.
b) 10 mg of the fine flour is mixed with 1 ml 0.1% aqueous Calcofluor.
c) 4 ml 96% ethanol is added and the sample is centrifuged (4000 rpm; 10 min).
d) The precipitate is washed three times with 5 ml 80% (v/v) aqueous ethanol.
e) 10 ml of 75% (w/w) aqueous glycerol is added, the suspension mixed, a subsample transferred to a cuvette and fluorescence measured at 360 nm excitation and 425 nm emission.

3.3 Enzymatic Methods

3.3.1 Methods Using Bacterial β-Glucanase

The most specific enzymatic method for β-glucan analysis is based upon a bacterial enzyme [(1→3)(1→4)-β-D-glucan 4-glucanohydrolase, EC 3.2.1.73)]produced by *Bacillus subtilis* (Anderson et al. 1978; Henry 1984; McCleary and Glennie-Holmes 1985). This enzyme only cleaves the (1→4)-β-linkages of 3-substituted-D-glucopyranosyl residues and is hence very specific for barley β-glucan. This β-glucanase is found as a contaminant in a commercial bacterial α-amylase (1000 S or 240 l from NOVO industries, Denmark) and the enzyme must therefore be purified prior to use. All starch-degrading enzymes, converting starch into glucose, would interfere and lead to an overestimation of the β-glucan content. The most simple purification procedure for the enzyme is described below.

Purification of β-Glucanase from *Bacillus subitilis* (Henry 1984)

a) 50 g 1000 S enzyme is mixed with 500 ml 50 m*M* phosphate buffer pH 8.0 and centrifuged (4000 *g* 5 min).
b) The supernatant is applied to a column of 50 g DEAE cellulose equilibrated with 50 m*M* of the above phosphate buffer.
c) The column is washed with 500 ml of the phosphate butter and the β-glucanase eluted with 100 m*M* acetate buffer pH 4.0.
d) The first 165 ml from the column is discarded and the next 400 ml is collected.
e) Residual contamination of α-amylase is removed by precipitation with starch. 0.5% $CaCl_2$, 40% (v/v) ethanol and 20% (w/v) starch is added to the above 400 ml fraction. The mixture is stirred and centrifuged (4000 *g* 10 min).
f) Step(e) is repeated.

The final preparation should contain no α-amylase. If the above procedure for β-glucanase purification is followed, most of the α-amylase activity present in the preparation is removed by the ion-exchange chromatography step b) and c). McCleary and Glennie-Holmes (1985) have described a simple procedure whereby 95% of the α-amylase is inactivated, or removed, prior to the ion-exchange chromatography step.

Inactivation of *Bacillus subtilis* α-Amylase (McCleary & Glennie-Holmes 1985)

a) 1 l bacterial amylase Novo 240 l (a liquid product similar to the 1000 S, which is a powdery material) is diluted to 2 l and the pH adjusted to 3.5 by careful addition of 1 *M* HCl with vigorous stirring.
b) The solution is kept at 4 °C for 2 h and then centrifuged (3500 *g* 30 min).
c) The pH is readjusted to 7.5 with 1 *M* NaOH after addition of 100 ml phosphate buffer pH 6.5.
d) The solution is again kept at 4 °C for 2 h and then centrifuged (3500 *g* 30 min) to remove a colloidal precipitate.

For extraction and measurement of β-glucan, different approaches have been taken, but all include the following general steps.

Analysis of β-Glucan by β-Glucanase from *Bacillus subtilis*

a) Inactivation of enzymes, which can be done by boiling flour in 80% ethanol (50% ethanol has been used by McCleary and Glennie-Holmes 1985). At the same time low molecular weight carbohydrates, that might otherwise interfere in the final determination, will be solubilized and can be removed.

b) Extraction of the β-glucan from the barley flour, e.g. extraction 0.5 g flour with 5 ml 100 m*M* phosphate buffer pH 6.5 with approx 10 U of the enzyme for 1 h at 40 °C. (U = reducing sugar per min per ml of the β-glucanase assayed at 40 °C at pH 6.5 with 1 mg/l purified β-glucan.)
c) Centrifugation and measurement of liberated degraded β-glucan.

Measurement of liberated β-glucan oligomers is done by a complete acid hydrolysis (Anderson et al. 1978) and measurement of glucose by a specific method or by measuring the oligosaccharides directly by p-hydroxybenzoic acid hydrazide (Henry 1984). The latter method requires a purified β-glucan which is used as a standard. By the recent procedure of McCleary and Glennie-Holmes (1985), the *Bacillus subtilis* β-glucanase (referred to as lichenase by the authors) has been used. After extraction and degradation of the β-glucan to oligomers, the β-glucan fragments are further degraded by a β-glucosidase to glucose, which is measured enzymatically. β-glucosidase is purified from NOVOZYM 188 (NOVO industries, Denmark), a detailed purification scheme has, however, not yet been published. As an aliquot of the β-glucanase, extracted flour can be run as a blank (without addition of β-glucosidase); this procedure does not require the removal of moderate concentrations of low molecular weight carbohydrates in the flour (e.g. glucose) prior to analysis. The method is commercially available through Biocon Ltd. and the analytical scheme is illustrated in Fig. 2.

3.3.2 Methods Based on Fungal Cellulases

Fungal cellulase systems, (1→4)-β-D-glucanases, which release glucose from β-glucan, are present in crude extracts of *Trichoderma reesei* and *Penicillum funicu-*

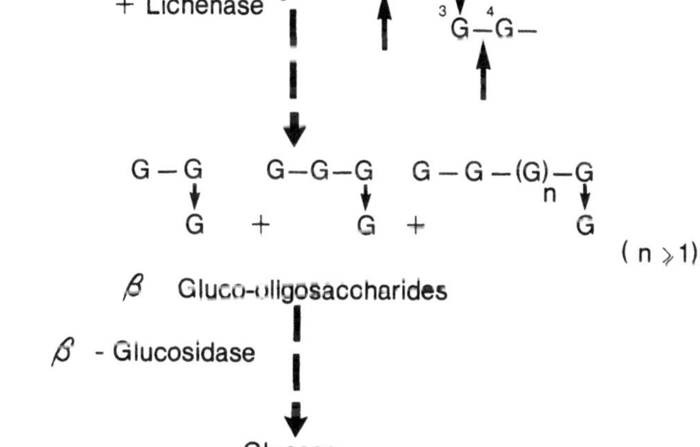

Fig. 2. Principle of enzymatic β-glucan analysis by the procedure of McCleary and Glennie-Holmes (1985) using *Bacillus subtilis* β-glucanase (lichenase) and β-glucosidase. (Courtesy of Dr. McCleary)

losum. The use of these enzymes for measurement of β-glucan has been described by Martin and Bamforth (1981), Prentice et al. (1980), Bamforth (1983), and Ahlwalia and Ellis (1984). Presumably because the enzymes also convert cellulose or (1→3)-β-D-glucan into glucose (not reported), the β-glucan has in all the cited works been extracted and removed from the cellulose material (e.g. husk and pericarp) of the barley grain, before these enzymes were used. During a complete chemical extraction of β-glucan, much or all of the starch will also be solubilized, which necessitates that the enzymes used do not contain any amylolytic activity. Prentice et al. (1980) describe a method based on ion-exchange chromatography to remove the starch-degrading enzymes from *T. reesei* extracts. The most common method has been to inactivate amyloglucosidase by heating. The first method of Martin and Bamforth (1981) was not adequate for all batches of the enzyme and the method was modified by Bourne et al. (1982). That method, however, sacrifices a larger amount of the β-glucanases. β-glucanases produced by *P. funiculosum* seem to overcome these problems, as they are more heat-stable. An inactivating scheme for starch degrading enzymes has been proposed.

Inactivation of amyloglucosidase activity in *P. funiculosum* extracts (Bamforth 1983)

a) 0.4 g cellulase CP (John & E. Sturge Ltd, UK) is added to 10 ml of 50 m*M* acetate-HCl pH 4.0.
b) The mixture is stirred thoroughly for 10 s at room temperature, kept 10 min at rest, stirred again and centrifuged.
c) The supernatant is heated at 70 °C for 60 min, and thereafter kept on ice for 2 min.
d) The solution is dialyzed at 4 °C for 16 h against 2 l of 50 m*M* acetate buffer pH 4.0 and centrifuged. The resultant solution is stable for 1 week when kept at 4 °C.

Prior to analysis by use of the enzyme solution obtained as above, the β-glucan can be extracted by hydrazinolysis as described by Martin and Bamforth (1981). This method requires an extensive dialysis of the extracts prior to enzyme digestion. A much simpler procedure has been described by Ahluwalia and Ellis (1984), which was also used for a complete extraction of starch:

Extraction and Measurement of Total β-Glucan Content by *P. funiculosum* β-Glucanase (Ahluwalia and Ellis 1984)

a) 20 mg barley flour is incubated with 4.0 ml 50 m*M* perchloric acid at 96 °C for 3 min.
b) The extract is cooled and centrifuged, the supernatant collected and the pellet washed with 1 ml cold 50 m*M* perchloric acid and recentrifuged. The supernatants are pooled and the volume made up to 10 ml with distilled water.
c) 250 μl of the extract is mixed with 650 μl of 0.2 *M* acetate buffer pH 5.4 and 100 μl heat-treated enzyme preparation is added.
d) The solution is incubated at 40 °C for 2 h and the released glucose measured enzymatically.

A somewhat different approach employing a β-glucanase of fungal origin has been described by Åman and Hesselman (1985). Although the method is time-consuming, it is interesting because crude, commercially available enzymes are used directly in the assay. By this method starch is first removed and thereafter the β-glucan is extracted by a β-glucanase present in an extract of *Rhizomucor pusillus*. The enzyme converts the β-glucan into oligomers.

β-Glucan Method of Åman and Hesselman (1985)

a) 200 mg flour is mixed with 5 ml 0.1 *M* acetate buffer pH 5.0 and 100 μl Termamyl 120 l, a very heat-stable α-amylase, (NOVO industries, Denmark) and incubated in boiling water for 1 h with occasionally stirring.
b) The solution is cooled to 40 °C and 100 μl amyloglucosidase suspension is added (from Boehringer Mannheim, 140 U/ml) and incubated for 2 h at 60 °C in a waterbath with shaker.
c) 20 ml absolute ethanol is added and soluble polymers precipitated at 4 °C overnight. The sample is centrifuged (1500 *g* 5 min), the supernatant decanted and the pellet washed twice with 25% (v/v) ethanol. The pellet is recovered by centrifugation, and excess ethanol is removed after the last wash by inverting the tubes on a filter paper.
d) 5 ml 0.05 *M* acetate buffer pH 4.8 and 50 μl β-glucanase from *R. pusillus* [Norsk Hydro, sold as HYDROZYME (6.4 EU/ml)] is added and the samples incubated for 2 h at 50 °C with shaking.
e) Soluble polymers are precipitated by addition of 20 ml absolute ethanol and the supernatant fraction isolated by centrifugation (1500 *g* 5 min). The pellet is resuspended in 20 ml 80% (v/v) ethanol and recovered by centrifugation. The supernatant fractions are pooled and diluted to 50 ml with 80% ethanol.
f) 1 ml of the supernatant is transferred to a 8-ml Pyrex glass fitted with a Teflon-lined screw cap and evaporated to dryness. 250 μl 2 *M* trifluoroacetic acid is added and the carbohydrates hydrolyzed for 2 h at 125 °C in capped tubes. The sample is evaporated as above, 4 ml water is added and glucose measured by an enzymatic method.

4 Quantification of β-Glucan in Malt, Wort and Beer

For quality control, the β-glucan content of malt is more interesting than the β-glucan content of barley. Malt is the raw material processed in brewing, and problems encountered due to β-glucan are directly reflected by the content in the malt.

Evaluation of a malt to be used for brewing always involves the production of a laboratory wort by a standardized mashing scheme, such as the European Brewery Convention "congress mashing", and hence a "congress wort" (Enari 1975). Other schemes following other brewing traditions are also used (see Wainwright and Buckee 1977). The laboratory wort is subjected to a range of different analyses, e.g. nitrogen content, extract content, fermentability, viscosity etc. The β-glucan content in this wort can be considered to be an estimate of the β-glucan in malt which is likely to be extracted in the brewing process.

Complications arise when attempting to use many of the methods for β-glucan analysis described for barley for the quantification of malt, wort or beer β-glucan. The following general complications can be summarized:

1. Malt contains a higher amount of enzymes (e.g. α-amylases, β-glucanases).

2. Malt, wort and beer contain low molecular weight carbohydrates of both starch and β-glucan origin, the highest content being found in wort, as almost all the starch is hydrolyzed into fermentable sugars and oligosaccharides. Although, in general, 70% of the sugars are converted into alcohol during fermentation, beer still contains considerable amounts of low molecular weight carbohydrates.

4.1 Total β-Glucan Content of Malt

4.1.1 Non-Enzymatic Methods

The method described by Jensen and Aastrup (1981) for measurement of β-glucan using complex formation with Calcofluor in fine flour suspensions (Sect. 3.2.2) has also been used for fine flours of malt. Calcofluor reacts only with the intact cell walls. Limitations in the technique are the same as noted for the quantification of the barley β-glucan.

4.1.2 Enzymatic Methods

Enzymes will attack small as well as large β-glucan molecules and since in most methods free glucose will interfere with the final quantification, procedures have been described for lowering the low molecular weight carbohydrate fraction. Although dialysis has been used by Martin and Bamforth (1981) after perchloric acid extraction, the most simple treatment seems to be an extraction with 50% ethanol in a steam bath (McCleary and Nurthen 1986). Henry and Blakeney (1986) have taken another approach. In their procedure, reducing sugars were reduced to their alditols by sodium borohybride prior to enzymatic extraction of the β-glucan, and therefore did not interfere in the final assay. By this procedure low molecular weight β-glucan is not removed, and it was also reported that values for malt β-glucan were somewhat higher than if pre-extraction with 80% ethanol was performed. Whether this is of any practical significance was not reported.

4.2 β-Glucan Content of Wort and Beer

4.2.1 Non-Enzymatic Methods

Traditionally, the high molecular weight β-glucan fraction in wort and beer is precipitated prior to analysis. Organic solvents are used (e.g. Crabb and Bathgate 1973), but more commonly β-glucan is selectively precipitated using 30% (w/v) ammonium sulphate (e.g. Erdal and Gjertsen 1967). Ammonium sulphate precipitation was originally designed as a part of an isolation procedure (Preeze and McKenzie 1952), and its specificity for an analytical technique has often been questioned (e.g. Bamforth 1982).

Determination of Wort and Beer β-Glucan (Erdal and Gjertsen 1967)

a) To 100 ml 0 °C wort or beer is added 30 g ammonium sulphate.
b) The β-glucan is precipitated at 0 °C for 1 h and the precipitate recovered by centrifugation (10,000 *g* 10 min).
c) The precipitate is washed twice with 20% (v/v) aqueous ethanol and recovered by centrifugation (10,000 *g* 10 min) after each wash.
d) The precipitate is redissolved in 1 N H_2SO_4 by boiling for 15 min.
e) Measurement of glucose is carried out with phenol and sulphuric acid (Dubois et al. 1956).

A method using the complex formation between Calcofluor and wort or beer β-glucan and precipitation of the complex formed was described by Jensen and Aastrup (1981). The method was further improved by Jørgensen (1983). By the latter method, a sample of wort or beer is mixed with a solution of Calcofluor in aqueous dimethylsulfoxide and the β-glucan Calcofluor complex formed is precipitated with ethanol. The precipitate is washed with ethanol and the fluorescence measured after a solubilization in an alkaline buffer. Major limitations were observed with this method, e.g. the β-glucan/Calcofluor complex is only quantitatively precipitated in the presence of other low molecular weight carbohydrates and the method should be calibrated to the above-mentioned method using ammonium sulphate precipitation.

4.2.2 Enzymatic Methods

Few methods are described for quantification of β-glucan in wort and beer by use of enzymatic methods. Probably this is due to difficulties in achieving a well-defined high molecular weight fraction prior to degradation and analysis by the enzymes. Bamforth and Martin (1981) have used dialysis and McCleary and Nurthen (1986) have used a precipitation with 50% ammonium sulphate, and the precipitate was washed twice with 50% ethanol before the β-glucan content was determined.

Determination of β-Glucan in Wort and Beer (McCleary and Nurthen 1986)

a) 5 ml of wort or decarbonated beer is transferred to a plastic centrifuge tube, 2.5 g ammonium sulphate is added and the salt carefully dissolved.
b) β-glucan is precipitated by placing the sample at 4 °C for 20 h.
c) The sample is centrifuged (3000 *g* 10 min) and the supernatant discarded.
d) The pellet is suspended in 1 ml of 50% (v/v) ethanol, further 10 ml of the ethanol is added and the washed precipitate recovered by centrifugation.
e) step(d) is repeated once more.
f) 5 ml of 20 m*M* phosphate buffer pH 6.5 is added and the β-glucan content determined as illustrated in Fig. 2.

5 An Automatic β-Glucan Analyzer System

5.1 Principle

The optical brightener Calcofluor white M2R New, the major active ingredient of which is the sodium salt of 4,4′-bis {{4-anilino-6-[bis(2-hydroxy-ethyl)aminol]-s-triazin-2-yl}amino}-2,2′-stilbenedisulfonic acid, (see Fig. 3 from Wood and Fulcher 1978) is a fluorescent dye used commercially as a whitening agent.

The dye complexes with glucose polymers linked by (1→3)-β- or (1→4)-β-linkages (Wood and Fulcher 1978; Wood 1980). Used as a dye in histochemical studies by fluorescence microscopy, however, Calcofluor was found to be extremely specific for the (1→3)(1→4)-β-D-glucan in cereals. This was concluded by the absence of fluorescence, when the samples were pre-treated with the β-glucanase purified from *Bacillus subtilis* (Wood et al. 1983). This specificity in reaction of Calcofluor towards β-glucan has been used by Wood and Weiz (1984), Jensen and Aastrup (1981), and Jørgensen (1983) for detection of β-glucan in barley, oats, malt, wort and beer, as described previously.

Another approach using the specificity of Calcofluor has recently been taken for the measurement of β-glucan in wort and beer (Jørgensen et al. 1985) and in barley and malt (Jørgensen 1986; Jørgensen and Aastrup 1986; Aastrup and Jørgensen 1987). Complex formation between Calcofluor and β-glucan is aqueous solutions, even at very low concentrations, induces an increase in the fluorescence intensity of the dye (Wood 1982; Wood et al. 1983). The exact mechanism for this reaction is not known, but it is probably caused by a hydrophobic interaction (Wood 1985), and only β-glucan larger than approximately 1×10^4–5×10^4 dalton will induce a full increase in the fluorescence intensity (Foldager and Jørgensen 1984). This is the basis for a simple detection principle for only the larger β-glucan molecules. This is ideal from the brewing point of view. One disadvantage is that Calcofluor in dilute concentrations is very sensitive to light (e.g. Wood 1982). However, by employing the reaction in a system based on Flow Injection Analysis (FIA) a reproducible determination of β-glucan can be made, as this system automatically assures a complete control of the time.

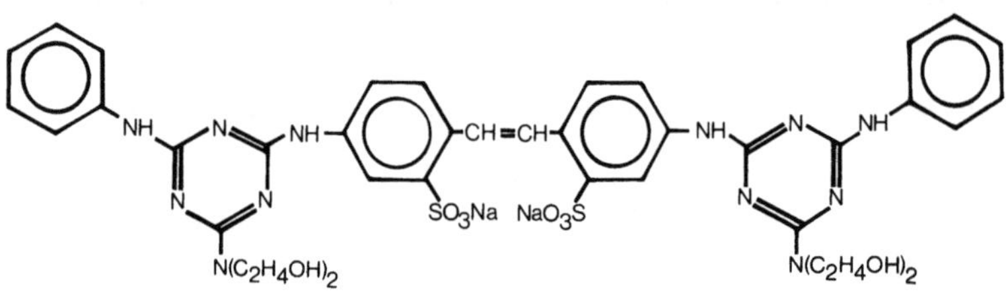

Fig. 3. Chemical structure of Calcofluor

5.2 Apparatus and Conditions

FIA is a well-described technique in the analytical literature (e.g. Ruzicka and Hansen 1981) and systems are commercially available. The principle is simple and easy to set up. In the above-mentioned text book, general rules of thumb and many examples are given.

An apparatus based on FIA for the determination of β-glucan in solution is shown in Fig. 4 (top).

The apparatus consists of: two peristaltic pumps (eight rollers to achieve a high pulsation rate), an injector valve mounted with a 10 μl injection loop, a flu-

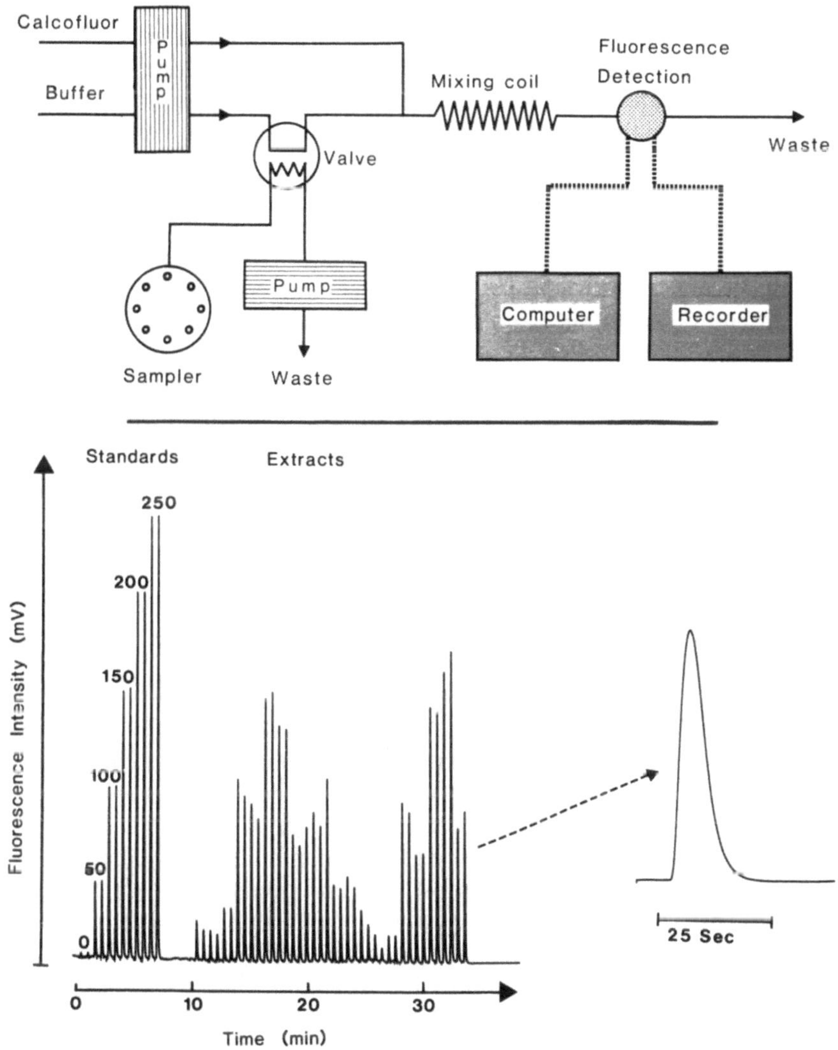

Fig. 4. (*Top*) Apparatus used for automatic β-glucan analysis. (*Bottom*) Responses from the automatic β-glucan analyzer. (Jørgensen 1986)

orescence detector set at 360 nm and 425 nm for excitation and emission respectively and mounted with an 8- μl flow-through cell, tubes of PVC of 0.5 mm ID, a mixing coil made of 1.5 m of the same PVC tubes and a mixing T-piece is made in a block of PVC with drilled 0.5-nm holes and connectors made of pieces of capillary steel glued into the block. The small sample volume used for injection combined with the mixing coil provides an approximately 15 times dilution of the sample before measurement. Sampling is made by a pump connected through the injector valve to an automatic sampler. A computer is used for control of time events (injections) and automatic data collection, calibration and calculation of the results. A recorder is also connected to the fluorescence detector, but only for the purpose of visualization of the reaction.

The system is run as follows:

The reagent consists of a $2.5 \times 10^{-3}\%$ (w/v) Calcofluor solution in 0.2 *M* glycine-NaOH puffer pH 10 and the carrier solution is the buffer alone. The reagent and buffer solutions are delivered at 2 ml/min. Sample rate is approximately 1 sample per 30 s, allowing more than 100 samples to be analyzed per hour [Fig. 4 (bottom)].

Measured as peak heights [Fig. 4 (bottom)], a linear relationship exists between increase in fluorescence intensity and concentration of a *β*-glucan standard in the range 0–250 mg/l. Unlike many other techniques, the system uses the undegraded polymer for calibration. *β*-glucan is commercially available (Biocon Ltd.), but different batches may vary in actual content (Jørgensen 1986). The purity should thus be checked prior to use, e.g. as glucose content after a complete acid hydrolysis. If other sources of *β*-glucan or "home-made" preparations are used, the content of *β*-glucan should be measured by a specific method, e.g. an enzymatic method.

5.3 Quantification of *β*-Glucan in Barley, Malt, Wort and Beer

A sample preparation scheme for barley, malt, wort and beer samples to be measured with the automatic system is illustrated in Fig. 5.

5.3.1 Sample Preparation for Wort and Beer

The system (Fig. 4) will directly measure the large *β*-glucan molecules in solution and the only sample preparation needed for wort and beer is a possible dilution in order to obtain a response within the working range of the system. Beer is in addition decarbonated prior to analysis, e.g. 10 min in an ultrasonic bath.

Although the *β*-glucan contents of most wort and beer samples are measured directly, special attention should be given to very dark samples (e.g. beer of stout types), as the colour causes a minor decrease in the observed fluorescence intensity. The effect of colour can be judged by measuring the absorbance of the sample at 360 nm after a proper dilution (that is, at the excitation wavelength used for fluorescence). If the absorbance is less than about 4 units, no correction is necessary, but if the absorbance is about 15 units (equivalent to a stout type beer), the fluorescence intensity is decreased by approximately 20%. A standard

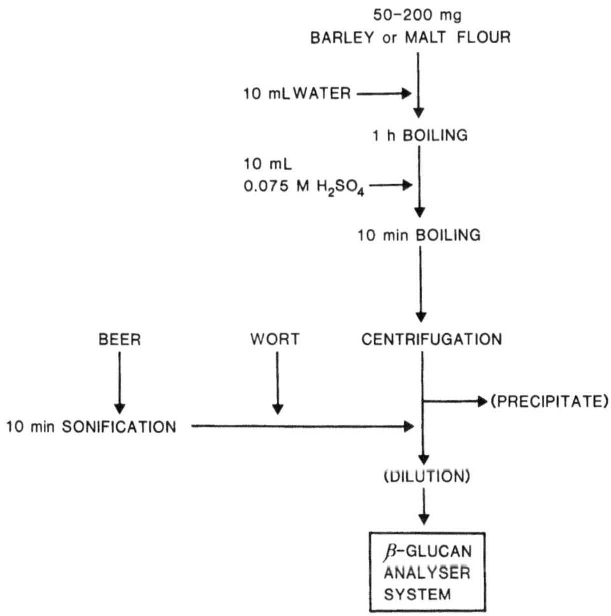

Fig. 5. Sample preparation scheme for barley, malt, wort and beer for determination of β-glucan with the automatic system. (Aastrup and Jørgensen 1987)

curve, where β-glucan is dissolved in solutions of beer caramel giving the absorbance value equal to the samples to be measured may be made to compensate for the colour.

5.3.2 Sample Preparation for Barley and Malt

The system shown in Fig. 4 will only measure β-glucan in solution. In order to determine the total β-glucan content of barley and malt, the water-insoluble β-glucan must be modified in order to be soluble. This is most easily done by a mild acid hydrolysis. By this procedure insoluble β-glucan is degraded to smaller soluble fragments. The hydrolysis however, does not degrade the β-glucan molecules to sizes that are unable to give a full complex formation and increase in fluorescence intensity of Calcofluor.

Solubilization of Total β-Glucan Content of Barley or Malt (Jørgensen and Aastrup 1986)

a) About 50 mg barley or 150 mg malt flour is poured into a 35 ml Pyrex test tube with Teflon-lined screw cap.
b) 10 ml of water is added, the suspension mixed, the cap closed and the sample place for 1 h in boiling water.
c) The sample is cooled, the cap removed and 10 ml of 0.075 M H_2SO_4 is added. The cap is closed and the sample incubated in boiling water for 10 min.
d) The sample is cooled, centrifuged and an aliquot is taken from the supernatant for analysis.

6 Evaluation of Methods

As demonstrated in this chapter, many offers have been made for the analysis of the β-glucan, most for the determination of the barley β-glucan and fewest for the wort and beer β-glucan.

When choosing an appropriate method several factors should be taken into consideration, e.g. speed, accuracy, cost, equipment needed etc. These factors must, however, be judged in respect to the purpose of the measurement. A method which is used for plant breeding must be very simple and fast and does not need to be very accurate. On the other hand, a method for quality control must be reproducible and should preferentially be capable of automatization. If only a limited number of samples are going to be measured, the analysis should be performed adequately with the apparatus available.

6.1 Specificity

A simple way to judge the specificity of different methods is to compare the results which have been reported in the literature. Considering the total β-glucan content of barley, these values differ with different barley varieties, the year of harvest, growing conditions etc. (e.g. Aastrup 1979 b). The differences are, however, relatively small compared with what could be expected for malt, wort and beer samples, as these are also dependent on the processing conditions. For wort and beer a comparison would also be further complicated by different methods differing in their molecular weight specificity. Few investigations have been done in this respect, but Jørgensen et al. (1985) have reported that 30% (w/v) ammonium sulphate precipitates β-glucan molecules larger than about 5×10^4 dalton and that β-glucan larger than about 1×10^4 induce an increase in the fluorescence intensity of Calcofluor. A method based upon precipitation of the β-glucan with 30% (w/v)

Table 1. β-glucan contents of barley

β-Glucan (%)	References
2.7 – 5.2	Jørgensen and Aastrup (1986)[a]
3.64– 6.44	Anderson et al. (1978)[b]
1.93– 6.81	Aastrup (1979)[b]
3.16– 5.78	Smith et al. (1980)[b]
3.80– 4.81	McCleary and Glennie-Holmes (1985)[b]
6.13–10.7	Bamforth and Martin (1981)[c]
4.3 – 6.0	Ahluwalia and Ellis (1984)[d]
3.5 – 4.0	Åman and Hesselman 1985[e]

[a] Calcofluor fluorescence and Flow Injection Analysis.
[b] Enzymatic method, *Bacillus subtilis* β-glucanase.
[c] Enzymatic method, *T. Reseei* cellulase system.
[d] Enzymatic method, *P. funiculosum* cellulase system.
[e] Enzymatic method, *Rhizomucor pusillus* β-glucanase.

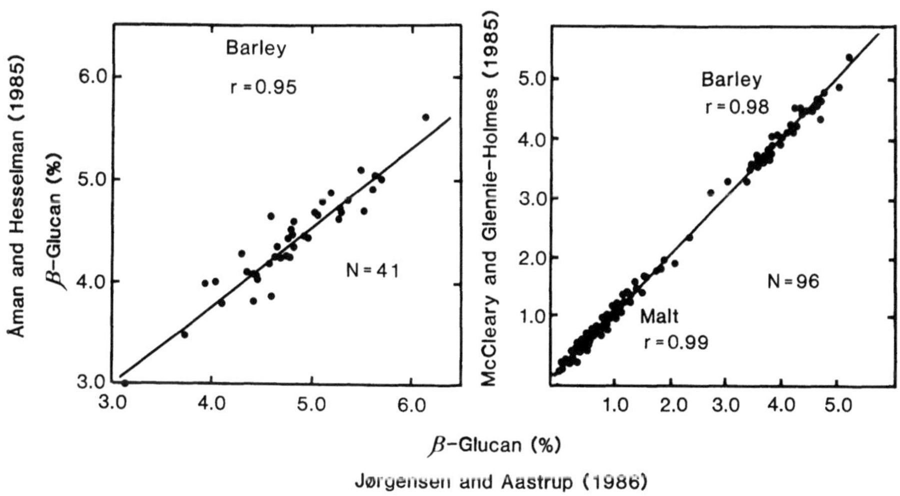

Fig. 6. Relationship between different β-glucan methods

(Sect. 4.2) will therefore give a lower value than the automatic β-glucan analyzer system (Sect. 5).

In Table 1, the results obtained for total barley β-glucan contents using different procedures are shown.

The enzymatic method of Bamforth and Martin (1981) seems to overestimate the β-glucan content, but the other methods give results at the same levels.

The most direct way for comparing different methods is by analysing the same samples with the methods. This has been done for three methods, using barley and malt samples. The results are illustrated in Fig. 6. The non-enzymatic method for Jørgensen and Aastrup (1986) is compared with two enzymatic methods, the method of Åman and Hesselman (1985) and of McCleary and Glennie-Holmes (1985), respectively. Although minor deviations are found, the methods seems to give similar results.

Acknowledgements. The authors wish to thank Dr. L. Munck and Dr. B. Ahrenst-Larsen, heads of Dept. of Biotechnology and Dept. of Brewing Chemistry, respectively for permission to publish this work. The advice of our colleagues Dr. P. Vaag, Dr. K. Erdal, and Dr. E. A. Jackson and the skilful drawing of the figures by Mr. K. Hansen is greatly appreciated.

References

Aastrup S (1979a) The relationship between the viscosity of an acid flour extract and its β-glucan content. Carlsberg Res Commun 44:289–304

Aastrup S (1979b) The effect of rain on β-glucan content in barley grains. Carlsberg Res Commun 44:381–393

Aastrup S (1983) Selection and characterization of low β-glucan mutant from barley. Carlsberg Res Commun 48:307–316

Aastrup S, Jørgensen KG (1986) Fluorimetric methods for the analysis of malt modification and pre-harvest sprouting. In: Linskens HF, Jackson JF (eds) Modern methods of plant analysis, new series, vol VII. Springer, Berlin Heidelberg New York, pp 56–66

Aastrup S, Jørgensen KG (1987) A new and simple method for determination of β-glucan. J Am Soc Brew Chem (in press)

Ahluwalia B, Ellis EE (1984) A rapid and simple method for the determination of starch and β-glucan in barley and malt. J Inst Brew 90:254–259

Åman P, Hesselman K (1985) An enzymatic method for total mixed-linkage β-glucans in cereal grains. J Cereal Sci 3:231–237

Anderson MA, Cook JA, Stone BA (1978) Enzymatic determination of 1,3:1,4-β-glucans in barley grain and other cereals. J Inst Brew 84:233–239

Bamforth CW (1982) Barley β-glucans, their role in malting and brewing. Brew Dig June:22–27, 35

Bamforth CW (1983) Penicillium funiculosum as a source of β-glucanase for the estimation of barley β-glucan. J Inst Brew 89:391–392

Bamforth CW, Martin HL (1981) β-glucan and β-glucan solubilase in malting and mashing. J Inst Brew 87:365–371

Bamforth CW, Martin HL (1983) The degradation of β-glucan during malting and mashing: the role of β-glucanase. J Inst Brew 89:303–307

Bamforth CW, Martin HL, Wainwright T (1979) A role for carboxypeptidase in the solubilisation of barley β-glucan. J Inst Brew 85:334–338

Bendelow WM (1975) Determination of non-starch polysaccharides in barley breeding programmes. J Inst Brew 81:127–130

Bourne DT, Powesland T, Wheeler RE (1982) The relationship between total β-glucan of malt and malt quality. J Inst Brew 88:371–375

Clarke AE, Stone BA (1963) Chemistry and biochemistry of 1→3-glucans. Rev Pure Appl Chem 13:134–156

Crabb D, Bathgate GN (1973) The influence of β-glucanase on effiency of wort separation. J Inst Brew 79:519–524

Dubois M, Gilles KA, Hamilton JK, Rebers PA, Smith F (1956) Colorimetric method for determination of sugars and related substances. Anal Chem 28:350–356

Enari TM (ed) (1975) Analytical EBC, 3rd edn. Schweiz Brau Rundsch

Erdal K, Gjertsen P (1967) β-glucans in malting and brewing, II. The fate of β-glucans during mashing. Proc Eur Brew Conv Congr, Madrid, pp 49–57

Erdal K, Gjertsen P (1971) β-glucans in malting and brewing, III. The action of endo-β-glucanases. Proc Eur Brew Conv Congr, Estoril, pp 49–57

Fincher GB (1975) Morphology and chemical composition of barley endosperm cell walls. J Inst Brew 81:116–122

Fleming M, Kawakami K (1977) Studies of the fine structure of β-D-glucans of barley extracted at different temperatures. Carbohydr Res 57:15–23

Fleming M, Manners DJ, Jackson RM, Cooke SC (1974) The estimation of β-glucan in barley. J Inst Brew 80:399–404

Foldager L, Jørgensen KG (1984) The molecular weight distribution of β-glucan in wort from malts of different barley varieties at different stages of malting. Carlsberg Res Commun 49:525–534

Forrest IS, Wainwright T (1977a) The mode of binding of β-glucans and pentosans in barley endosperm cell walls. J Inst Brew 83:279–286

Forrest IS, Wainwright T (1977b) Differentiation between desirable and troublesome β-glucans. Proc Eur Brew Conv Congr, Amsterdam, pp 401–413

Gjertsen P (1966) β-glucans in malting and brewing. I. Influence of β-glucans on the filtration of strong beer. Proc Am Soc Brew Chem, pp 113–120

Greenberg DC, Whitmore ET (1974) A rapid method for estimating the viscosity of barley extracts. J Inst Brew 80:31–33

Henry RJ (1984) A simplified enzymic method for the determination of (1→3)(1→4)-β-glucans in barley. J Inst Brew 90:178–180

Henry RJ (1985) Near infrared reflectance analysis of carbohydrates and its applications to the determination of (1→3)(1→4)-β-D-glucan in barley. Carbohydr Chem 141:13–19

Henry RJ, Blakeney AB (1986) Determination of total β-glucan in malt. J Inst Brew 92:354–356
Jensen SA, Aastrup S (1981) A fluorimetric method for measuring 1,3:1,4-β-glucan in beer, wort, malt and barley by use of Calcofluor. Carlsberg Res Commun 46:87–95
Jørgensen KG (1983) An improved method for the determination of β-glucan in wort and beer by use of Calcofluor. Carlsberg Res Commun 48:505–516
Jørgensen KG (1986) Quantification of (1→3)(1→4)-β-D-glucan using Calcofluor complex formation and flow injection analysis. I. Analytical principle. Carlsberg Res Commun (in press)
Jørgensen KG, Aastrup S (1986) Quantification of (1→3) (1→4)-β-D-glucan using Calcofluor complex formation and flow injection analysis. II. Determination of total β-glucan content of barley and malt. Carlsberg Res Commun (in press)
Jørgensen KG, Jensen SA, Hartlev P, Munck L (1985) The analysis of β-glucan in wort and beer using Calcofluor. Proc Eur Brew Conv Congr, Helsinki, pp 403–410
MacDonald RS (1986) Review: infrared spectroscopy. Anal Chem 58:1906–1925
Madacsi JP, Parrish FW, Roberts EJ (1983) Non-enzymic method for the determination of beta-glucan in the presence of starch. J Am Soc Brew Chem 41(4):161–162
Martens HA (1985) Multivariate calibration. Dr. Tech Thesis, Technical University of Norway, Trondheim
Martin HL, Bamforth CW (1981) An enzymatic method for the measurement of total and water-soluble β-glucan in barley. J Inst Brew 87–88–91
McCleary BV, Glennie-Holmes M (1985) Enzymic quantification of (1→3)(1→4)-β-D-glucan in barley and malt. J Inst Brew 91:285–295
McCleary BV, Nurthen E (1986) Measurement of (1→3)(1→4)-β-D-glucan in malt, wort and beer. J Inst Brew 92:168–173
Morgan AH, Gothard PG (1977) A rapid, simple viscometric technique for indirect estimation of soluble β-glucan content of raw barley. J Inst Brew 83:37–38
Morrall P, Briggs DE (1978) Changes in cell wall polysaccharides of germinating barley grains. Phytochemistry 17:1495–1502
Preece IA, MacKenzie KG (1952) Non-starchy polysaccharides of cereal grains. II. Distribution of water soluble gum-like materials in cereals. J Inst Brew 58:457–464
Prentice N, Babler S, Faber S (1980) Enzymic analysis of β-D-glucans in cereal grains. Cereal Chem 57(3):198–202
Ruzicka J, Hansen EH (1981) Flow injection analysis. In: Elving PJ, Wineforner JD, Kolthoff IM (eds) Chemical analysis, vol 62. Viley, New York
Schuster K, Narziss L, Kumada J (1967) Über die Gummistoffe der Gerste und ihr Verhalten während der Malz- und Bierbereitung. I. Problemstellung und Analysemethoden. Brauwissenschaft 20:125–135
Smith DB, Morgan AG, Aastrup S (1980) Variation in the biochemical composition of acid extracts from barley of contrasting malting quality. J Inst Brew 86:277–283
Staudte RG, Woodward JR, Fincher GB, Stone BA (1983) Water soluble (1→3)(1→4)-β-D-glucans from barley (Hordeum Vulgare) endosperm. III. Distribution of cellotriosyl and cellotetraosyl residues. Carbohydr Polym 3:299–312
Wainwright T, Buckee GK (1977) Barley and malt analysis – a review. J Inst Brew 83:325–347
Wood PJ (1980) Specificity in the interaction of direct dyes with polysaccharides. Carbohydr Res 85:271–287
Wood PJ (1982) Factors affecting precipitation and spectral changes associated with complex formation between dyes and β-D-glucans. Carbohydr Res 102:283–293
Wood PJ (1984) Physiochemical properties and technological and nutritional significance of cereal β-glucans. In: Rasper VF (ed) Polysaccharides in technology and nutrition. Am Acad Cer Chem Symp, St. Paul, pp 35–77
Wood PJ (1985) Dye-polysaccharide interactions – recent research and applications. In: Hill RD, Munck L (eds) New approaches to the research in cereal carbohydrates. Elsevier, Amsterdam (Progress in Biotechnology, vol 1)
Wood PJ, Fulcher RG (1978) Interaction of some dyes with cereal β-glucans. Cereal Chem 55(6):952–966

Wood PJ, Weiz J (1984) Use of Calcofluor in analysis of oat β-D-glucan. Cereal Chem 61(1):73–75
Wood PJ, Paton D, Siddiqui IR (1977) Determination of β-glucan in oats and barley. Cereal Chem 54(3):524–533
Wood PJ, Fulcher RG, Stone BA (1983) Studies on the specificity of interaction of cereal cell wall components with Congored and Calcofluor. Specific detection and histochemistry of (1→3)(1→4)-β-D-glucan. J Cereal Sci 1:95–110
Woodward JR, Fincher GB (1983) Water-soluble barley β-glucans. Fine structure solution behaviour and organization in the cell wall. Brew Dig May:28–32

Analysis of Proanthocyanidins and Phenolic Acids in Barley, Malt, Hops and Beer

B. JENDE-STRID

1 Introduction

Barley grains and hop cones contain a great number of phenolic compounds, such as proanthocyanidins, catechins, phenolic acids and coumarins. A variety of methods for separation and identification of these substances in beer and its raw materials have been developed, and many investigations concerning the role of different phenolic compounds as flavour components of beer and possible constituents of beer haze have been carried out. In this chapter some of the most frequently used analytical methods are described with the emphasis on high-performance liquid chromatography (HPLC).

The proanthocyanidins have attracted much interest because of their ability to precipitate proteins in beer, resulting in non-biological haze (Vancraenenbroek et al. 1979; Outtrup and Erdal 1983; Delcour et al. 1984). The proanthocyanidins were rather early recognized as major phenolic components of beer haze (McFarlane et al. 1955).

In the literature there is some confusion concerning the proper designation of proanthocyanidins. Brewery chemists often call them anthocyanogens. This term was introduced by Harris and Ricketts (1959) for all substances in beer and wort that yield red anthocyanidins on heating with hydrochloric acid. Sometimes catechins and proanthocyanidins are referred to as flavanoids because of their more saturated nature when compared, for instance, to flavones and flavonols, which have an additional double bond between C2 and C3 in their heterocyclic ring. Other frequently used terms for proanthocyanidins and catechins are polyphenols and tannins.

The proanthocyanidins of barley consist mainly of dimers and trimers of (+)-catechin and (+)-gallocatechin, which also occur as monomers. The latter is only reported by Brandon et al. (1982), and it was found in small quantities, compared to (+)-catechin (1:6). The structures of two dimers (procyanidin B-3 and prodelphinidin B-3), one trimer (procyanidin C-2) and three prodelphinidin trimers were established by Mulkay et al. (1981 a, b) and Outtrup and Schaumburg (1981) by use of ^{1}H and ^{13}C NMR spectroscopy (see Fig. 1). Brandon et al. (1982) isolated a trimer consisting only of (+)-gallocatechin units from grains of Gwylan barley. The structure was based on degradation studies with phloroglucinol and hydrochloric acid.

Around 500 proanthocyanidin-free barley mutants have been induced at the Carlsberg Laboratory. The biosynthesis of catechin and proanthocyanidins is blocked in the grains of these mutants, but other flavonoids and phenolic acids are synthesized (von Wettstein et al. 1985).

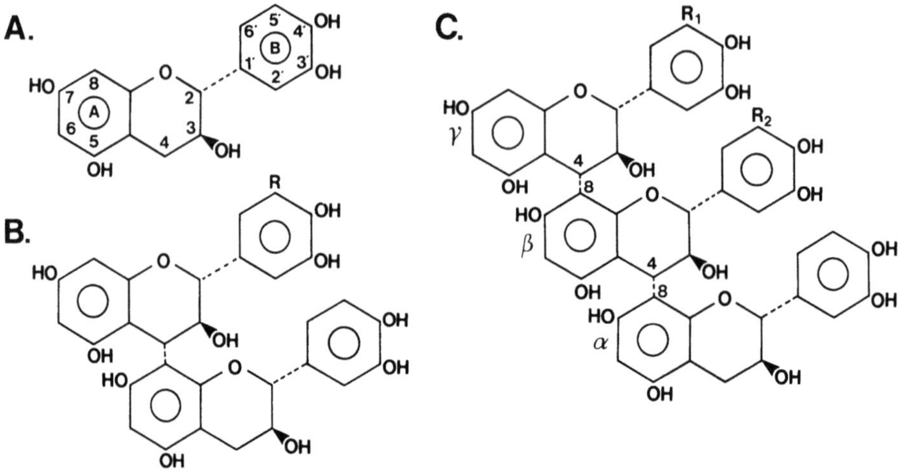

Fig. 1 A–C. Structures of barley flavanoids. **A** (+)-catechin. **B** *1* R=H procyanidin B-3; *2* R=OH prodelphinidin B-3. **C** *1* R_1=H, R_2=H procyanidin C-2; *2* R_1=OH, R_2=H prodelphinidin trimer (α C_{8-4}, β C_{8-4}, γGC); *3* R_1=H, R_2=OH prodelphinidin trimer (αC_{8-4}, β GC_{8-4}, γC); *4* R_1=OH, R_2=OH prodelphinidin trimer (α C_{8-4}, β GC_{8-4}, γ GC). C=(+)-catechin, GC=(+)-gallocatechin. The catechin units are linked in positions *8-4*

The concentration of proanthocyanidins in hop cones is considerably higher than in barley grains, and there is a great variation between different hop varieties. A content as high as 5% (dry weight) has been reported by Erdal et al. (1985). The hop proanthocyanidins consist mainly of (+)-catechin and (−)-epicatechin units. These are also found as monomers. According to McMurrough et al. (1983), the major part of the hop proanthocyanidins is more highly polymerized than is the case in barley. The structures of these higher polymers are not known. The structures of procyanidins B-1, B-3, and B-4, isolated from hops, were elucidated by Gorissen (1969) by ^{1}H NMR spectroscopy. McMurrough (1981) detected the procyanidin dimers B-1, B-2, B-3, and B-4, and also reported the presence of delphinidin and cyanidin in hydrolysates of hop extracts. Erdal et al. (1985) found the procyanidin dimers B-2, B-3, and B-4, the procyanidin trimer C-1 and one dimer and one trimer of prodelphinidin in hop samples analyzed by HPLC.

In addition to proanthocyanidins, hops contain great amounts of flavonol glucosides. The most abundant flavonols are mono- and diglucosides of kaempferol and quercetin. Also triglucosides are present (Vancraenenbroek et al. 1980).

McMurrough (1981) analyzed ten varieties of hops and found an average of 1.20 mg/g of kaempferol and 0.92 mg/g of quercetin, measured as aglycones. Flavonols are not believed to be an important factor in haze formation.

The different proanthocyanidins from barley and hops and the hop flavonols are recovered in the beer. In addition, a prodelphinidin B-9 [(+)-epigallocatechin – (+)-catechin] has been isolated from beer by Delcour and Tuytens (1984). The structure was determined by ^{1}H NMR and circular dichroism spectroscopy.

Around 80% of the beer proanthocyanidins originate from barley and the rest from hops (Knorr 1977). Moll et al. (1984) followed the changes in polyphenol content from barley and malt to finished beer. They found that malting strongly reduces the content of (+)-catechin, whereas procyanidin B-3 remains constant. During malting and wort boiling the dimeric proanthocyanidins decrease in concentration, and (+)-catechin and (−)-epicatechin, as well as complexed proanthocyanidins, are formed. Some of these complexed proanthocyanidins are removed with the break, but the remaining soluble part, together with free proanthocyanidins and catechins, will combine with proteins and amino acids and give rise to permanent haze during the storage of beer. The composition of the complexed polyphenols is difficult to investigate and has not been studied in detail.

Phenolic acids are widespread in the plant kingdom and several of them have been identified in barley (Urion et al. 1963; McMurrough et al. 1984; Jende-Strid 1985) The phenolic acids seldom occur in free form, but usually as glucosides or esters. The hydroxycinnamic acids are important precursors in the biosynthesis of flavonoids and lignins, and some of the phenolic carboxylic acids are present in lignins as esters (Harborne and Simmonds 1964). The chemical structures of some phenolic acids are shown in Fig. 2. The following phenolic acids have been

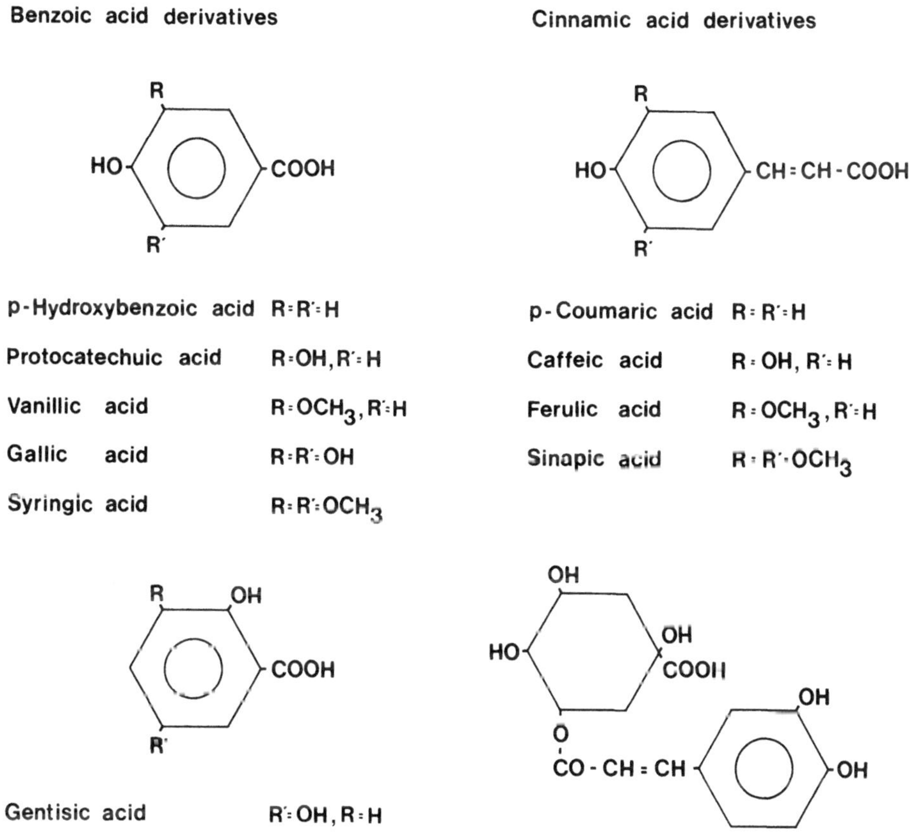

Fig. 2. Structures of phenolic acids

found in barley grains: gallic acid, protocatechuic acid, gentisic acid, p-hydroxy-bensoic acid, chlorogenic acid, vanillic acid, caffeic acid, syringic acid, p-coumaric acid, ferulic acid, sinapic acid and cinnamic acid. The composition of phenolic acids varies, depending on the type of hydrolysis used in the preparation of the barley extract. No qualitative differences between wild-type barley and proanthocyanidin-free mutants in different genes have been detected.

The amount of phenolic acids in barley grains is low compared to the amount of proanthocyanidin and catechin. Typically wild-type barley varieties contain around 100–160 mg proanthocyanidin and catechin per 100 g (dry weight) as measured by HPLC (Jende-Strid and Møller 1981; Jerumanis 1985), and the total amount of acetone extractable phenolic acids in wild-type barley and proanthocyanidin-free mutants is around 10–20 mg per 100 g, i.e. ca. 10% of the amount of the flavanoids (Jende-Strid 1985). Simple phenolic acids are not considered as significant flavour constituents of beer or major components of beer haze (Dadic and Belleau 1980; McMurrough et al. 1984).

2 Choice of Analytical Methods

The different methods available for analysis of phenolic compounds in barley, malt, hops and beer can be divided into two main categories: chromatographic methods and colourimetric methods.

The chromatographic methods include: column chromatography, paper chromatography (PC), thin layer chromatography (TLC), gas-liquid chromatography (GLC) and high-performance liquid chromatography (HPLC). Early investigations using PC and TLC showed that beer and its raw materials contained very complex mixtures of phenolic compounds. It was impossible to achieve good separation of proanthocyanidins with these techniques, and quantitative analyses of different substances were not accurate. Separation and quantitative analysis of phenolic compounds has improved greatly with the application of the HPLC technique. Analysis by HPLC has become the method of choice for most scientists working with phenolic compounds in plant material. It is a fast procedure compared to other chromatographic methods, and its resolving capacity and accuracy is unsurpassed (see Fig. 3–5). It is an expensive method, however. GLC has been used for separation and quantitative analysis of phenolic compounds (Salomonsson et al. 1978; Vancraenenbroek et al. 1979). Compared to HPLC, GLC is a more cumbersome technique to use, because derivatization of the phenolic compounds prior to analysis is necessary, and the high molecular weight of dimers and higher oligomers of proanthocyanidins make them less volatile. Column chromatography is mainly used for fractionation and purification of phenolic samples. It is an important method when large samples are processed, but usually too slow a procedure to be included in routine analysis of barley and beer samples.

Many colourimetric methods for quantitative analysis of proanthocyanidins and catechins in barley, malt, hops and beer have been worked out. Analyses employing colourimetric methods are fast and cheap compared to HPLC analysis,

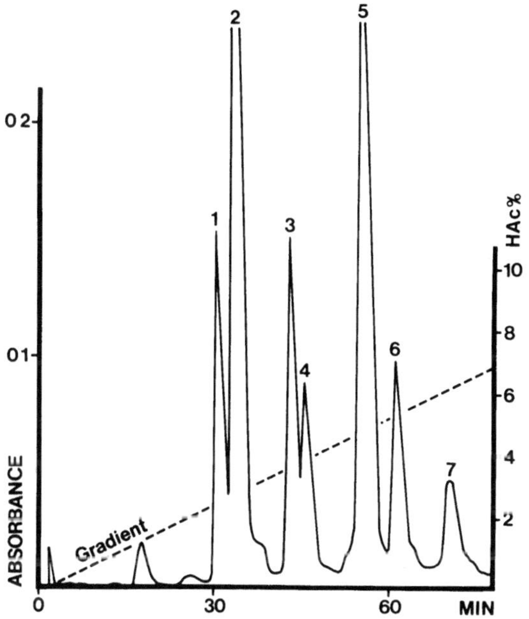

Fig. 3. Separation of flavanoids of barley (var. Eldorado) by HPLC. Column: RSil C_{18} HL 10 µ, eluent : water : acetic acid, flow rate 1 ml/min, absorbance 280 nm. *1* Prodelphinidin trimer (α C_{8-4}, βGC_{8-4}, γ GC); *2* prodelphinidin B-3, *3* prodelphinidin trimer (α C_{8-4}, β C_{8-4}, γ GC); *4* prodelphinidin trimer (α C_{8-4}, β GC_{8-4}, γ C); *5* procyanidin B-3; *6* procyanidin C-2; *7* (+)-catechin. (Jerumanis 1985)

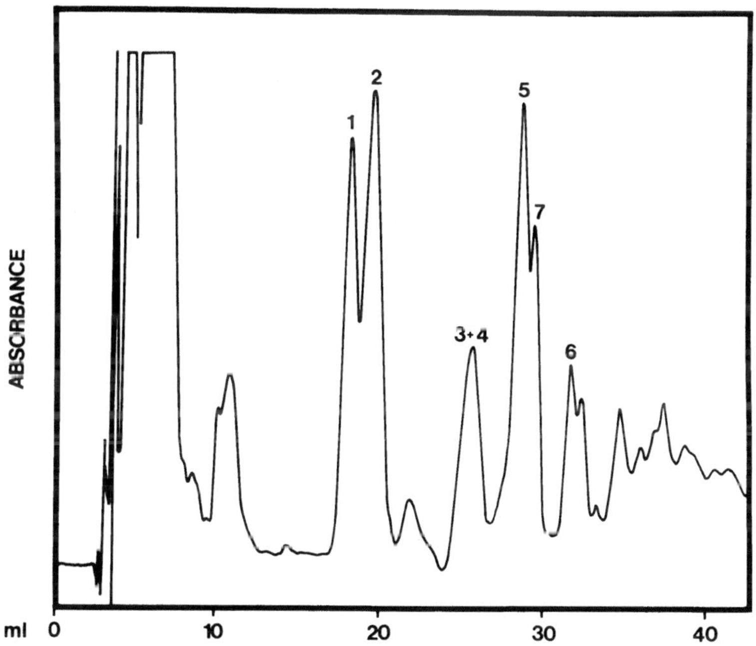

Fig. 4. Separation of flavanoids in barley (var. Bonus) by HPLC. Column μ-Bondapak Phenyl, eluent : water : acetic acid, flow rate 1 ml/min, absorbance 280 nm. Peak numbers refer to the same compounds as in Fig. 3.

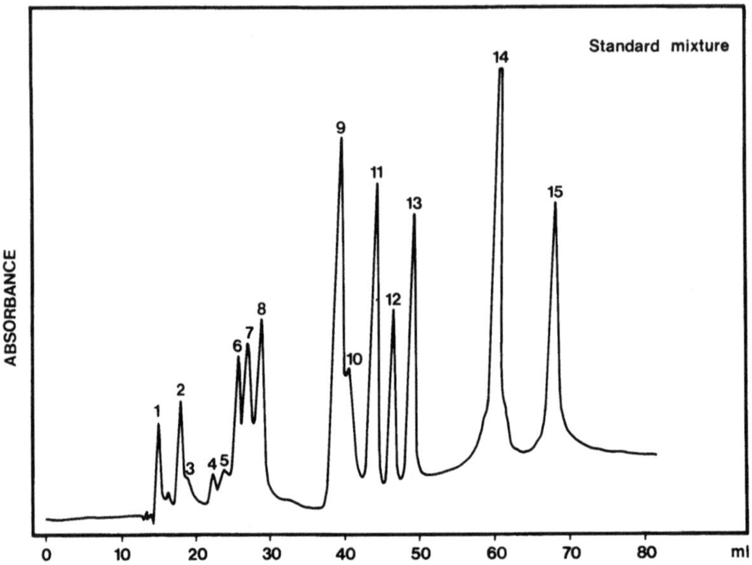

Fig. 5. Separation of 16 phenolic standards by HPLC. Column: μ-Bondapak C_{18}, eluent: methanol : acetic acid : water, flow rate: 2 ml/min, absorbance 280 nm. *1* Gallic acid; *2* protocatechuic acid; *3* (+)-catechin; *4* gentisic acid + p-hydroxybenzoic acid; *5* chlorogenic acid; *6* vanillic acid; *7* caffeic acid; *8* syringic acid; *9* p-coumaric acid; *10* dihydroquercetin; *11* ferulic acid; *12* sinapic acid; *13* o-coumaric acid; *14* cinnamic acid; *15* naringenin. 0.25 µg of each sample was injected

but the amounts of individual flavanoids cannot be determined using these methods. However, if routine analyses of great numbers of samples are performed, it is often sufficient to know the total amount of flavanoids in a sample.

Several methods have been based on the conversion of proanthocyanidins to red anthocyanidins upon heating in acidic solution. Unfortunately, this conversion is not quantitative, and the yield of anthocyanidin has been reported to vary from 10% to 27%. In alcoholic solution the yield is around 18% (Dadic 1971). The proanthocyanidins are adsorbed on a polyamide resin (Nylon 66, Divergan or Polyclar AT) followed by heating of the adsorbate in alcoholic hydrochloric acid and spectrophotometric measurement of the resulting anthocyanidins (Harris and Ricketts 1959; McFarlane 1961). The amount of proanthocyanidins in a sample is calculated from a standard curve. These methods do not determine free catechins. McFarlane et al. (1964) and Dadic (1971) have developed methods for determination of catechin as well as proanthocyanidin in the same sample. McFarlane et al. separated the catechins from the proanthocyanidins by chromatography on Polyclar AT. Dadic based his method on binary analysis of the absorbances at two suitable wavelengths (455 and 545 nm) after heating the sample in alcoholic hydrochloric acid. All these methods are too time-consuming to be used for routine analysis of large numbers of samples.

Jerumanis' method (1968) is frequently used to measure total polyphenol content in beer and wort. The polyphenols are reacted with ammonium ferric citrate in alkaline solution. A red-coloured complex is formed and the extinction measured on a spectrophotometer. Only compounds having an o-diphenol structure will react under these conditions. The proanthocyanidins and catechins present in beer are o-diphenols, but some of the flavonols and phenolic acids also have this structure. This makes the method less specific for flavanoids than the other colourimetric methods described here.

Vanillin and p-dimethylaminocinnamaldehyde in acidic solution are two specific reagents for proanthocyanidins and catechins. Flavanoids form red-coloured complexes with vanillin in acidic solution and green ones with p-dimethylaminocinnamaldehyde. A number of methods have been based on the vanillin reaction. An improved vanillin method has been worked out by Broadhurst and Jones (1978). Delcour and de Varebeke (1985) have developed an assay for analysis of proanthocyanidins and catechins in beer founded on the p-dimethylaminocinnamaldehyde reaction. The two methods are described later in this chapter (p 124).

If the same material is analyzed with different colourimetric methods, the results are not commensurable. Delcour and de Varebeke (1985) analyzed beer with different methods and found a wide variation in polyphenol content in the same sample, depending on the method used. For instance, one beer sample had a polyphenol content of 148 mg/1, when analyzed with Jerumanis' method, 117 mg/1 (vanillin method), 21.6 mg/1 (p-dimethylaminocinnamaldehyd reagent) and 30.4 mg/1 (anthocyanogen method). Another sample contained 86, 124, 16.4, and 13.3 mg/1 respectively, when analyzed with the same methods. But the same method used on different samples wil provide adequate comparisons of the contents of proanthocyanidins and catechins, even if the results are not as accurate as those obtained by HPLC analysis.

3 Chromatographic Methods

3.1 High Performance Liquid Chromatography (HPLC)

HPLC equipment of standard type can be used for analysis of proanthocyanidins and phenolic acids.

3.1.1 Preparation of Barley and Malt Samples

Grains of barley or malt are ground into a fine powder, which is shaken with a suitable solvent for 30–60 min. An Omnimixer can also be used for the extraction. The proportion of barley or malt powder to solvent should be 1 : 5 to 1 : 10 (w/v). Acetone-water (3 : 1, v/v) is the most widely used solvent, and it gives a higher yield of proanthocyanidins than aqueous ethanolic or aqueous methanolic solvents do (Jones et al. 1976). Dimethylformamide has also been used for extraction, but should be avoided as it is very poisonous. The yield of phenolic acids and proanthocyanidins is increased, if the extraction is repeated two or three times. About 99% of the acetone extractable compounds are obtained in the first three extractions (Jende-Strid and Møller 1981; Jende-Strid 1985). The acetone-water extracts contain many UV-absorbing substances, and purification of the extracts is needed before analysis by HPLC or by other chromatographic methods.

Lipids are removed from the combined acetone-water extracts by shaking with petroleum benzin or petroleum ether. If needed, centrifugation at low speed (3000 *g*) will improve the separation of the petroleum benzine phase from the aqueous phase. The acetone is evaporated on a Rotary evaporator.

The aqueous residue can be treated in different ways, depending on its further use. The phenolic compounds can be fractionated on a Sephadex LH-20 column or extracted from the water phase with ethylacetate. The extraction should be repeated three to four times to ensure maximum yield. The combined ethylacetate extracts are evaporated to dryness and the residue dissolved in a small volume of ethanol or methanol. After centrifugation or filtration, the sample is ready to be injected into the HPLC.

Ethylacetate extraction is a method commonly used to purify and concentrate samples containing phenolic compounds. The disadvantage of this method is that polar compounds, such as prodelphinidins, gallic acid and phenolic glucosides are not extracted quantitatively into the ethylacetate. The same problem will arise if hexan is used instead of ethylacetate (McMurrough et al. 1984). If phenolic acids are analyzed, it is advantageous to hydrolyze the water phase prior to ethylacetate extraction. Phenolic acids occur mainly as glucosides or esters, and the yield of free acids is greatly increased if the samples are subjected to either acidic or alkaline hydrolysis (Salomonsson et al. 1978; Andersen and Pedersen 1983). Acidic hydrolysis can be performed in 2N-HCl at 100 °C for 20 min, and alkaline hydrolysis can be done in 1*M*-NaOH with N_2 bubbling at room temperature over night (Jende-Strid 1985). Ethylacetate extraction should be omitted if proanthocyanidins are analyzed quantitatively. Jende-Strid and Møller (1981) found that the total yield of proanthocyanidins in a sample is less than 50%, and there is specific loss of the more polar proanthocyanidins if an ethylacetate extraction step is included in the preparation of samples for HPLC analysis.

Jerumanis (1985) has devised a method for quantitative analysis of flavanoids in barley, hops and beer by high-performance liquid chromatography. He recommends the following procedure for preparation of barley and malt samples:

Fifteen g of finely ground barley or malt flour is extracted in an Omnimixer with 100 ml of 75% (v/v) acetone for 2 × 2 min with an interval of 15 min. The mixture is filtrated and the first 15 ml of filtrate discarded. 25 ml of filtrate and 10 ml of chloroform are shaken in an centrifuge tube and then centrifuged at 7000 *g* for 15 min. The content is poured into a separating funnel and the bottom phase, which is chloroform, is removed. The upper phase is poured into a 100-ml flask. Centrifuge tube and funnel are washed with water and the washings are poured into the flask. 25 mg of ascorbic acid is added to prevent oxidation, and the volume is adjusted to 100 ml with distilled water.

A glass column (25 × 2 cm), filled with Polyamid-6 (1.2 g in distilled water), is washed with 75% acetone and distilled water. 40 ml of barley extract is mixed with 0.5 g of Polyamid-6 and poured into the column. After adsorption of the polyphenols, the column is rinsed with distilled water and eluted with 2 × 10 ml and 1 × 5 ml of 75% acetone. The eluent is evaporated to dryness and dissolved in 0.6 ml of 50% (v/v) methanol. Aliquots of 50 μl are injected into the HPLC. This quantity corresponds to 125 mg of barley or malt.

3.1.2 Preparation of Hop Samples

Hop extracts suitable for HPLC analysis of proanthocyanidins and phenolic acids can be produced using the same methods as described for barley and malt.

McMurrough (1981) prepared hop extracts for HPLC analysis of flavonol aglycones by refluxing the finely ground hop flour in methanol-4N-HCl (3 : 1, v/v) for 2 h. After filtration and washing with methanol, the combined filtrate and washings are shaken with hexane. The hexane phase is discarded, and a sample of the methanol filtrate is centrifuged and analyzed by HPLC.

Erdal et al. (1985) extracted hop flour with 60% (v/v) ethanol containing 1% ascorbic acid. After evaporation to dryness, the residue is taken up in water and used for HPLC analysis.

3.1.3 Preparation of Beer Samples

Beer samples should always be decarbonated prior to analysis. In beer and wort proanthocyanidins and phenolic acids are already in liquid phase, and acetone extraction is not necessary. Centrifugation may be needed if the beer and wort samples are cloudy. Qureshi et al. (1979) analyzed polyphenols and other potential flavouring compounds in wort and beer by HPLC. They injected decarbonated beer samples and centrifuged wort samples directly into the HPLC without any prior treatment. Other workers have been unable to obtain good separation of polyphenols in untreated beer samples, and McMurrough et al. (1984) recommended shaking of the beer with isooctane after the pH has been adjusted to 2.0 with HCl. The remaining aqueous phase is then extracted four times with ethylacetate. The combined ethylacetate extracts are evaporated in vacuo to 1–2 ml and diluted with methanol. Samples of the methanol solution are analyzed by HPLC. Jerumanis (1985) prepared decarbonated and centrifuged beer and wort samples in the same way as acetone extracts of barley or malt. Moll et al. (1984) have developed a system where by the polyphenols from the beer and wort samples are adsorbed on Nylon 66 and then eluted with 75% acetone. The acetone extract is treated as described for barley and malt extracts.

3.1.4 Choice of HPLC Columns

The best separations of phenolic compounds are achieved on reversed phase HPLC columns. The compounds are eluted from a reversed phase column in order of decreasing polarity. Most workers have used μ-Bondapak C_{18} columns (Waters Assoc.), but good separations have also been obtained on other types of reversed phase columns, for instance: R Sil C_{18} (Jerumanis 1979), Li-Chrosorb RP 18 (Van de Casteele et al. 1983) and Nucleosil C_{18} (Andersen and Pedersen 1983). A particle size of 10 μ of the column packing material is recommended for analytical work, whereas a particle size of 30 μ can be used for preparative purposes. A precolumn will prolong the life of the analytical column.

3.1.5 Mobile Phase

Good separation of polar phenolic compounds, such as proanthocyanidins can be achieved using a gradient system consisting of acetic acid and water.

Two examples:

1. A linear gradient from 0% to 10% acetic acid (v/v) for 2 h. Flow rate 1 ml/min. [Separation of all proanthocyanidins in barley, Jerumanis (1985, see Fig. 3).]

2. A linear gradient from 2% to 10% acetic acid (v/v) for 30 min. Flow rate 2 ml/min. [Separation of (+)-catechin, leucocyanidins, and proanthocyanidins, Kristiansen (1984, see Table 1).]

For separation of other types of flavonoids and phenolic acids, methanol should be included in the solvent mixtures. Methanol will decrease the retention times of the less polar compounds, and the proportion of methanol in the mixtures can be chosen to ensure reasonable retention times and good separation of the different compounds. Acetonitrile is used by some workers instead of methanol. It gives at least as good a separation as methanol, but is more poisonous.

Examples of solvent systems:

1. Solvent A (methanol : acetic acid : water, 20 : 5 : 75, v/v/v) for 10 min, then a linear gradient to solvent B (methanol : acetic acid : water, 40 : 5 : 55, v/v/v) for 10 min and maintained at solvent B for 20 min. Flow rate 2 ml/min. [Separation of phenolic acids, Jende-Strid (1985, see Fig. 5 and Table 1).]
2. A linear gradient from 2.5% aqueous acetic acid (v/v) to 50% methanol in 2.5% aqueous acetic acid (v/v) for 30 min. Flow rate 2 ml/min. [Separation of phenolic acids and proanthocyanidins, McMurrough et al. (1984).]
3. A linear gradient from solvent A (water : formic acid, 95 : 5, v/v) to solvent B (methanol, 100%) for 34 min. Flow rate 2.5 ml/min. [Separation of anthocyanins, proanthocyanidins, phenolic acids and other phenolic compounds, Van de Casteele et al. (1983).]
4. Isocratic elution with water : acetonitrile : acetic acid (88 : 10 : 2). Flow rate 2 ml/min. [Separation of phenolic acids, coumarins and flavonols, Carpentier and Cowles (1981).]

280 nm is a suitable wavelength for peak detection when proanthocyanidins and phenolic acids are analyzed.

Elution volumes of some phenolic acids and proanthocyanidins are given in Table 1.

3.1.6 Standards

The peaks on the HPLC chromatograms can be identified by use of standard compounds. The retention times or elution volumes of the unknown peaks and those of the standard compounds are compared. This is, however, not an absolutely safe way to identify a peak on a complex chromatogram of a beer or a barley sample, because two different compounds can have the same or very similar retention times under the conditions used for the analysis, and thus appear as one peak on the chromatogram. In order to investigate that, the eluate containing the peak can be collected, concentrated and analyzed by TLC. Standard phenolic

Table 1. Elution volumes with HPLC and R_f values with TLC of phenolic acids and proanthocyanidins. Details of the chromatographic systems are given in Sects. 3.1.5 and 3.2. (Kristiansen 1984; Jende-Strid 1985)

	Elution volume (ml)		R_f value		s BAWC
	1	2	CAW	15% CH_3COOH	
Gallic acid	13.5	–	0.17	0.53	–
Protocatechuic acid	16.7	–	0.44	0.58	–
Gentisic acid	21.2	–	0.60	0.76	–
p-Hydroxybenzoic acid	21.2	–	0.77	0.71	–
Chlorogenic acid	22.7	–	0.20	0.72	–
Vanillic acid	24.3	–	0.89	0.68	–
Caffeic acid	25.5	–	0.48	0.34	–
Syringic acid	27.3	–	0.91	0.63	–
p-Coumaric acid	37.4	–	0.82	0.54	–
Ferulic acid	42.4	–	0.90	0.50	–
Sinapic acid	44.9	–	0.92	0.42	–
o-Coumaric acid	47.4	–	0.85	0.58	–
Cinnamic acid	58.4	–	0.94	0.64	–
Dihydroquercetin	38.8	49.9	0.33	0.53	0.86
2,3-trans-3,4-trans-leucocyanidin	–	21.6	–	–	0.61
2,3-trans-3,4-cis-leucocyanidin	–	14.0	–	–	0.48
(+)-Catechin	17.2	24.4	0.11	0.51	0.66
Procyanidin B-3	–	19.9	–	–	0.44
Prodelphinidin B-3	–	12.6	–	–	0.32
Procyanidin C-2	–	21.3	–	–	0.33
Prodelphinidin trimer (α C_{8-4}, β GC_{8-4}, γ GC)	–	11.3	–	–	0.14
Prodelphinidin trimer (α C_{8-4}, β C_{8-4}, γ GC)	–	15.5	–	–	0.24

acids can be bought commercially, and it is also possible to buy preparations of some flavonoids, for instance dihydroquercetin, rutin, (+)-catechin and naringenin. Unfortunately, it is not possible to buy proanthocyanidin standards, and many investigations on proanthocyanidin content or composition of plant material have only been made possible by gifts of standards from other laboratories. It is very time-consuming to isolate or synthesize, purify and identify proanthocyanidins if the only purpose is to use them as standards for other investigations.

Outtrup (1981) isolated procyanidins B-3 and C-2, prodelphinidin B-3 and three prodelphinidin trimers from barley using the following method: Barley grains are finely ground and the flour is separated by sieving into two fractions, larger and smaller than 250 μ particle size. 35 g of the fine fraction flour is extracted in an Omnimixer with 3 × 200 ml of 75% (v/v) acetone containing 1% (w/v) ascorbic acid. The combined acetone extracts are fractionated on a Sephadex LH-20 column (85 × 2.5 cm). The column is eluted with 60% (v/v) ethanol at a flow rate of 50 ml/h. The eluate is collected as 25-ml fractions, and presence of proanthocyanidins can be detected by TLC or acidic hydrolysis of aliquots of the fractions. Fractions eluted within 20–30 h contain procyanidin B-3 and prodelphinidin B-3, whereas fractions eluted within 30–50 h contain the four trimers. The proanthocyanidins are purified by repeated Sephadex LH-20 chromatography.

Further separation and purification is accomplished by semipreparative and analytical HPLC. For semipreparative HPLC a R Sil C_{18} HL 30 μ column (30 × 0.8 cm) is eluted with a linear gradient from 10 to 20% methanol (v/v) for 60 min. Flow rate 2 ml/min. Loading capacity is 5–10 mg. For analytical HPLC a Sil C_{18} HL 10 μ column (30 × 0.46 cm) is eluted with a linear gradient from 2 to 10% acetic acid for 100 min. Flow rate 1 ml/min. Loading capacity is 100–150 μg. Fractions eluted from the HPLC containing the proanthocyanidin peaks can be collected and freeze-dried. The structures of unknown proanthocyanidins can be established by means of NMR spectroscopy or mass spectroscopy. For these techniques the reader is referred to Vol. 2 of this series. Specific references for proanthocyanidins are as follows:

1. Proanthocyanidins of barley: separation and identification (Mulkay et al. 1981 a).
2. Structure elucidation of some proanthocyanidins in barley by ^{1}H 270 MHz NMR spectroscopy (Outtrup and Schaumburg 1981).
3. Proanthocyanidins of barley and sorghum: Composition as a function of maturity of barley ears (Brandon et al. 1982).
4. Synthesis of condensed tannins. Part 9. The condensation sequence of leucocyanidin with (+)-catechin and with the resultant procyanidins (Delcour et al. 1983).
5. Structure elucidation of three dimeric proanthocyanidins isolated from a commercial Belgian pilsner beer (Delcour and Tuytens 1984).

3.1.7 Quantitation

The quantities of different phenolic compounds in a sample can be calculated from the HPLC chromatograms. The peak area of a certain compound is compared with that of a known amount of standard. Concentration curves of proanthocyanidins and phenolic acids can be made if samples of standard compounds are run on the HPLC in three to four different concentrations. A linear relationship between peak area and concentration has been observed for proanthocyanidins as well as phenolic acids (see Fig. 6 and 7). The reproducibility of the peak areas is around ±2%. The slopes of the standard curves are dependent on the solvent system used for elution of the column and on the quality of the column itself. Analysis of the same compound on two different columns of the same type, eluted under the same conditions, can give quite different retention times, and a new standard curve should be made up every time the conditions of the HPLC analysis are changed.

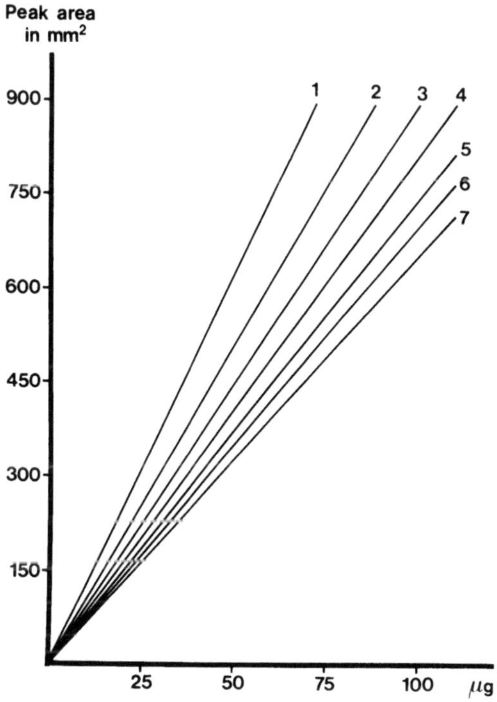

Fig. 6. Standard curves for HPLC of flavanoids. *1* (+)-Catechin; *2* procyanidin B-3; *3* procyanidin C-2; *4* prodelphinidin B-3; *5* prodelphinidin trimer (α C_{8-4}, β GC_{8-4}, γ C); *6* prodelphinidin trimer (α C_{8-4}, β C_{8-4}, γ GC); *7* prodelphinidin trimer (α C_{8-4}, β GC_{8-4}, γ GC). (Jerumanis 1985)

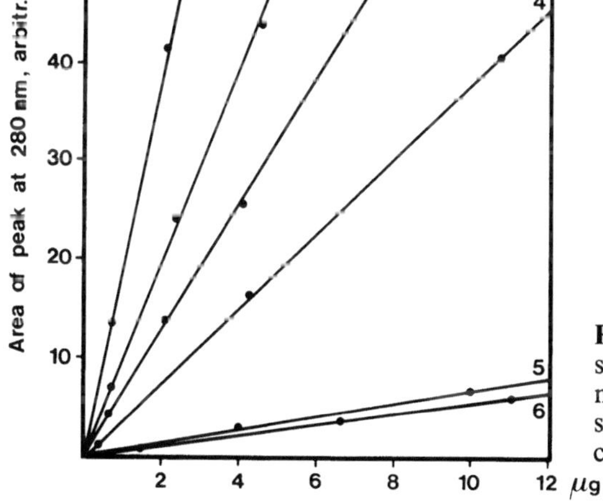

Fig. 7. Standard curves for HPLC of selected phenolic compounds. *1* Cinnamic acid; *2* dihydroquercetin, *3* syringic acid; *4* gallic acid; *5* (+)-catechin; *6* p-hydroxybenzoic acid

3.2 Thin Layer Chromatography (TLC)

Phenolic compounds are usually separated on cellulose plates. Many different solvent system have been used for development of the thin layer plates. Some examples are given:

1. BAW (butanol: acetic acid: water, 4:1:5, v/v/v, upper phase).
2. Different concentrations of aqueous acetic acid (for instance 6%, 15% and 30% v/v).
3. CAW (chloroform: acetic acid: water, 10:9:1, v/v/v). A good system for separation of phenolic acids (Table 1).
4. BzAW (benzene: acetic acid: water, 125:72:3, v/v/v).
5. sBAWC (s-butanol: acetic acid: water: chloroform, 7:1:2:1, v/v/v/v, Kristiansen 1984). A good system for separation of proanthocyanidins, catechin and leucocyanidins (Table 1).

The most used system for two-dimensional thin layer chromatography is BAW in the first dimension and aqueous acetic acid in the second dimension.

The phenolic compounds are localized under UV light. Treatment with NH_3 fumes will in some instances improve the visibility of the spots. Spraying the plates with different dyes can make the flavonoids and phenolic acids visible in daylight. A few examples of spray reagents:

1. A mixture of 1% (w/v) $FeCl_3 \cdot 6H_2O$ and 1% (w/v) $K_3Fe(CN)_6$ (1:1, v/v) produces a blue colour with most phenolic compounds.
2. 5% (w/v) vanillin in ethanol and conc. HCl (4:1, v/v) gives a red colour with proanthocyanidins and catechins.
3. 1% (w/v) $FeCl_2$ in ethanol produces different colours with different phenolic acids (Urion et al. 1963).
4. 2% (w/v) 2,6-dichloroquinonechlorimide in ethanol (Gibbs reagent) makes phenolic acids and proanthocyanidins appear blue or purple.
5. p-dimethylaminocinnamaldehyde (100 mg in 100 ml of methanol: conc. HCl, 3:1, v/v) gives a green colour with proanthocyanidins and catechins (McMurrough and McDovell 1978).
6. 0.02–0.2% (w/v) diazotized p-nitroaniline or diazotized sulphanilic acid with 20% (w/v) Na_2CO_3 overspray can be used for all phenolic compounds.

3.3 Column Chromatography

Chromatography on Sephadex LH-20 with its high affinity for phenolic compounds is the most widely used procedure for purification and fractionation of phenolic compounds. Sephadex-25 Superfine columns have also been used for separation of proanthocyanidins and catechins from beer and its raw materials (McMurrough 1979).

The columns can be eluted with aqueous methanol or ethanol and fractions containing phenolic compounds are detected by UV absorbance.

4 Colourimetric Methods

4.1 Determination of Proanthocyanidins in Barley, Malt, Wort and Beer

The procedure is modified from Harris and Ricketts' method (1959). 5 g of freshly ground barley or malt flour is extracted, while rotating, with 50 ml of 60% (v/v) ethanol containing 1% (w/v) ascorbic acid for 1 h. After centrifugation at 4000 *g*, the flour is re-extracted in the same way three times. The volume of the combined extracts is adjusted to 250 ml with 60% ethanol. An aliquot of 10 ml of extract, wort or decarbonated beer is shaken with 0.5 g of Divergan polyamide powder in a 25 ml graduated centrifuge tube for 40 min and then centrifuged at 3000 *g* for 5 min. The pellet is washed with distilled water and recentrifuged. 20 ml of butanol:conc. HCl (5:1, v/v) is added to the polyamide pellet with the adsorbed proanthocyanidins, and the sample is placed in a boiling water bath for 30 min. The proanthocyanidins are converted to red anthocyanidins, and after cooling and adjustment of the volume to 25 ml, the anthocyanidin concentration is measured on a spectrophotometer at 550 nm against a blank prepared from distilled water treated in the same way as the sample. The proanthocyanidin content of the sample is calculated from a calibration curve obtained with known amounts of cyanidin chloride.

4.2 Measurement of Total Polyphenols in Beer and Wort

Jerumanis' method for measurement of polyphenol content in beer and wort (1968) has been adopted by the European Brewery Convention (Analytica-EBC 1975). The beer is decarbonated by shaking. Hazy beers and worts are centrifuged. 10 ml of beer and 8 ml of a 1% (w/v) solution of sodiumcarboxymethylcellulose (low viscosity) with the addition of 0.2% (w/v) disodium ethylene diamine tetra acetate (CMC/EDTA) are pipetted into a 25-ml graduated flask and mixed. 0.5 ml of green ammonium ferric citrate solution (3.5%, w/v) is added and mixed and after that 0.5 ml of NH_3 ($NH_3:H_2O$, 1:2, v/v) is added and the solution thoroughly mixed again. The volume is adjusted to 25 ml with distilled water and after 10 min the extinction is measured on a spectrophotometer at 525 nm against a blank prepared from beer, treated in the same way as the sample but with the ferric reagent omitted. The polyphenols in beer and wort react with ferric ions in alkaline solution and a red-coloured complex is produced. Concentration of polyphenols in ppm is given by: $E_{525} \times 630$.

4.3 Analysis of Proanthocyanidins and Catechins Using Acidified Vanillin

This method has been developed by Broadhurst and Jones (1978). 0.5 ml of degassed beer or extract of barley or malt flour, prepared as described in 4.1, is pipetted into a test tube wrapped in aluminium foil. 3.0 ml of 2.4% (w/v) vanillin in methanol is added followed by 1.5 ml of concentrated hydrochloric acid. The solution is thoroughly mixed and allowed to stand for 15 min at room temperature. The flavanols react with vanillin in acidic solutions and a red coloured product with the absorbance maximum at 500 nm is formed. The absorbance is measured on a spectrophotometer at 500 nm against a blank made up with water instead of sample.

The total concentration of proanthocyanidins and catechins is calculated from a standard curve obtained with (+)-catechin as a standard.

4.4 Determination of Proanthocyanidins and Catechins in Beer with p-Dimethylaminocinnamaldehyde as a Reagent

The procedure is worked out by Delcour and de Varebeke (1985), and is based on the formation of a green complex when p-dimethylaminocinnamaldehyde reacts with the A rings of proanthocyanidins and catechins. 10 ml of decarbonated beer is diluted to 100 ml with water. 1.0 ml of this solution is pipetted into a test tube and 5.0 ml of a reagent, consisting of p-dimethylaminocinnamaldehyde (1.0 g) in 1000 ml of methanol and concentrated hydrochloric acid (3:1, v/v), is added. After 10 min the absorbance is measured on a spectrophotometer at 640 nm against a blank made up with water instead of beer.

The flavanoid content can be calculated from the following equation: Flavanoids as catechin equivalents, in mg/1 $= 335 \times E_{640}$.

5 Vanillin Test of Barley Grains

A qualititative test for presence or absence of proanthocyanidins in barley grains or malt has been developed by Aastrup (1985). It is a useful method for control of proanthocyanidin-free barley lines or varieties and samples of malt at breeding stations, malt houses and breweries.

The grains are prepared using the Carlsberg Seed Fixation System (Heltved et al. 1982; Jensen and Aastrup 1985) as follows: From a representative sample 100 grains are placed in the undermatrix containing depressions, using an overmatrix with round holes (Fig. 8 A, B). The depressions in the undermatrix have a depth of half a grain length. The grains are fixed in a block of termoplastic clay, Cernit, by means of a specially designed press, in such a way that half of each grain is embedded in the Cernit clay (Fig. 8 C, D).

One quarter of each grain is removed using a disc sander (Fig. 8 E, F). The sanded grains are stained with a freshly prepared solution of 1% (w/v) vanillin in 6N-HCl. The Cernit blocks are placed in the solution for 30 min. A distinct red colour develops in the testa layer of the proanthocyanidin containing grains, while the proanthocyanidin-free grains lack the red colour (Fig. 8 G, H). The colour remains stable for at least 30 min.

With minor modifications the vanillin test procedure can be used in a non-destructive way for screening for proanthocyanidin-free mutants in large materials of mutagenized barley: An under- and overmatrix is used in which the depressions are arranged in such a way that the barley grains are lying on their side. Thus the

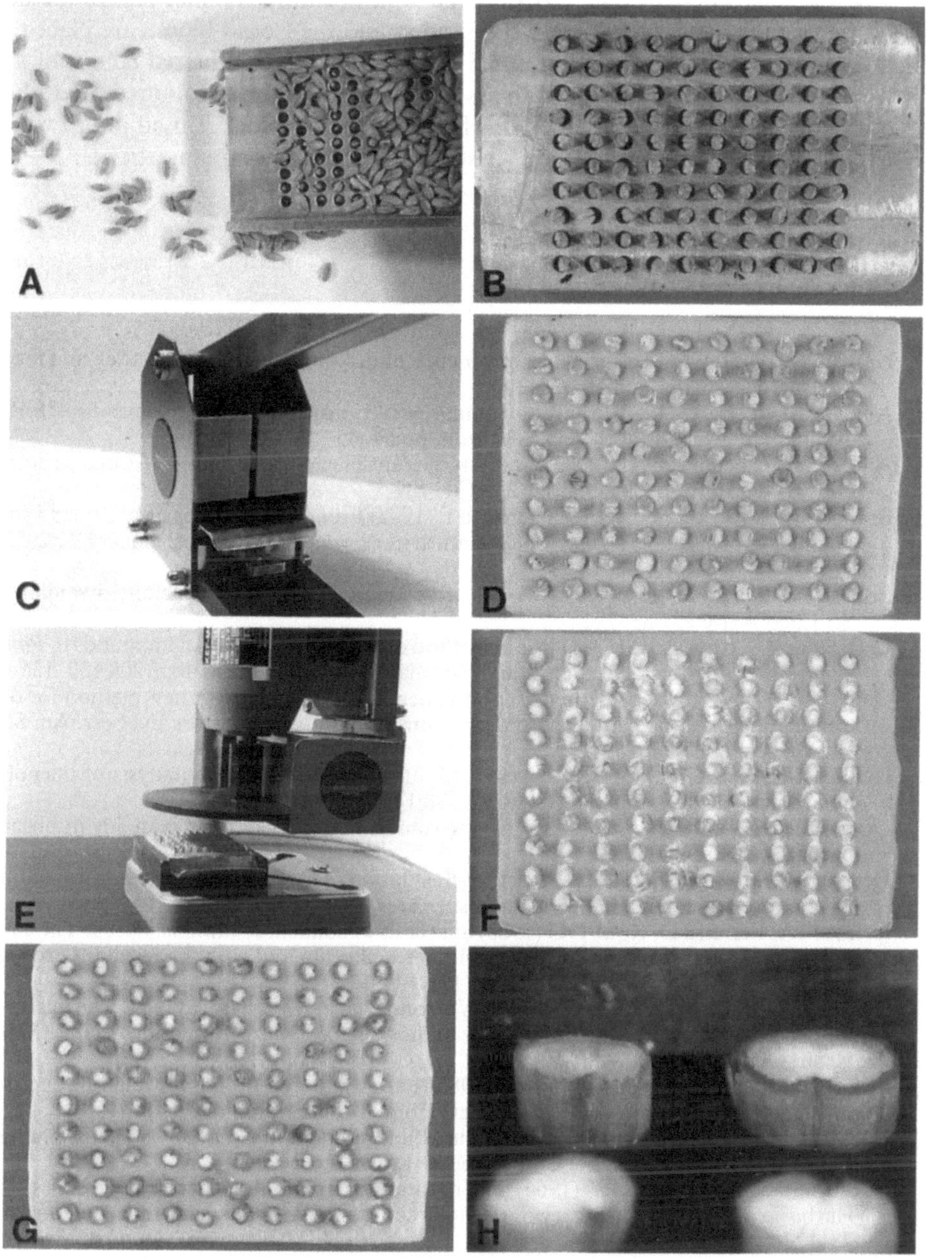

Fig. 8 A–H. Pictorial description of vanillin test of barley grains. 100 grains are oriented by placing them on an overmatrix (**A**) and shaking them into position in the undermatrix (**B**). The undermatrix with the grains in position is placed in the press unit (**C**) and the grains pressed into thermoplastic clay, Cernit (**D**). A disc sander (**E**) is used for removing one quarter of each of the grains (**F**). After staining (**G** and **H**) the proanthocyanidin containing grains show a distinct red ring, while no red colour is observed in the proanthocyanidin-free grains (Aastrup 1985)

grains are fixed in the Cernit blocks in the same position. A minor part of the side of the grain is removed with the disc sander, and the Cernit blocks are placed in a solution of 1% (w/v) vanillin in 4.5 N-HCl for 20 min. The red colour of the proanthocyanidin-containing grains is still visible, and the uncoloured proanthocyanidinfree grains can be taken out from the Cernit block, rinsed in tap water overnight and planted (Henriette Kristensen, personal communication).

References

Aastrup S (1985) A test for presence or absence of proanthocyanidins in barley and malt. Carlsberg Res Commun 50:37–42

Analytica-EBC (1975–1983) Analysis Committee of European Brewery Convention. Enari T-M (ed) Schweiz Brau Rundsch 3rd edn, pp 64–65

Andersen JM, Pedersen WB (1983) Analysis of plant phenolics by high-performance liquid chromatography. J. Chromatogr 259:131–139

Brandon MJ, Foo LY, Porter LJ, Meredith P (1982) Proanthocyanidins of barley and sorghum: composition as a function of maturity of barley ears. Phytochemistry 21:2953–2957

Broadhurst RB, Jones TJ (1978) Analysis of condensed tannins using acidified vanillin. J Sci Food Agric 29:788–794

Carpentier BA, Cowles JR (1981) Rapid method of analysing phenolic compounds in *Pinus elliotti* using high-performance liquid chromatography. J Chromatogr 208:132–136

Dadic M (1971) Analytical methods for polyphenols in brewing. II. A new method for determination of anthocyanogens and catechins (tanninogens) in wort and beer. Am Soc Brew Chem Proc, pp 159–170

Dadic M, Belleau G (1980) Beer hazes. I. Isolation and preliminary analysis of phenolic and carbohydrate components. ASBC J 38:145–158

Delcour JA, de Varebeke JD (1985) A new colourimetric assay for flavanoids in pilsner beers. J Inst Brew 91:37–40

Delcour JA, Tuytens GM (1984) Structure elucidation of three dimeric proanthocyanidins isolated from a commercial belgian pilsner beer. J Inst Brew 90:153–161

Delcour JA, Ferreira D, Roux DG (1983) Synthesis of condensed tannins. Part 9. The condensation sequence of leucocyanidin with (+)-catechin and with the resultant procyanidins. J Chem Soc Perkin Trans I, pp 1711–1717

Delcour JA, Schoetens MM, Meysman RW, Dondeyne P, Moerman E (1984) The intrinsic influence of catechin and proanthocyanidins on beer haze formation. J Inst Brew 90:381–384

Erdal K, Outtrup H, Ahrenst-Larsen B (1985) Barley and hop proanthocyanidins and the colloidal stability of beer. Proc Eur Brew Conv Congr, Helsinki, pp 459–466

Gorissen H (1969) Les composés phénoliques du houblon (*Humulus lupulus* L.). Dissertation de Doctorat, Université Catholique de Louvain.

Harborne JB, Simmonds NW (1964) The natural distribution of phenolic aglycones. In: Harborne JB (ed) Biochemistry of phenolic compounds. Academic Press, London, pp 78–127

Harris G, Ricketts RM (1959) Beer haze–the problem and a solution. Proc Eur Brew Conv. Congr, Rome, pp 290–302

Heltved FS, Aastrup S, Jensen O, Gibbons G, Munck L (1982) Preparation of seeds for mass screening. Carlsberg Res Commun 47:291–296

Jende-Strid B (1985) Phenolic acids in grains of wild-type barley and proanthocyanidin-free mutants. Carlsberg Res Commun 50:1–14

Jende-Strid B, Møller BL (1981) Analysis of proanthocyanidins in wildtype and mutant barley (*Hordeum vulgare* L.). Carlsberg Res Commun 46:53–64

Jensen S Aa, Aastrup S (1985) Determination of malt modification. Cerevisia 10:113–123

Jerumanis J (1968) Simplification et amélioration de la méthode de dosage de polyphenols totaux en brasserie. Bull Ass Anc Etud Ec Super Brass Univ Louv 64:223–240

Jerumanis J (1979) Séparation et identification de flavanoides par chromatographie liquide à haute performance (HPLC). Proc Eur Brew Conv Congr, Berlin, pp 309–319

Jerumanis J (1985) Quantitative analysis of flavanoids in barley, hops and beer by high-performance liquid chromatography (HPLC). J Inst Brew 91:250–252

Jones WT, Broadhurst RB, Lyttleton JW (1976) The condensed tannins of pasture legume species. Phytochemistry 15:1407–1409

Knorr F (1977) Polyphenole im Brauprozeß. Brauindustrie 62:1017–1028

Kristiansen KN (1984) Biosynthesis of proanthocyanidins in barley: genetic control of the conversion of dihydroquercetin to catechin and procyanidins. Carlsberg Res Commun 49:503–524

McFarlane WD (1961) Determination of anthocyanogens. J Inst Brew 67:502–506

McFarlane WD, Wye E, Grant HL (1955) Studies on the flavonoid compounds in beer. Proc Eur Brew Conv Congr, Baden-Baden, pp 298–308

McFarlane WD, Sword PT, Blinoff G (1964) Catechins-tannins in brewing, Proc Eur Brew Conv Congr, Brussels, pp 174–181

McMurrough I (1979) Chromatographic procedure for measuring polyphenol haze precursors. Proc Eur Brew Conv Congr Berlin, pp 321–334

McMurrough I (1981) High-performance liquid chromatography of flavonoids in barley and hops. J Chromatogr 218:683–693

McMurrough I, McDovell J (1978) Chromatographic separation and automated analysis of flavanols. Anal Biochem 91:92–100

McMurrough I, Hennigan GP, Loughrey MJ (1983) Contents of simple, polymeric and complexed flavanols in beers and their relationship to haze formation. J Inst Brew 89:15–23

McMurrough I, Roche GP, Cleary KG (1984) Phenolic acids in beers and worts. J Inst Brew 90:181–187

Moll M, Fonknechten G, Carnielo M, Flayeux R (1984) Changes in polyphenols from raw materials to finished beer. MBAA Tech Q 21:79–87

Mulkay P, Toillaux R, Jerumanis J (1981 a) Proanthocyanidins of barley: Separation and identifikation. J Chromatogr 208:419–423

Mulkay P, Touillaux R, Jerumanis J (1981 b) Les prodelphinidines de l'orge: séparation, identification, et influence sur la stabilité colloidale de la bière. Cerevisia 6:29–33

Outtrup H (1981) Structure of prodelphinidins in barley. Proc Eur Brew Conv Congr, Copenhagen, pp 323–333

Outtrup H, Erdal K (1983) Haze-forming potential of proanthocyanidins. Proc Eur Brew Conv Congr, London, pp 307–314

Outtrup H, Schaumburg K (1981) Structure elucidation of some proanthocyanidins in barley by ^{1}H 270 MHz NMR spectroscopy. Carlsberg Res Commun 46:43–52

Qureshi A, Burger WC, Prentice N (1979) Quantitation of potential flavoring compounds in worts and beers by HPLC. ASBC J 37:153–160

Salomonsson AC, Theander O, Åman P (1978) Quantitative determination by GLC of phenolic acids as ethyl derivatives in cereal straws. J Agric Food Chem 26:830–834

Urion E, Metche M, Haluk JP (1963) Bestimmung der Phenolsäuren der Gerste mittels Polyamid-Dünnschicht-Chromatographie. Brauwissenschaft 16:211–215

Vancraenenbroek R,Kara-Zaitri M, Devreux A (1979) Influence de la teneur en proanthocyanidin dimères et la catéchines de la bière sur la stabilité colloidale. Proc Eur Brew Conv Congr, Berlin, pp 293–307

Vancraenenbroek R, Callewaert W, Lontie R (1980) Identification de glycosides de flavonols dans le houblon. Cerevisia 2:69–73

van de Casteele K, Geiger H, De Loose R, Van Sumere CF (1983) Separation of some anthocyanidins, anthocyanins, proanthocyanidins and related substances by reversed phase high-performance liquid chromatography. J Chromatogr 259:291–300

Von Wettstein D, Nilan RA, Ahrenst-Larsen B, Erdal K, Ingversen J, Jende-Strid B, Kristiansen K, Larsen J, Outtrup H, Ullrich S (1985) Proanthocyanidin-free barley for brewing: Progress in breeding for high yield and research tool in polyphenol chemistry. MBAA Tech Q 22:41–52

Yeast

Breeding and Characterization of Brewer's Yeast Including Single Chromosome Transfer

C. Gjermansen

1 Introduction

Man has brewed beer for thousands of years. Improvements have been achieved in the course of time, mainly by amendment of raw materials and brewing technologies. Genetic improvement of the yeast *Saccharomyces*, the living organism responsible for conversion of wort to beer, has attracted relatively little attention. By Hansen's selection of pure yeast cultures in 1883 (Hansen 1888), it became obvious that the brewing performance of a yeast depends on its genetic background. Since that time yeast breeding has mainly been carried out by selection for good brewing performance. The properties selected for have been good attenuation of the wort, fermentation rate, flocculation, cropping rate and taste. Because of the need for constant properties in each strain, there may have been selection against efficient sexual reproduction; brewing strains in general are non-maters, sporulate poorly, if at all, and exhibit low spore viability (Johnston 1965; Fowell 1969; Anderson and Martin 1975). This makes breeding and genetic analysis difficult.

Another class of *Saccharomyces* strains are used in laboratories all over the world in fundamental genetic research. They may be referred to as laboratory strains. In contrast to brewer's yeast their derivation has involved constant selection for regular sexual reproduction including high spore viability. These strains are very well characterized. Unfortunately they cannot be used in the brewery since they produce an unpalatable beer, but they can be taken advantage of by the breeder as model strains. All of the methods worked out for manipulation of laboratory yeast are potentially applicable for breeding of brewer's yeast. In spite of the aberrations in sexual reproduction, genetic improvement of brewer's yeast has a great potential: genes responsible for degradation of dextrins and starch can be transferred from one strain to another by crossing (Emeis 1971) and transformation (Barney et al. 1980; Yamashita and Fukui 1983; Pretorius et al. 1986). Genes involved in flocculation have been identified (reviewed by Johnston and Reader 1983). Construction of yeast strains that do not produce diacetyl has been carried out (Ramos-Jeunehomme and Masschelein 1977). A gene causing phenolic off-flavour has been identified (Goodey and Tubb 1981). Possibilities and difficulties in manipulating the genome of brewer's yeast have been rewiewed (Kielland-Brandt 1981; Spencer and Spencer 1983; von Wettstein et al. 1984).

The key words in all breeding programmes are: characterization, breeding and testing. If the breeder does not know the genetic constitution of his organism and the detailed genetic processes that take place, he will soon run into severe problems, many of which would be trivial if the strain was well characterized. Since most genetic improvements emerge stepwise, it is of great importance to be able to test the newly constructed strains and to identify promising cell lines for the next experiments. In this chapter, methods of genetic characterization and breeding of brewers yeast are described. Sigsgaard (this volume) discusses methods for evaluating new potential brewing yeasts.

2 Sporulation, Isolation of Meiotic Progeny and Construction of Hybrids

Like other lager yeasts, the Carlsberg production strain 244 does not undergo a regular meiosis. It was considered to be non-sporulating at 30 °C, which is the standard sporulation temperature for laboratory strains. However, at 22 °C and lower temperature it is able to produce asci with 1, 2, and 3 spores (Gjermansen and Sigsgaard 1981). The sporulation frequency as well as the spore viability is low. Only 1% of the spores give rise to colonies upon microdissection and most of the clones are non-maters. However, by mass spore isolation it is possible to obtain an unlimited number of different clones derived from the parental strain by meiosis.

2.1 Mass Spore Isolation

Cells are streaked in a thin layer to cover the surface of a sporulation plate consisting of 1% potassium acetate and 0.1% Difco Bacto Yeast Extract, solidified with 2% Difco Bacto Agar. After incubation for 5–8 days at 22 °C the cells are washed off, spun down, resuspended in 5 ml of freshly prepared 75 m*M* thioglycolate, 0.1 *M* Tris-HCl, pH 8.8 and incubated for 15 min. The cell and ascus suspension is washed twice and treated with 2% (w/w) β-glucuronidase, Sigma No. G-0751, in 2.5 ml 1*M* sorbitol for 15–30 min. Hydrolytic enzymes from snails, present in this product, attack some of the structural components of the walls of asci and vegetative cells. After washing twice with 1*M* sorbitol the spores are released from the asci by lysis in 10 ml of water. To kill vegetative cells, the spore suspension is frozen over night. After thawing, a drop of Triton X-100 is added and the spores are separated by sonication. Different dilutions are plated on YPD (1% Difco Bacto Yeast Extract, 2% Difco Bacto Peptone, 2% glucose) solidified with agar. Incubation should not be at higher temperatures than 22 °C since some of the progeny are temperature-sensitive for growth.

2.2 Isolation of Colonies and Spore Staining

After a week of growth the spore-derived colonies are usually large enough to permit isolation in a replica pattern on new YPD plates. The colonies vary in size and morphology, which is in accordance with the known extensive, structural heterozygosity of the parental strain (Nilsson-Tillgren et al. 1981, 1986; Casey 1986b). All types of colonies should be picked up during isolation. The isolated clones are tested for sporulation by replica plating to sporulation medium and incubated as described in Sect. 2.1. Non-sporulators are considered to be potential mating strains. Since many of the clones sporulate poorly, staining is necessary to detect sporulation. A modification of the malachite green-safranine staining method (Schaeffer and Fulton 1933) can be used: Clones in a replica pattern on sporulation medium are replicated with velvet onto a 10 × 10 cm glass plate. After air-drying, the glass plate is heated gently on a hot plate and transferred to a 5% malachite green oxalate solution at 70 °C–80 °C for 1 min. After rinsing with tap water, the plate is transferred to a 0.5% safranine O solution for 1 min, whereafter it is again rinsed. The wet glass plate can be examined under a microscope, using cover slips. Spores are stained green while unsporulated cells are red. The plates should be inspected within 10 min as the colors fade slowly away.

2.3 Determination of Mating Type

Expression of mating type of the non-sporulating clones is determined by mixing a loopful of cells on YPD plates with cells of tester strains of mating types **a** and **α**. After 2 days of growth the plates are replica-plated to sporulation medium. A clone is designated as non-mater when no sporulation occurs, as mating type **a** when sporulation occurs after mating with the **α** tester and as **α** when sporulation occurs after mixing with the **a** tester. While it is conceivable that some maters might escape this screen, it allows the handling of meiotic progeny in large numbers.

2.4 Construction of Hybrids by Mating of Meiotic Segregants

Meiotic segregants of opposite mating type can be crossed by mixing cells on YPD medium. In contrast to laboratory strains, where mating usually can be observed after a few hours, meiotic offspring of lager yeast often need longer time to perform efficient mating. It is thus recommended to incubate the mating mixture at least 24 h. The efficiency of expression of mating type is highly strain-dependent and varies from zero to normal as compared to laboratory strains. Construction of hybrids are complicated by the absence of good selective markers. However, the hybrids grow faster and form larger colonies than the maters. This allows enrichment for hybrids. Cells from the mating mixture are transferred to liquid YPD. After several generations of growth, dilutions of cells are plated on YPD and large colonies are isolated. Hybrids do not express mating type and should be able to sporulate. In some cases temperature sensitivity for growth can be used for selections. Crossing of high temperature (34 °C) sensitive strains can yield hybrids which are able to grow at the restrictive temperature. Hybrids of meiotic offspring of the same parental strain differ in many characteristics important to the brewer, including flocculation, growth rate, and cropping rate. Thus it is by conventional genetic methods possible to utilize the extensive heterozygosis of the parental strain and to create genetic variation by reshuffling chromosomes.

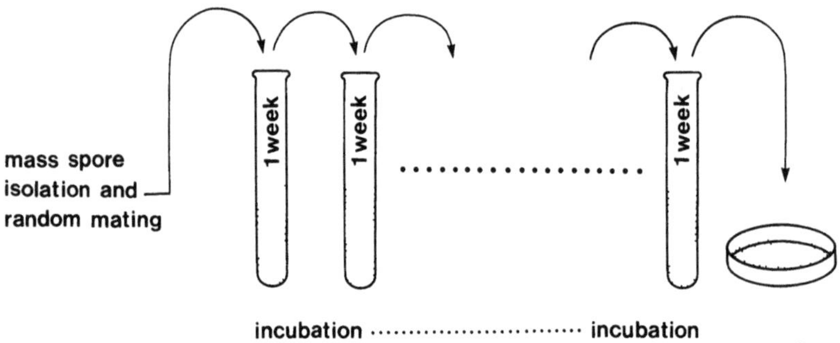

Fig. 1. Selection for rapid growth of randomly produced hybrids

2.5 Random Mating

By the method outlined above, hybrids can be constructed from identified maters, which can be subsequently used in other crosses if it is deemed expedient. Hybrids can also be constructed at random by mass-mating of germinated spores, isolated as described in Sect. 2.1. This results in a genetically heterogeneous population of hybrids. By exposing the cells to specific selection pressures, one can enrich for cell lines showing improved characteristics. Upon selection for rapid growth under fermentative conditions at 10 °C, hybrids of meiotic segregants of the Tuborg lager production strain 224 were isolated. Some of the hybrids showed increased fermentation rate and increased yeast yield (Sigsgaard, Fig. 12, this volume). The procedure for random mating and subsequent selection for rapid growth was the following:

Mass spore isolation was performed as described in Sect. 2.1. The spore suspension from one sporulation plate was centrifuged and the spores were resuspended in 5 ml of YPD. After shaking for 2 days at 22 °C, germination of the spores and random mating was performed. The cells were spun down, resuspended in 0.2 ml of H_2O and plated on YPD. Following incubation for 2 days, the cells were washed off, spun down and resuspended in 10 ml of wort in a test tube. The culture was incubated without agitation at 10 °C for 7 days. Selection for rapid growth was carried out by subsequently transferring 0.2 ml of suspended cells of the culture to 10 ml of fresh wort. Dilution and fermentative growth was repeated ten times (Fig. 1). From the last fermentation tube dilutions were plated on YPD. After incubation, large colonies were isolated and tested in EBC-tubes.

3 Genetics of Diacetyl Formation

Two vicinal diketones, diacetyl and 2,3-pentanedione are off-flavour components produced by normal brewing strains. In the production of lager beer, their amount is reduced during the secondary fermentation (lagering). This is the main

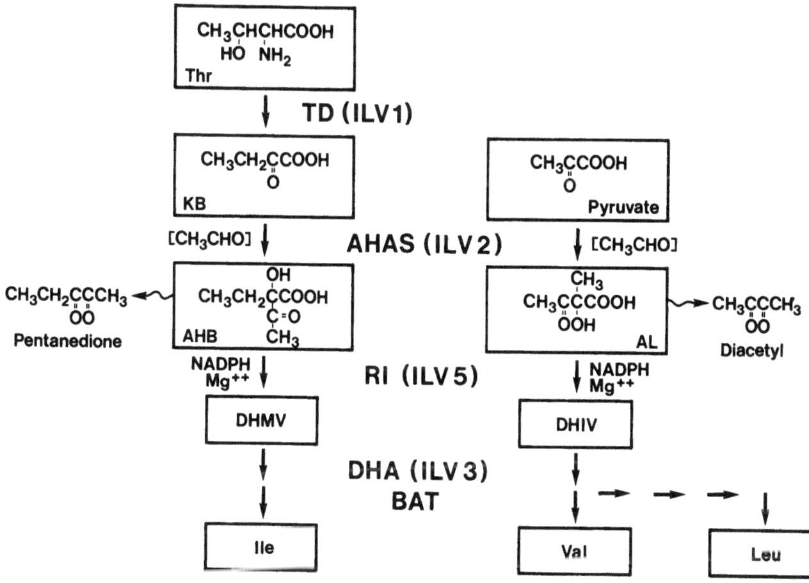

Fig. 2. The biosynthesis of isoleucine and valine in yeast and the formation of vicinal diketones. Four of the enzymes, threonine deaminase (*TD*), acetohydroxy acid synthase (*AHAS*), acetohydroxy acid reductoisomerase (*RI*), dihydroxy acid dehydratase (*DHA*), and branched-chain amino acid aminotransferase (*BAT*) are shared by the two pathways. (*thr* threonine; *ile* isoleucine; *val* valine; *leu* leucine; *KB* α-ketobutyrate; *AHB* α-aceto-α-hydroxybutyrate; *AL* α-acetolactate; *DHMV* α,β-dihydroxy-β-metylvalerate; *DHIV* α,β-dihydroxy isovalerate)

function of lagering, since maturation of beer can be effected enzymatically in 24 h by addition of α-acetolactate decarboxylase (Godtfredsen and Ottesen 1982; Godtfredsen et al. 1983). The vicinal diketones are side products of the isoleucine valine pathway (Fig. 2). Diacetyl is derived from α-acetolactate while 2,3-pentanedione is formed from α-acetohydroxy butyrate (Fig. 2). The structural genes responsible for the pathway, *ILV1, ILV2, ILV3,* and *ILV5,* are situated on chromosomes V, XIII, X, and XII, respectively. They have been cloned from laboratory strains of *S. cerevisiae,* as well as from strain 244 of *S. carlsbergensis* (Petersen et al. 1983a; Polaina 1984; Casey 1986a; Nilsson-Tillgren et al. 1986), except *ILV5,* which has only been cloned from *S. cerevisiae.*

Ramos-Jeunehomme and Masschelein (1977) have demonstrated that *ilv2* mutants produce neither diacetyl nor 2,3-pentanedione during the main fermentation. Unfortunately, it has not been possible to induce these mutations in strain 244 or its meiotic segregants. The tested meiotic segregants are disomic for chromosome V as well as chromosome XIII, and this may explain why recessive mutations in these chromosomes are not easily obtained. Although mutations are probably induced with the same frequency as in haploids, recessives can usually not be found because of the presence of more than one dominant wild-type allele. Kielland-Brandt et al. (1979) isolated mutants of the Carlsberg lager strain 244 which are resistant to the amino acid analogue thiaisoleucine. These mutants are dominant alleles of the *ILV1* gene. Falco and Dumas (1985) have shown that a class of mutants resistant to the herbicide sulfomethuron methyl (SM) are dominant alleles of *ILV2*. The drug inhibits α-acetohydroxy acid synthase, thereby making the cells unable to synthesize isoleucine and valine. In the presence of SM, wild-type cells cannot grow on medium lacking the two amino acids. Since the mutants produce an enzyme with altered properties compared to the wild type, they may in some cases produce lower amounts of diacetyl.

3.1 Isolation of Spontaneous Mutants Resistant to Sulfometuron Methyl

With the possibility of a positive selection, the isolation of spontaneous mutations are often preferred over mutagenesis since the latter increases the chance of altering other genes. Spontaneous mutants resistant to SM can be isolated by the following procedure:

Cells are grown over night in liquid YPD. After washing twice with water, 10^7 cells are plated on medium lacking isoleucine and valine, supplemented with sulfometuron methyl (10–30 mg/l). Typically 10–20 colonies will appear after 7 days of incubation. Table 1 shows diacetyl values after trial fermentations with eight SM-resistant mutants of strain 244. The fermentations were carried out in EBC-tubes as described by Sigsgaard, this volume. Notice that six of the mutants produce less amounts of diacetyl as compared to the wild-type strain. To decrease the diacetyl formation further, homozygosity for the mutant gene may be expedient. This may be obtained by cross-breeding, as outlined in Sect. 2.

Table 1. The amount of vicinal diketones and their precursors in racked beer from EBC-tube fermentations after 7 days of fermentations at 12 °C. Fermentations were carried out in duplo five times, whereby bottom yeast from one fermentation was used as pitching yeast in the following fermentation. Only the results from the first and the fifth fermentation are shown

Strain	Diacetyl (ppm)		2,3-Pentanedione (ppm)	
	1st generation	5th generation	1st generation	5th generation
244	1.16	0.80	0.67	0.62
CG1309	0.55	0.46	0.38	0.37
CG1310	0.70	0.53	0.48	0.47
CG1311	1.20	0.90	0.60	0.78
CG1316	0.58	0.68	0.46	0.63
CG1317	0.79	0.62	0.50	0.56
CG1318	0.70	0.64	0.44	0.57
CG1319	0.64	0.77	0.58	0.64
CG1320	1.19	0.89	0.72	0.70

4 Transformation

Development of techniques of genetic transformation has expanded the possibilities of directed breeding of brewer's yeast. Genes from *Saccharomyces* and other organisms can be introduced in the genome by treating the cells with purified DNA. Thus a gene for flocculence *FLO1* (Barney et al. 1980) as well as genes allowing utilization of dextrins and starch (Yamashita and Fukui 1983; Pretorius et al. 1986) have been reported transferred by transformation. Recently, yeast strains able to synthesize and excrete β-glucanase from barley have been constructed (Jackson et al. 1986). In transformation experiments only a minority of the cells are able to take up DNA. It is thus nessesary to have a selection system allowing only transformants to grow. Most genes of interest to the brewer cannot be selected for directly. They must be coupled to another gene which can be positively selected for. Such genes are usually wild-type alleles of genes allowing the cells to grow without the supplement of a specific nutrient to the medium. Cells carrying a mutation in the gene are auxotrophic, but can be transformed to prototrophy by DNA with the wild-type allele. A protocol for transformation of yeast is given by Keiding (this volume).

A meiotic segregant from the Carlsberg lager strain 244 has been mutagenized with ethyl methanesulphonate. Among other mutants a leucine auxotroph *(leu2)* was isolated. This mutant was transformed to leucine prototrophy with a plasmid carrying the *LEU2* gene from *S. cerevisiae* (Gjermansen 1983). Any isolated gene can be inserted into vectors carrying the *LEU2* wild-type gene. Thus the *leu2* mutation allows the introduction of desired genes into lager yeast by transformation.

Transformation systems independent of mutations in the recipient strain have been developed in recent years. R. Yocum and R. Daves have constructed sophisticated vectors carrying a gene for geneticin sulphate (G418) resistance from the bacterial transposon Tn903. Since wild-type cells are sensitive to G418, transformants can be selected by supplementing the medium with the drug. The transformation and selection procedures are described by Keiding (this volume). In a model experiment, the lager strain 244 was transformed with the integration vector, pDY3 (Eur. Pat. Appl. 85303625.9, 1985), linearized with restriction endonuclease KpnI. 2250 transformants were obtained by plating 2×10^8 cells, treated with 10 µg DNA, on G418 medium (0.5 mg/ml).

5 Single Chromosome Transfer

In order to solve specific breeding aims in brewer's yeast, a genetic characterization of the involved chromosomes and genes is needed. This has been hampered by the aberrant sexual behaviour of the strains. The technique of single chromosome transfer (Nilsson-Tillgren et al. 1980; Dutcher 1981) has helped to overcome the problems. This technique makes use of the *kar1* mutation (Conde and Fink 1976), which greatly reduces the frequency of karyogamy after mating, even if present in only one of the parents. Such a cross will yield progeny with the nuclear genotype of either of the parents and cytoplasmic elements from both. Importantly it also yields, at low frequency, progeny that have the nuclear genotype of one parent plus an extra chromosome of the other. Thus a mater derived from a brewing strain can be mated with a laboratory strain carrying the *kar1* mutation as well as appropriate auxotrophic and antibiotic resistance markers. Selection of cells in which a chromosome from the brewing strain has been transferred to the genetic background of the laboratory strain can be carried out on appropriate media. Occasionally the *S. cerevisiae* chromosome is lost, yielding chromosome substitution lines. They are *S. cerevisiae* strains in which a chromosome has been replaced by a functionally homologous chromosome of *S. carlsbergensis*. By crossing these strains to appropriate genetically marked laboratory strains, diploids giving four-spored asci are obtained. The spore viability is high, allowing tetrad analysis. Using this technique it has been possible to isolate and analyse in *S. cerevisiae* strains, chromosomes III, V, X, XII, and XIII from brewing strain 244 of *S. carlsbergensis* (Nilsson-Tillgren et al. 1981; Petersen et al. 1983b; Casey 1986b; Nilsson-Tillgren et al. 1986). As an example the isolation and characterization of chromosome V is described.

5.1 Transfer of Chromosome V

A prototrophic α-mater derived from strain 244 was crossed to a laboratory strain, C80-II-17 of *S. cerevisiae* with the genotype, *MATa can1 ura3 his1 ilv1 rad3 ade2 cyh2 kar1* (*can1, ura3, his1, ilv1,* and *rad3* are situated on chromosome V while the other genes are carried by other chromosomes). Colonies of C80-II-17 are red due to the *ade2* mutation and resistant to canavanine and cycloheximide. Efficient mating was accomplished as follows: log phase cells in YPD of the two strains to be mated were harvested by centrifugation, washed and resuspended in fresh medium to a density of $2\text{–}4 \times 10^8$ cells/ml. Equal numbers of cells of the two strains (total volume 0.4–0.6 ml) were mixed and spread on YPD plates. When high frequency of mating could be observed by microscopy, the cells were rinsed off, washed twice with distilled water and plated on medium supplemented with cycloheximide and lacking uracil, histidine, isoleucine, and valine. This medium does not allow growth of the parental strains. Neither will it allow growth of normal hybrids, since the gene for cycloheximide resistance *(cyh2)* is recessive. White as well as red colonies appeared with low frequency. Probably the white colonies represent normal hybrids that have undergone mitotic recombination resulting in homozygotization of the *cyh2* allele. The red colonies have received the *URA3, HIS1,* and *ILV1* genes from the brewing strain but not the genes *ADE2* and *CYH2*. They are disomic for chromosome V (addition lines). Reversion to auxotrophy for one of the three markers *ilv1, ura3,* and *his1* was always accompanied by simultaneous reversion of the three others, indicating that these genes are located on the same chromosome in *S. carlsbergensis* and that the reversion was due to chromosome loss. Chromosome substitution strains which had lost the *S. cerevisiae* chromosome V (after growth on YPD) could be recognized by replica plat-

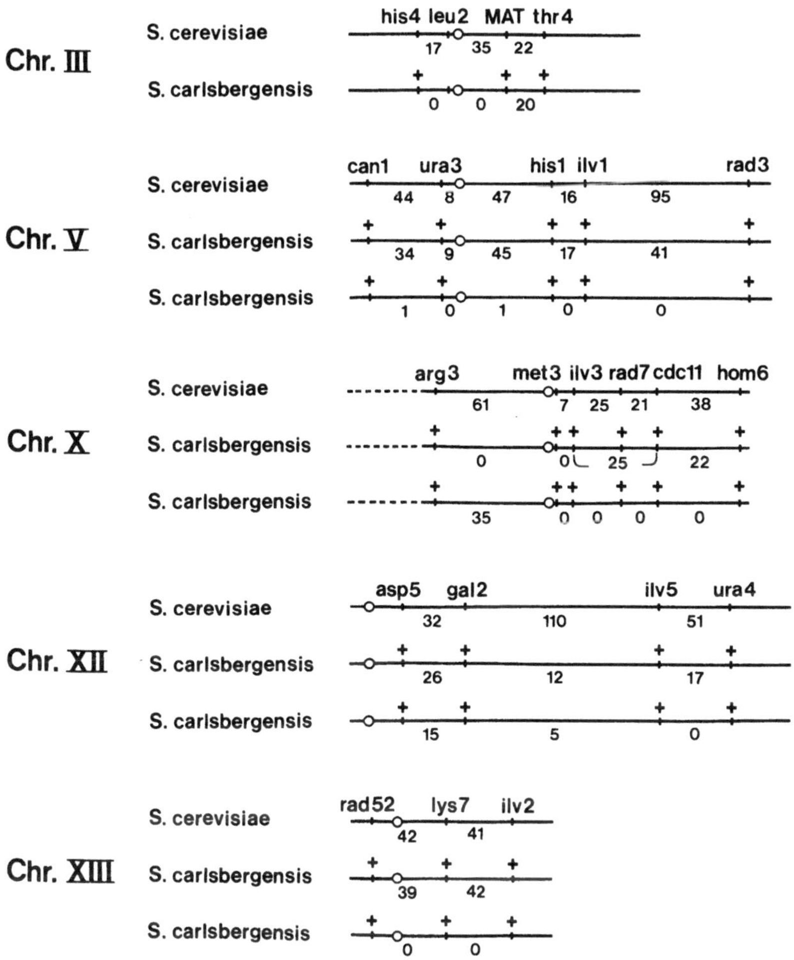

Fig. 3. Genetic analysis of chromosomes III, V, X, XII and XIII. Beneath the map of each marked *S. cerevisiae* chromosome are shown the two types of functionally homologous *S. carlsbergensis* chromosomes. The recombination values (centimorgans) between the markers indicate the recombination frequencies observed in crosses with the *S. cerevisiae* chromosome

ing on medium containing canavanine. Colonies that did not give resistant papillae could be shown to carry only a chromosome V from the brewing strain.

To investigate the recombination behaviour of the transferred chromosomes, chromosome substitution lines as well as addition lines were crossed to appropriately marked tester strains. The resulting diploids were sporulated and subjected to tetrad analysis. It appears that in each case the *S. carlsbergensis* strain has donated either of two different chromosomes V. One of them is not able to recombine at proper levels with the *S. cerevisiae* chromosome V while the other recombines at normal levels. Similar results have been found for chromosomes III, X, XII, and XIII (Fig. 3). Put in a simplified way, the brewing strain 244 contains two different sets of chromosomes, respectively chromosome regions, one which recombines with the corresponding chromosome region in *S. cerevisiae*, and one which does not.

5.2 Characterization of Chromosome V by Restriction Endonuclease Analysis

The observed lack of recombination between *S. carlsbergensis* and *S. cerevisiae* chromosomes was expected to correlate with DNA sequence inhomology. The technique of molecular hybridization according to Southern (1975) is a convenient tool to study the extent of sequence homology between well defined regions of the two chromosomes. For chromosomes V the homology between four differ-

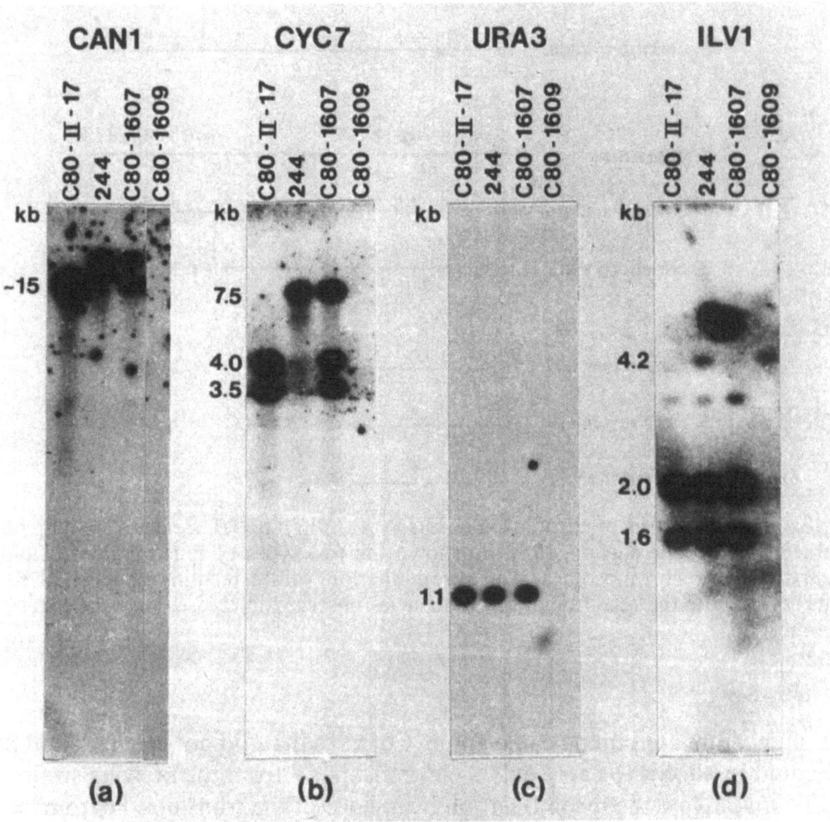

Fig. 4. Molecular hybridization analysis of four chromosome V regions. DNA from strains 244 (*S. carlsbergensis*), C80-II-17 (recipient *S. cerevisiae*), C80-1607 (chromosome addition strain with recombining chromosome) and C80-1609 (chromosome substitution strain with non-recombining chromosome) was digested with restriction enzymes EcoRI (a and d) or HindIII (b and c). The resulting fragments were separated by electrophoresis, transferred to nitrocellulose filters, and hybridized to ^{32}P-labelled DNA carrying the genes *ILV1*, *URA3*, *CYC7*, and *CAN1*. (Nilsson-Tillgren et al. 1986)

ent loci has been investigated. DNA was prepared from the recipient laboratory strain C80-II-17, the brewing strain 244, the substitution strain with a non-recombining chromosome, C80-1609, and the chromosome addition strain C80-1607 containing a recombining *S. carlsbergensis* chromosome. The DNA was cut with the restriction enzymes EcoRI, and HindIII. After gel electrophoresis, the fragments were transferred to a nitrocellulose filter and hybridized to ^{32}P-labelled DNA carrying the genes *ILV1*, *URA3*, *CYC7*, and *CAN1* from *S. cerevisiae*. The sizes of the cleaved fragments that bind to the probes can be investigated by au-

```
                    1
S. carlsbergensis         T                  GA   G   C     A   G       TC          A    G  AGT
S. cerevisiae       ATG TCA GCT ACT CTA CTA AAG CAA CCA TTA TGT ACG GTT GTT CGG CAA GGT AAA CAG
S. cerevisiae       Met Ser Ala Thr Leu Leu Lys Gln Pro Leu Cys Thr Val Val Arg Gln Gly Lys Gln
S. carlsbergensis                           Arg             Ser Ala     Ser         Ser Arg Ser
                  58
                     G   G    C         C T              A                A   A   T       C   T
                    TCC AAA GTG TCT GGA TTG AAC CTT TTG AGA CTA AAG GCT CAT TTG CAC AGA CAA CAC
                    Ser Lys Val Ser Gly Leu Asn Leu Leu Arg Leu Lys Ala His Leu His Arg Gln His
                        Arg                             Lys             Gln             His
                 115
                    T A           T   A                   C   G         C                     C
                    CTG TCA CCT TCC TTG ATA AAA CTA CAC TCT GAA TTG AAA TTG GAT GAG CTG CAA ACT
                    Leu Ser Pro Ser Leu Ile Lys Leu His Ser Glu Leu Lys Leu Asp Glu Leu Gln Thr

                 172
                              T   G   C   T       C   G   A   G   A   G   C   G       C   A
                    GAT AAC ACC CCT GAT TAC GTC CGT TTA GTT TTA AGG TCC TCT GTA TAC GAT GTT ATT
                    Asp Asn Thr Pro Asp Tyr Val Arg Leu Val Leu Arg Ser Ser Val Tyr Asp Val Ile

                 229
                              A                           C                 T G
                    AAT GAA TCT CCA ATC TCT CAA GGT GTA GGT TTG TCT TCC CGT CTA AAC ACG AAT GTC
                    Asn Glu Ser Pro Ile Ser Gln Gly Val Gly Leu Ser Ser Arg Leu Asn Thr Asn Val

                 286
                      T               G       T           G                               T
                    ATC TTG AAA AGA GAA GAT CTA TTG CCT GTT TTC TCT TTC AAG CTT CGT GGT GCC TAT
                    Ile Leu Lys Arg Glu Asp Leu Leu Pro Val Phe Ser Phe Lys Leu Arg Gly Ala Tyr

                 343
                                        C A                         A
                    AAC ATG ATT GCC AAG TTG GAC GAT TCT CAA AGA AAC CAG GGT GTT ATT GCC TGT TCA
                    Asn Met Ile Ala Lys Leu Asp Asp Ser Gln Arg Asn Gln Gly Val Ile Ala Cys Ser

                 400
                      A   A   C   C                   T       A   C       T       G   C   G
                    GCT GGG AAT CAT GCC CAA GGT GTG GCC TTT GCT GCT AAA CAC TTG AAA ATA CCT GCT
                    Ala Gly Asn His Ala Gln Gly Val Ala Phe Ala Ala Lys His Leu Lys Ile Pro Ala

                 457
                      C       C           C                   C               C       C       C
                    ACT ATC GTT ATG CCT GTT TGT ACA CCA TCT ATT AAG TAT CAA AAT GTC TCG AGA TTA
                    Thr Ile Val Met Pro Val Cys Thr Pro Ser Ile Lys Tyr Gln Asn Val Ser Arg Leu

                 514
                                  T   T   T  T        T               A   C   A   A
                    GGG TCT CAA GTC GTC CTA TAT GGT AAC GAT TTT GAC GAG GCT AAG GCT GAA TGT GCC
                    Gly Ser Gln Val Val Leu Tyr Gly Asn Asp Phe Asp Glu Ala Lys Ala Glu Cys Ala
                                            Phe
                 571
                              A       A G       C A   A                                   C
                    AAA TTG GCT GAA CAC CGT GGC TTG ACG AAC ATT CCT CCT TTC GAT CAT CCT TAT GTC
                    Lys Leu Ala Glu Glu Arg Gly Leu Thr Asn Ile Pro Pro Phe Asp His Pro Tyr Val
                                        Gly
```

Fig. 5. Comparison of *S. cerevisiae* and *S. carlsbergensis ILV1* coding sequences. The first 627 nucleotides of the coding region, and the deduced amino acid sequence are shown. For *S. carlsbergensis* the differences from *S. cerevisiae* are depicted

toradiography. From Fig. 4 it is seen that only the *ILV1* probe hybridizes to DNA from the substitution strain carrying the non-recombining chromosome V. revealing a faint band at 4.2 kb. In contrast strong hybridization is obtained with DNA from the three other strains. The pattern observed for strain 244 probed with *ILV1* DNA contains bands identical to those of the recipient in addition to the faint band specific for the non-recombining *S. carlsbergensis* chromosome. In the chromosome addition strain only the *S. cerevisiae*-like pattern is observed. By probing with *URA3*, no hybridization to DNA from the substitution strain is seen. In the case of the *CYC7* and *CAN1* loci a slightly altered nucleotide sequence in the *S. cerevisiae*-like chromosome of the donor strain is revealed (Fig. 4a, b). It can be seen that a HindIII site is absent in the *CYC7* region, creating a larger DNA fragment. The chromosome addition strain shows, as expected, the pattern of the recipient as well as that of *S. carlsbergensis*. Thus the results from the molecular analysis are in agreement with the genetic data showing the presence of two structurally different chromosomes V in the brewing strain 244 of *S. carlsbergensis*.

To obtain detailed information about the observed nucleotide sequence differences within a single gene, the *ILV1* alleles from *S. carlsbergensis* and *S. cerevisiae* have been cloned (Petersen et al. 1983b; Nilsson-Tillgren et al. 1986). The *ILV1* gene from the non-recombining chromosome V has been sequenced (Gjermansen in preparation) and compard to the corresponding sequence from *S. cerevisiae* (Kielland-Brandt et al. 1984). Figure 5 shows a comparison of the first 627 nucleotides of the coding region of *S. carlsbergensis* and *S. cerevisiae* alleles. From these sequences the corresponding amino acid sequences have been deduced. The inhomology of the shown sequence is 15%. This resulted in 93 codon changes which gave only 13 amino acid substitutions.

Acknowledgements. I thank Steen Holmberg, Jens Litske Petersen, and Torsten Nillson-Tillgren for valuable discussions, Morten Kielland-Brandt and Diter von Wettstein for critical reading of the manuscript.

References

Anderson E, Martin PA (1975) The sporulation and mating of brewing yeasts. J Inst Brew 81:242–247

Barney MC, Jansen GP, Helbert JR (1980) Use of genetic transformation for the introduction of flocculence into yeast. Am Soc Brew Chem J 38:71–74

Casey GP (1986a) Cloning and analysis of two alleles of the *ILV3* gene from *Saccharomyces carlsbergensis*. Carlsberg Res Commun 51:327–341

Casey GP (1986b) Molecular and genetic analysis of chromosomes X in *Saccharomyces carlsbergensis*. Carlsberg Res Commun 51:343–362

Conde J, Fink GR (1976) A mutant of *Saccharomyces cerevisisae* defective for nuclear fusion. Proc Natl Acad Sci USA 73:3651–3655

Dutcher SK (1981) Internuclear transfer of genetic information in *kar1-1/KAR1* heterokaryons in *Saccharomyces cerevisiae*. Mol Cell Biol 1:245–253

Emeis C-C (1971) A new hybrid yeast for the fermentation of wort dextrins. Am Soc Brew Chem Proc, pp 58–62

Falco SC, Dumas KS (1985) Genetic analysis of mutants of *Saccharomyces cerevisiae* resistant to the herbicide sulfometuron methyl. Genetics 109:21–35
Fowell RR (1969) Sporulation and hybridization of yeast. In: Rose AH, Harrison JR (eds) The yeasts vol I. Academic, London, pp 303–383
Gjermansen C (1983) Mutagenesis and genetic transformation of meiotic segregants of lager yeast. Carlsberg Res Commun 48:557–565
Gjermansen C, Sigsgaard P (1981) Construction of a hybrid brewing strain of *Saccharomyces carlsbergensis* by mating of meiotic segregants. Carlsberg Res Commun 46:1–11
Godtfredsen SE, Ottesen M (1982) Maturation of beer with α-acetolactate decarboxylase. Carlsberg Res Commun 47:93–102
Godtfredsen SE, Rasmussen AM, Ottesen M, Sigsgaard P, Erdal K, Mathissen T, Ahrenst-Larsen B (1983) Use of α-acetolactate decarboxylase for accelerated maturation of beer. Proc Eur Brew Conv Congr, London, pp 161–168
Goodey AR, Tubb RS (1981) A yeast gene responsible for the production of 4-vinyl guaiacol from ferulic acid. Soc Gen Microbiol. Q 8:103 (abstr)
Hansen EC (1888) Undersøgelser fra gjaeringsindustriens praxis. Meddelser fra Carlsberg Laboratoriet 2:257–322
Jackson EA, Ballance GM, Thomsen KK (1986) Construction of a yeast vector directing the synthesis and release of barley (1,3; 1,4)-β-glucanase. Carlsberg Res Commun 51:445–458
Johnston JR (1965) Breeding yeast for brewing. I. Isolation of breeding strains. J Inst Brew 71:130–135
Johnston JR, Reader HP (1983) Genetic control of flocculantion. In: Spencer JFT, Spencer DM, Smith ARW (eds) Yeast genetics. Fundamental and applied aspects. Springer, Berlin Heidelberg New York
Kielland-Brandt MC (1981) The genetics of brewer's yeast. Proc Eur Brew Conv Congr, Copenhagen, pp 263–276
Kielland-Brandt MC, Petersen JGL, Mikkelsen JD (1979) Mutants in the biosynthesis of isoleucine in a non-mating, non-sporulating brewing strain of *Saccharomyces carlsbergensis*. Carlsberg Res Commun 44:27–36
Kielland-Brandt MC, Holmberg S, Petersen JGL, Nilsson-Tillgren T (1984) Nucleotide sequence of the gene for threonine deaminase *(ILV1)* of *Saccharomyces cerevisiae*. Carlsberg Res Commun 49:567–575
Nilsson-Tillgren T, Petersen JGL, Holmberg S, Kielland-Brandt MC (1980) Transfer of chromosome III during *kar*-mediated cytoduction in yeast. Carlsberg Res Commun 45:113–117
Nilsson-Tillgren T, Gjermansen C, Kielland-Brandt MC, Petersen JGL, Holmberg S (1981) Genetic differences between *Saccharomyces carlsbergensis* and *S. cerevisiae*. Analysis of chromosome III by single chromosome transfer. Carlsberg Res Commun 46:65–76
Nilsson-Tillgren T, Gjermansen C, Holmberg S, Petersen JGL, Kielland-Brandt MC (1986) Analysis of chromosome V and the *ILV1* gene from *Saccharomyces carlsbergensis*. Carlsberg Res Commun 51:309–326
Petersen JGL, Holmberg S, Nilsson-Tillgren, Kielland-Brandt MC, Gjermansen C (1983 a) Approaches to genetic improvement of brewer's yeast. Cloning of the isoleucine-valine genes. E.B.C.-Symp Biotech Nuttfield, GB, Nov 1983
Petersen JGL, Holmberg S, Nilsson-Tillgren T, Kielland-Brandt MC (1983 b) Molecular cloning and characterization of the threonine deaminase (*ILV1*) gene of *Saccharomyces cerevisiae*. Carlsberg Res Commun 48:149–159
Polaina J (1984) Cloning of the *ILV2, ILV3,* and *ILV5* genes of *Saccharomyces cerevisiae*. Carlsberg Res Commun 49:577–584
Pretorius IS, Chow T, Modena D, Marmur J (1986) Molecular cloning and characterization of the *STA2* glucoamylase gene of *Saccharomyces diastaticus*. Mol Gen Genet 203:29–35
Ramos-Jeunehomme C, Masschelein CA (1977) Contrôle génétique de la formation des dicetones vicinals chez *Saccharomyces cerevisiae*. Proc Eur Brew Conv Congr, Amsterdam, pp 267–283

Schaeffer AB, Fulton M (1933) A simplified method of staining of endospores. Science 77:194
Southern EM (1975) Detection of specific sequences among DNA fragments separated by gel electrophoresis. J Mol Biol 98:503–517
Spencer JFT, Spencer DM (1983) Genetic improvement of industrial yeasts Ann Rev Microbiol 37:121–142
Yamashita I, Fukui S (1983) Molecular cloning of a glucoamylase producing gene in the yeast *Saccharomyces*. Agric Biol Chem 47:2689–2692
Von Wettstein D, Gjermansen C, Holmberg S, Kielland-Brandt MC, Nilsson-Tillgren T, Pedersen MB, Petersen JGL, Sigsgaard P (1984) Genetic engineering in the improvement of brewers' yeast. MBAA Tech Q 21:45–61

Genetic Analysis and Transformation of Distiller's Yeast

A. K. KEIDING

1 Introduction

Industrial yeasts are as yet poorly characterized from a genetic point of view, which is in contrast to the strains of *Saccharomyces cerevisiae* on which our knowledge of yeast genetics is based.

Sporulation sufficient for microdissection and tetrad analysis has been reported for some industrial yeasts, such as baker's yeasts (Gunge 1966; Keiding 1985) and wine yeasts (Ismail and Ali 1971; Benítez et al. 1983). Contrary to this, most brewing strains, whether classified as *S. cerevisiae* or *S. carlsbergensis*, show no or poor mating ability, poor sporulation and low spore viability (Winge 1944; Fowell 1969; Anderson and Martin 1975; Gjermansen and Sigsgaard 1981). As a consequence, it is impossible to perform tetrad analysis on brewing strains directly.

In this chapter the strategy and the methods employed in a genetic characterization of the industrial yeast, strain A from the Danish Distilleries Ltd. is described. Several of the conclusions are based on experiments that involve tetrad analysis, which is useful for investigators of industrial strains predominantly producing asci with three to four viable spores. In strains which never produce asci with four viable spores, single chromsome transfer in combination with molecular characterization is expedient for genetic characterization of the strain. This method is described by Gjermansen (this volume, pp 131–144).

Two transformation systems for obtaining unstable and stable transformants are described. One utilizes complementation of auxotrophic mutants of an industrial strain and the other a dominant marker selectable in wild-type strains. Particularly the latter procedure will be important for breeders of industrial yeasts.

2 Genetic Analyses

2.1 Sporulation and Mating Ability

To initiate a genetic characterization of an unknown strain, sporulation and mating abilities are the first traits which should be analyzed. If the strain in question upon sporulation yields four-spored asci, and one to four spores per ascus germinate, microdissection is worthwhile. Otherwise, isolation of spores can be performed by mass spore isolation. Isolated clones derived from spores should be tested for growth requirements, mating type and sporulation ability.

2.1.1 Sporulation Test

After 2 days at 30 °C on PS plates (pre-sporulation medium, consisting of 0.8% Difco yeast extract, 0.3% Difco peptone, 10% glucose, 2% Bacto agar for solidification), yeast cells are transfered to sporulation plates (1% potassium acetate, 0.1% yeast extract, 0.05% glucose, 2% agar) and incubated at 10 °C, 20 °, 30 °, and 37 °C for 14, 7, 7, and 5 days, respectively. Sporulation is detected by a modification (Gjermansen and Sigsgaard 1981) of the malachite green-safranin staining procedure described by Schaeffer and Fulton (1933): The colonies are replica-plated onto glass plates and fixed on a hotplate (until the glass is too hot to touch, 10–15 s), incubated for 1 min in 5% malachite green oxalate (Merck No. 1398) at 70 °–80 °C, washed in water, incubated 1 min in 0.5% safranin 0 (Gurr No. 27300) (room temperature) and again washed in water. Examination in the microscope reveals green spores and red somatic cells.

2.1.2 Spore Isolation

Asci to be microdissected are incubated at room temperature with *β*-glucuronidase (Sigma No. G-0715) on thin slabs of 4% Bacto noble agar until appropriate degradation of ascus wall material has taken place. For strain A 10 µl *β*-glucuronidase solution (40000 U/ml in H_2O) and cells are well spread on a limited part of the slab and incubated for 30 min at 20 °C. The slab on a glass plate is placed in a micromanipulator. Several micromanipulators are commercially available, see Sherman (1975). The needle is made from a Pasteur pipette by heating the thin part of the pipette in a gas flame and stretching it out. The obtained microneedle stem is cut about 7 cm from the thick part of the Pasteur pipette. Two stems are fused in the flame and stretched out to get microneedles perpendicular to the stems. They are cut about 1 cm from the stems and pumiced.

The mass spore isolation is designed to open asci and to kill non-sporulated cells. The washed cell and ascus suspension in 1M sorbitol is treated with 2% (w/w) *β*-glucuronidase at 30 °C until microscopy reveals that cells have been protoplasted and thus lyse in 10% (w/v) sodium dodecyl sulphate (SDS). The cell mixture is suspended in water and frozen overnight at −18 °C. The spores are separated by sonication and a drop of Triton X-100 (Serva No. 37238) is added to prevent them from sticking together. The spores are plated on appropriate medium.

2.1.3 Mating Type Tests

2.1.3.1 Induction of Respiratory-Deficient Mutations

Strains to be tested are made respiratory deficient, rho^-, by streaking on Ethidium bromide-containing (EtBr) medium (25 µg EtBr/ml) (Slonimski et al. 1968) and subsequent growth in the dark.

2.1.3.2 Determination of Mating Type

Single rho^- colonies derived from the industrial strain are mixed with **a** or *α* tester strains with auxotrophic requirements on YPD plates (1% yeast extract, 2% peptone, 2% glucose, 2% agar). Tester strains used in these experiments are D273-11A *(MATα ade1 his1)* and D286-2A (*MATa ade1 his1,* both Cold Spring Harbor Lab.). After 2 or 5 days at 30 °C, the cell mass is replica plated onto NF plates (non-fermentable minimal medium consisting of 0.67% Difco yeast nitrogen base without amino acids, 1% succinic acid, 0.6% NaOH, 1% glycerol, 1% glutamic acid, 1% aspartic acid, 1% sodium lactate, 2% agar) and the ability to grow is tested after 3–5 days at 30 °C. Only zygotes will give rise to growth on this medium, revealing mating between the industrial strain and the tester strain.

2.1.3.3 Rare Mating on Solid Medium

Lawns of rho$^-$ cells (5×10^6–5×10^7 cells × plate^{-1}) are replica plated onto YPD plates together with a tester strain of the same density. Tester strains used in these experiments are K396-22B (*MATα spo11 ura3 ade1 his1 leu2 lys7 met3 trp5,* Klapholz and Esposito 1982) and K396-11A (*MATa spo11 ura3 ade1 his1 leu2 lys7 met3 trp5*, Klapholz and Esposito 1982). After 2 days of growth, replica plating is carried out onto NF plates and growth is tested after 5 days at 30 °C. Only rare zygotes from rare mating events will give rise to colonies.

2.1.3.4 Rare Mating in Liquid Medium

Rho$^-$ cells and tester strains are grown in liquid YPD to log phase. All cultures are concentrated to 10^7 cells/ml by centrifugation, and equal amounts of rho$^-$ cells and tester strain cells are mixed. Incubation is performed by shaking at 30 °C for 1 h, whereafter the suspension is diluted tenfold and growth allowed to continue for an additional 18 h. A sample from each mixture is plated on YPG (complex non-fermentable medium consisting of 1% yeast extract, 2% peptone, 3% glycerol, 2% agar), SD (synthetic minimal medium consisting of 0.67% yeast nitrogen base without amino acids, 2% glucose, 2% agar) and NF plates. Growth is tested after 2–5 days at 30 °C.

2.1.4 Characterization of Auxotrophic Spore Clones

Meiotic segregants with growth requirement are characterized by complementation analysis: Mutant spores are crossed with genetically marked strains of *S. cerevisiae*. A mutation is determined to be allelic to a specific gene in a tester strain when complementation on SD medium failed. Rare mating may reveal complementation for spores characterized as non-maters.

2.1.5 Sporulation and Mating Ability of Strain A

B. E. Christensen and A. Sahlgren (personal communication) have sporulated strain A which yielded asci containing one to four spores. The four-spored asci constituted 40% of the total. Onehundred and ninety asci with four spores were dissected and 317 spores germinated (42%). In only four asci were all four spores viable. These 16 spores will hereafter be referred to as the four tetrads. In subsequent experiments 284 meiotic segregants were available and tested for growth

Table 1. Summary of sporulation and mating types of clones derived from dissected spores of strain A

Sporulation	Mating type	Sum
–	–	1
–	α	15
–	**a**	94
–	Weak bisexual	1
+	–	27
+	α	70
+	Bisexual	57
+	**a**	19

requirements, mating type and sporulation (Keiding 1985). A summary of the encountered combinations of mating phenotypes and sporulation ability is listed in Table 1. The phenotypes of the four tetrads are listed in Table 2. Many of the spores showed sporulation ability, also coincident with one or both mating types. Among non-sporulating spores, **a** maters predominant. These characteristics could be due to the genetic constitution *MAT**a** MAT**a** MAT*α of strain A and independent chromosome losses in individual cells of spore-derived clones.

Three of the clones had a requirement for adenine and turned out to be *ade3* mutants. The occurrence of these allelic mutants shows that strain A is heterozygous for an *ade3* mutation. Since the *ade3* mutant allele segregated with low frequency it can be assumed that strain A has more than two copies of chromosome VII, unless there is a recessive lethal mutation linked to the *ade3* allele.

2.2 Mutagenesis

Mutants, including auxotrophic mutants, may have value directly in breeding. Mutation breeding could seem a rational approach in cases where a large number of performance parameters are to be kept constant while one is changed. They have further use, however namely in the selection of progeny in a variety of breeding techniques such as *kar1* crosses, protoplast fusion, rare mating and transformation. In addition, the survival curve of a yeast strain after mutagen treatment could give hints about the ploidy of the strain (Parry and Cox 1972).

Appearance of recessive mutants from industrial strains is hampered by the polyploid nature of these yeasts. Mutations are probably induced with the same frequency in diploids or polyploids as they are in haploids, but they usually do not reveal themselves because of the presence of dominant wild type alleles. Secondary events that lead to the expression of a recessive mutant phenotype in a diploid or polyploid yeast could be chromosome loss or mitotic recombination. Appearance of recessive mutants could also be due to monosomy of the chromosome in question. As a consequence of the need for secondary events, a mutagenesis procedure including two subsequent mutagen treatments is expected to result in more recessive mutants than a procedure with only one mutagenesis. It is recommendable to use two different doses: a low dose (15%–65% survival) is chosen to obtain single-point mutations without extensive genetic damage which could affect the growth rate and/or the industrial qualities. A high dose (0.5%–1.5% survival) is preferential for induction of mitotic recombination and chromosome loss. The two doses could be combined to one's specific needs for mutants, as the mutagen source. (For review see Borstel and Mehta 1976.) Below is described mutagenesis by UV radiation and ethyl methanesulphonate (EMS) treatment.

2.2.1 Procedure

Cells are grown in liquid YPD to mid-log phase or early stationary phase. For UV-irradiation 5 ml of 10^7 cells/ml in H_2O are treated with a dose rate of 9 J/s/m^2. This can be obtained with the sterilization lamp TUV 30 W (GTE Sylvania Incorporated, Danvers, MA 01923, USA) at a distance of 30 cm. For strain A_{91} UV-irradiation time is 3 s and 40 s, which give doses of 27 J/m^2 and 360 J/m^2 and survival of 15% and 1.5%, respectively. Subsequent growth over-night in liquid YPG is carried out in the dark to eliminate the action of photorepair (Haynes and Kunz 1981). For EMS mutagenesis 10^7 cells/ml in H_2O are incubated with 1% and 4% EMS (Sigma No. M-0880), respectively, by shaking for 2 h at 30 °C to 65% and 0.5% survival, respectively. The cells are washed twice in 5% sodium thiosulfate and once in water before over-night incubation in liquid YPG. For a second mutagenesis either of the procedures may be repeated. Enrichment for auxotrophic mutants is performed by heat killing of exponentially growing cells essentially, as described by Walton et al. (1979). A stationary phase culture in YPG is washed in water and diluted tenfold in SD. After 4 h of shaking, 5 ml of the culture are heat-treated, which kills growing cells while permitting most stationary phase cells to survive. The conditions had to be optimized for each strain, for instance 4 min at 52 °C has been used for A_{91}. The treated cells are grown to stationary phase in YPG and subjected to a second cycle of heat killing. After another growth period, cells are plated on YPD. Auxotrophic mutants are found by replica plating onto SD. Colonies which grow on YPD, but failed to grow on SD, are isolated and the growth requirements are determined on SC (synthetic complete medium, SD+amino acids) lacking individual nutrients.

2.2.2 Mutants Obtained from Spore Clones of Strain A

Mutagenesis has been performed on the spore-derived strains A_{91} and A_{209} (Keiding 1985). From A_{91}, mutants appeared which were allelic to *ade1, ade3, lys2, met13, trp1, ura1,* and *ura4,* whereas from A_{209} only *ade3* mutants appeared. Since strain A has previously been shown to contain the *ade3* mutation, the appearance of *ade3* mutants upon mutagenesis is likely to be due to recombination, chromosomes loss, ect., rather than gene mutation. It can be concluded that different spore clones from the same parental strain may exhibit different sensitivity to mutagen treatment.

2.3 Crossing and Tetrad Analysis

Crosses between meiotic progeny of the industrial strain in question and genetically marked haploid strains could be one way to attempt a determination of the ploidy level of the strain (Fincham 1983; Roman et al. 1955). The resulting hybrids should be sporulated, and tetrad analysis should be performed (Hawthorne and Mortimer 1960). However, such analysis has little value if there are reduced levels of recombination between many of the chromosome from the industrial yeast and the standard strains of yeast, since improper disjunction will cause the frequency of spore germination to be too low. In these cases the crosses should be done with *kar1* mutants to allow transfer and analysis of single chromosomes (Nilsson-Tillgren et al. 1981; cf. Gjermansen, this volume, p 137 ff.). The segregation patterns in tetrads of a polyploid strain might also demonstrate preferential pairing of chromosomes and could indicate structural differences between the homeologous chromosomes from the industrial yeast and the tester strain, too small to affect recombination (James and Inhaber 1975).

Two crosses were made between spore clones of strain A and laboratory strains (a gift from G. Kawasaki): A_{91} *(MATα)* × *N248-1A (MAT**a** ade1 leu1 ura3 trp1 met14 his2)* and A_{209} *(MAT**a**)* × N248-1C *(MATα ade1 leu1 ura3 trp1 met14 his2)* (Keiding 1985). The first five mutations are all centromere-linked and located on chromosomes I, VII, V, IV, and IX, respectively. The mutation *his2* is located on chromosome VI, 21 cM from the centromere. Upon analysis of a triploid strain with the genotype +/+/−, predominantly 4:0 segregation of + : − in each ascus will appear. The ratio between 2:2, 3:1, and 4:0 segregation will be dependent on the distance of the gene from the centromere (Mortimer and Hawthorne 1973). For most of the markers a remarkably large fraction of the tetrads segregated 4:0, the only exception being the segregation of *TRP1/trp1* in the cross A_{91} × N248-1A, where regular 2:2 segregation was observed. Judging from these findings it appears most likely that A_{91} and A_{209} contain two copies of each of the chromosomes I, V, VI, VII, and XI. In addition A_{209} probably contains two copies of chromosome IV, whereas A_{91} contains a single chromosome IV. A structual difference between the chromosomes of the distiller's yeast and those of the tester strains is indicated by the observation that a larger part of the tetrads segregated 4:0 than expected for random disjunction in a cross of the type +/+ × − (Riley and Manney 1978; Shaffer et al. 1971). This may be due to a preferential pairing of the chromosomes originating from the industrial strain, vs. pairing of chromosomes from the industrial strain and the tester strain, analogous to what has been observed by James and Inhaber (1975).

3 Molecular Characterization

Molecular characterization of an industrial strain may provide information about its genetic constitution which is otherwise difficult to obtain. Nucleotide sequence divergence from the standard strains (laboratory strains) can be revealed, whether caused by base substitutions or loss of a chromosome segment. For strain A chromosome III was chosen for analysis because the good knowledge of the mating type locus *(MAT)* can be utilized in future breeding. In addition to the *MAT* locus, chromosome III contains the *HIS4* and the *LEU2* genes as well as two silent mating type loci (*HML* and *HMR*).

3.1 Procedure

The technique is based on sequence specific enzymatic scissions with restriction endonucleases of total DNA extracted from the yeast strain to be investigated. Total DNA is isolated from 1-l cultures by the method of Cameron et al. (1977). After electrophoretic separation by size of the many different sorts of DNA fragments formed, a separate preparation of radioactive DNA (Rigby et al. 1977), with yeast nucleotide sequence of the region in question only, is hybridized to the DNA in the electropherogram (Southern 1975). The fragments in the electropherogram that are formed from the region in question will bind molecules of the radioactive DNA and thereby reveal themselves, among the thousands of sizes of fragments, as black bands when the electropherogram is placed on a photographic film. The result is a region-characteristic pattern of bands. All methods are described in detail by Pedersen (this volume, pp 180–194) and by Maniatis et al. (1982).

3.2 Analysis of Chromosomes III of Strain A

Upon hybridization of the *HIS4* gene to restriction endonuclease digested DNA from strain A and the four tetrads, band patterns were found identical to that of *S. cerevisiae* strain S288C (*MATα mal gal2*, ATCC 26108) (Nilsson-Tillgren et al. 1981) (results not shown). Thus the nucleotide sequences in the *HIS4* region of S288C and strain A are similar.

Hybridization of the *LEU2* gene to digested DNA of strain A and the four tetrads show band patterns different from those of the *S. cerevisiae* strains S288C and A9275C (***MATa*** *his4-29 HOL1-1 can*^{s}, a gift from G. Fink) (Chinault and Carbon 1979; Storms et al. 1979; Dobson et al. 1981). The fragment patterns of strain A and the meiotic products A_{211}–A_{214} after digestion with restriction endonuclease PstI are shown in Fig. 1: from strain A three hybridizing bands were observed. The fastest migrating band results from the shortest fragment. This band showed a stronger hybridization intensity than the two other bands. This indicates either that this fragment is present twice or that the hybridization of the two other fragments to the probe is weaker due to reduced homology. An independent segregation of the bands was found in the four meiotic products. Three of the strains (A_{211}, A_{212}, A_{213}, lane d, e, f) contained two bands, the fourth strain (A_{214}, lane g) contained only one band which hybridized strongly. By assuming that this band contains twice the amount of DNA (e.g. the strain contains two copies of the gene) one can interpret this result as the segregation of the four fragments. Analyses of the three other tetrads and with other restriction endonucleases (results not shown) also showed segregation of four independent bands, of which two are of identical size (Keiding 1985). The difference in size of these

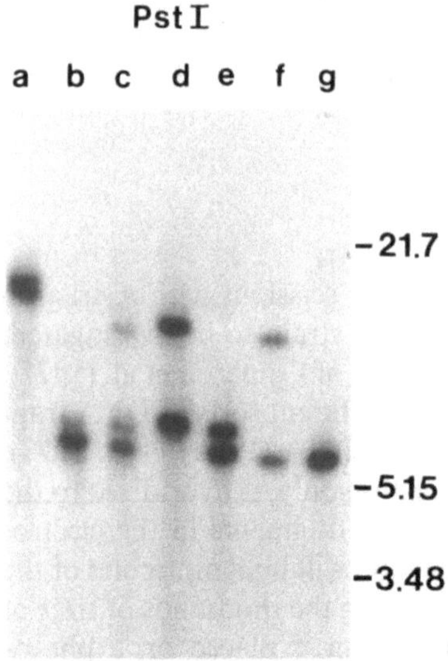

Fig. 1. Restriction endonuclease fragment analysis of the *LEU2* region in strain A and one of its tetrads consisting of strains A_{211}–A_{214}.
DNA from strains S288C (*a*), A9275C (*b*), A (*c*) and A_{211}–A_{214} (*d–g*) was cleaved with PstI, subjected to electrophoresis in 0.7% agarose and transferred to a nitrocellulose filter. The DNA was hybridized to a ^{32}P-labelled 830 bp EcoRI-SalI fragment containing part of the *LEU2* gene (Storms et al. 1979) from the plasmid pC512 (S. Holmberg, personal communication). Sizes are stated in kb

bands originates from three different nucleotide sequences centromere-distal from the gene caused by different transposable elements (for review see Williamson 1983). Strain A thus contains four copies of the *LEU2* gene, named $LEU2^1$, which is present in two copies, $LEU2^2$ and $LEU2^3$, both present in one copy. the distribution of the three alleles in the meiotic progeny is listed in Table 2.

Molecular analyses of the mating type loci *MAT, HML* and *HMR* were made by hybridization with the entire *MATα* gene from *S. cerevisiae* on a 4.3 kb HindIII fragment (Nasmyth and Tatchell 1980, cf. Holmberg 1982). The probe hybridizes to the silent mating type cassette *HML* with same intensity as to the *MAT* gene and to *HMR* with less intensity (Nasmyth and Tatchell 1980; Strathern et al. 1980). Hereby and by several restriction endonuclease digestions it was found that one *HML* locus, three *HMR* loci and four *MAT* loci segregated in the spores. The sequences adjacent to the *HMR* genes and the *MAT* genes all show divergence from each other. Furthermore the mating genotype of each spore could be determined (Table 2).

The molecular analysis indicates that strain A contains four chromosomes III, and the analyzed meiotic progeny strains contain two chromosomes III. An unexpected presence of a total of five *MAT**a*** and three *MATα* alleles in each tetrad is found. This finding indicates the presence of the *HO* (homothallism) gene, which can cause the homothallic shift of both *MAT**a** MAT**a*** and *MATα MATα* to *MAT**a** MATα*, in the presence of *HMα* and *HM**a***, respectively. In the four spore-derived strains A_{214}, A_{279}, A_{326}, and A_{355} the presence of two *MAT**a*** alleles coincides with the absence of the *HMLα* gene. This combination prevents a homothallic shift from *MAT**a** MAT**a*** to *MAT**a** MATα*. It can be assumed that the genotype of strain A is *MAT**a** MAT**a** MAT**a** MATα*, where MAT^{I}, MAT^{II}, and MAT^{IV} contain *MAT**a*** information while MAT^{III} contains *MATα*. On the basis of the segregation of genes in the four tetrads (see Table 2) calculations have been made to determine which of the different *MAT* and *LEU2* regions are

Table 2. Distribution of *MAT*, *MATa*, *MATα*, *HML*, *HMR*, and *LEU2* alleles in the spore-derived strains studied

	Mating phenotype	*MAT*				*MATa*	*MATα*	*HMR*			*HML*	*LEU2*		
		I	II	III	IV			I	II	III		1	2	3
A_{211}	nm			+	+	+	+		+		+		+	+
A_{212}	nm	+	+			+	+	+		+	+	+	+	
A_{213}	nm		+	+		+	+		+			+		+
A_{214}	**a**	+			+	++		+		+		++		
A_{277}	bs	+	+			+	+	+			+	++		
A_{278}	nm			+	+	+	+	+		+	+		+	+
A_{278}	**a**	+	+			++			++			+	+	
A_{280}	nm			+	+	+	+			+		+		+
A_{325}	bs		+		+	+	+			+	+	+	+	
A_{326}	**a**	+			+	++		+	+			+		+
A_{327}	nm	+		+		+	+	+	+			+		+
A_{328}	nm		+	+		+	+			+	+	+	+	
A_{353}	nm	+	+			+	+	+		+	+	++		
A_{354}	bs			+	+	+	+		+		+		+	+
A_{355}	**a**	+	+			++			+	+		++		
A_{356}	nm			+	+	+	+	+					+	+

Each + represents one copy. nm: non-mating; bs: bisexual.

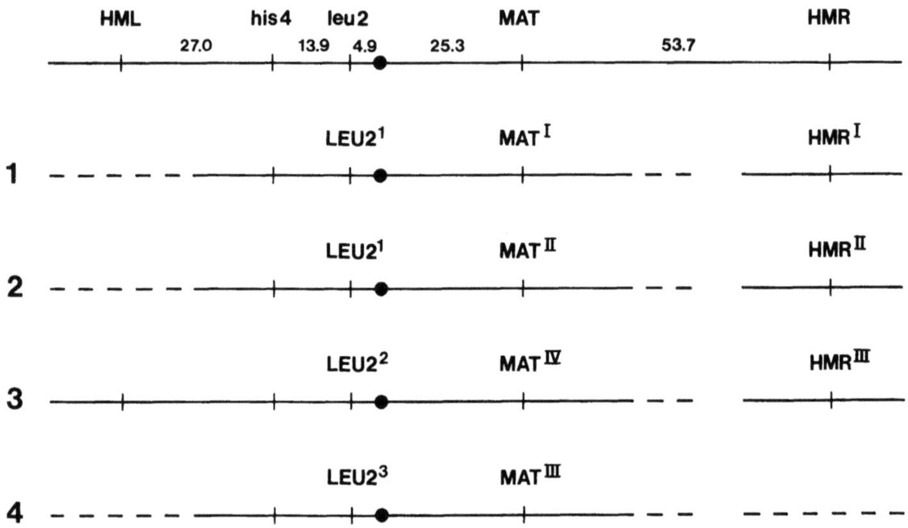

Fig. 2. Model of the four chromosomes III in strain A with linkage of *MAT, LEU2,* and *HML*. A map of relevant chromosome III markers of standard strains of *S. cerevisiae* is shown at the top. The map distances, expressed in cM, are from Mortimer and Schild (1982). It is not possible to decide about the linkage between *MAT* and *HMR* from the available tetrads as indicated by the *dashed lines*. It is not known if the incomplete chromosomes are deficient only in *HML* and/or *HMR*. This is also indicated by *dashed lines*

linked. Partial genetic maps of the four chromosomes III in strain A, including *HML*, are shown in Fig. 2. No deductions can be made about the linkage between the *HMR* regions and *MAT* regions without additional markers as the distance is expected to be more than 50 cM (centi Morgan) (Mortimer and Schild 1982).

In this kind of analysis, the availability of just a few full tetrads is a great advantage. However, the study of a strain which upon dissection results in two or three viable spores may well give information about the number of alleles of a gene in question and the ploidy level of the strain. Further use of these techniques is the identification of industrial strains by specific combinations of band patterns by hybridization to selected gene probes as described by Pedersen (this volume, p 180 ff.).

4 Genetic Transformation of an Industrial Yeast Strain

Genetic transformation is the change of the genetic setup of an organism by the introduction of purified DNA. There are many potential and actual uses for industrial yeast strains expressing specific genes and these may be introduced by transformation. Positive selection of transformants is only possible if the gene or other coupled genes are dominant. Different types of DNA can be used for transformation in *S. cerevisiae*. These can be circular molecules that have been constructed in vitro by placing a gene selectable in yeast on a bacterial plasmid together with (Hinnen et al. 1978) or without a yeast origin of replication (Beggs 1978; Struhl et al. 1979). In both cases the transformants will, in addition to the yeast gene(s) introduced, contain genes from foreign species. However, if proper constructs are used in plasmid integration, it is possible to remove all foreign DNA sequences from the yeast that has been transformed, leaving behind just the desired gene (Scherer and Davis 1979; Yocum 1984). Circular in vitro constructed molecules which only involve yeast DNA could also be used (Kielland-Brandt et al. 1979; Holmberg et al. 1979; Kielland-Brandt et al. 1981). Linear DNA can be useful, such as total native yeast DNA (Jansen et al. 1979). When the goal is interruption of a gene, transformation could be performed with that gene in linear form, interrupted with a selectable marker (Rothstein 1983).

Strain A is to a large extent genetically similar to characterized standard strains (Keiding 1985) and thus it can be anicipated that any *S. cerevisiae* gene harboured on a transformation vector can be expressed in it. Two transformation systems were set up with the aim of introducing and integrating foreign genes in industrial strains. In the first system auxotrophic mutants are needed while the second system can be used on wild-type strains.

4.1 Selection of Transformants by Acquired Prototrophy

Transformants are directly selectable if the recipient strain contains an auxotrophic recessive mutation and the plasmid contains the wild-type dominant allele, resulting in prototrophy upon transformation. In addition to the dominant *S. cerevisiae* marker, the vector may carry an origin of replication from yeast and a bacterial part.

One of the plasmids employed in this study, pMS12 (Thomsen 1983), is shown in Fig. 3. *TRP1* is the selectable gene and the 2 kb EcoRI-PstI fragment of the 2

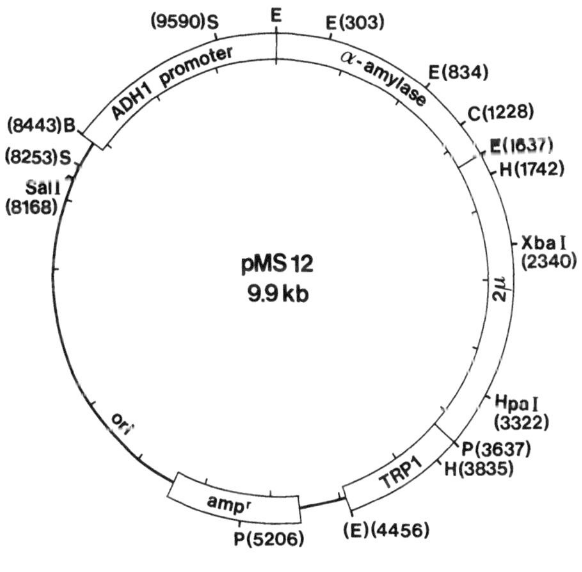

Fig. 3. Partial restriction endonuclease site map of the plasmid pMS12 (Thomsen 1983). *ADH1* promoter, 2 μ and *TRP1* are all from *S. cerevisiae*. The line including *amp*^{r} (ampicillin resistance) and *ori* (origin of replication) is from pBR322. *B* BamHI; *C* ClaI; *E* EcoRI; *H* HindIII; *P* PstI; *S* SphI; () = former site, destroyed during construction. Positions of sites are shown in bp. Marks are drawn every 500 bp

micron (2 μ) plasmid from yeast (in b form) contains an origin of replication. As a heterologous gene is included a cDNA gene for α-amylase from mouse, inserted behind the yeast promoter *ADH1* (alcohol dehydrogenase I). In the plasmid pMS12(ΔSphI) the small SphI-SphI fragment of pMS12 is delected.

4.1.1 Transformation Procedure

The transformation is performed by the procedure of Ito et al. (1983) using LiOAc: A loopful of cells is inoculated in 10 ml YPD and incubated at 30 °C overnight. The culture is washed in water, afterwards in TE+0.1 M LiOAc before resuspending the cells in TE+0.1 M LiOAc to a final concentration of 10^8 cells/ml. (10×TE stock solution: 0.1 M Tris; 0.01 M EDTA; pH 8.0.) 5 μl of salmon sperm DNA (10 mg/ml), 10 μg of plasmid DNA and 0.7 ml 40% PEG (polyethylene glycol) are added to 0.1 ml of this suspension and mixed gently. Incubation is performed with careful but gentle shaking for 60 min at 30 °C. After 5 min at 42 °C, the mixture is centrifuged for about 3 s in the Eppendorf centrifuge and the supernatant is removed with a pipette. Cells resuspended in water are spread on selective medium, i.e. plates without tryptophan. From strain KK7, a stable *trp1* mutant of A_{91} (a spore-derived clone of strain A), 16 transformants per μg of pMS12 were obtained, whereas the laboratory strain DB 746 *(MATα his3Δ1 leu2-3 leu2-111 ura3-52 trp1-289)* gave 100 transformants per μg of pMS12.

4.1.2 Screening for Amylase Activity

After 3 to 7 days of growth, transformants are picked up and screened for the ability to secrete active amylase to the surrounding medium. This is done by spotting loopfuls of yeast on medium containing 0.4% soluble starch. The plates are incubated for 6 to 18 h and then stained with iodine vapour by placing the Petri dish bottom up with the medium and the cells on top of the inverted Petri dish lid containing solid iodine. After 2 to 10 min colonies synthesizing and secreting amylase are surrounded by a clear zone–a halo–on the dark iodine-stained background (Fig. 4). As can be seen, more amylase is secreted from transformants of KK7 than from those of the laboratory strain. The reason for this is not yet known.

4.1.3 Stability Test of Transformants

The stability of tryptophan prototrophy is tested by growth for 17 to 20 generations in non-selective medium, i.e. YPD, followed by plating on YPD and replica plating onto SD and YPD+0.4% starch. For transformants with pMS12, about 20% of the colonies have lost both tryptophan prototrophy and amylase secretion. The corresponding value for pMS12(ΔSphI) is 7%. In other transformants of industrial yeast strains and their derivatives, plasmid loss frequencies of 50% (Gjermansen 1983), and 5% (Hinchliffe and Daubney 1986) have been reported under similar conditions.

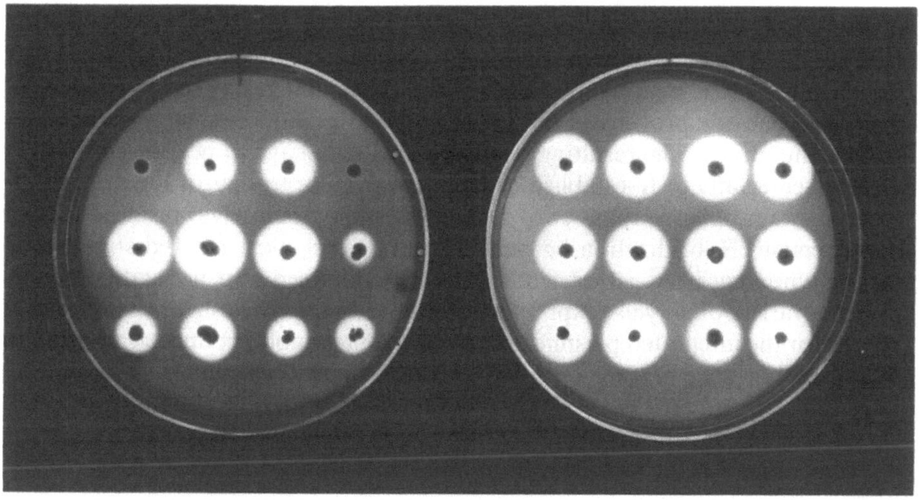

Fig. 4. Screening of transformed cells for secretion of α-amylase activity. On the *left plate*, starting with the upper row *from left to right* DB746 untransformed, DB746+pMS12, DB746+pMS12(ΔSphI), KK7 untransformed. *Middle and lower rows* KK7+pMS12, KK7+pMS12(ΔSphI), e7, three clones of cross e7 × A_6, 2 clones of cross e7 × A_{326}. A_6 and A_{326} are meiotic products of strain A. On the *right plate*, starting with the upper row from *left to right* transformants of KK7 with pMS12 partially digested with HindIII: e1–e11, e17

4.1.4 Integration of a Self-Replicating Plasmid

In attempts to obtain higher stability through integration of the plasmid in the genome of KK7, pMS12 was linearized through partial digestion with restriction endonuclease HindIII before transformation. The free ends obtained in the *TRP1* gene are expected to stimulate integration of pMS12 into the *trp1* gene of KK7 (Orr-Weaver et al. 1983). By this procedure the frequency of transformants was lowered by a factor of 2. Stability tests revealed that out of 12 transformants one strain, e7, had no loss of tryptophan prototrophy or amylase secretion after 60 generations in non-selective medium. Upon gel electrophoresis of restricted DNA, Southern transfer and hybridization to radioactive probe (1.6 kb PstI fragment of pMS12) it was demonstrated that e7 has pMS12 integrated in the *trp1* gene (results not shown).

Through integration of a plasmid containing a sequence homologous to one in the transformed strain partial stabilization can be obtained. By subsequent excision complete stabilization for the homologous sequence can be obtained. Also chromosome breakage and chromosome loss has been reported (Falco et al. 1982; Holmberg et al. 1982) due to unequal crossing-over between chromatids containing 2 μ sequences during mitosis. This will result in lower viability, just as crossing-over between integrated and unintegrated 2 μ may do.

In e7, some of the isolated clones show stability in both tryptophan prototrophy and amylase secretion. Other clones are stable in tryptophan prototrophy, unstable in amylase secretion and exihibit a reduced growth rate compared to KK7, whether grown in selective or unselective conditions.

4.1.5 Breeding Through Transformation of Spore-Derived Strains

One cannot expect to obtain auxotrophic mutants from polyploid strains by simple mutagenesis treatments. Plasmids with dominant yeast genes may be introduced into meiotic progeny of industrial yeasts after induction of relevant auxotrophic mutations. The resulting transformants may have to be crossed to other meiotic segregants to re-establish characteristics close to that of the original industrial strain. A large number of such crosses may themselves result in further improvement. Hybridization between e7 and several spore-derived strains with mating type **a** has been performed. The hybrids obtained were not stable and displayed a variety of amylase expression levels, see Fig. 4. These hybrids are heterozygous for the integrated plasmid which can lead to viable products with deletion of the amylase gene in some cells due to chromosome breakage.

4.2 Integration Vector with Drug Resistance as Selective Marker

While selection of antibiotic-resistant transformants is commonplace in bacterial transformation systems, such selection has only recently become available for yeast. *S. cerevisiae* is sensitive to geneticin (G418 sulphate, Jimenez and Davies 1980), which is an aminoglycoside antibiotic structually related to kanamycin and gentamycin. The transposon *Tn903* codes for resistance to G418.

The intergration vector pDY3 (see Fig. 5, Yocum 1984) was kindly provided by R. Yocum and R. Daves, Biotechnica International, Inc., Cambridge, MA, USA. It is constructed to introduce a desired gene in yeast and to jettison all vector sequences resulting in stable transformants.

The transposon *Tn903* is combined with *amp*^{r} and *ori* from pBR322 and *lacZ* (from *E. coli*) which is placed unter yeast *GAL* control. Furthermore, the vector contains a sequence which is homologous with a sequence (a "target" sequence)

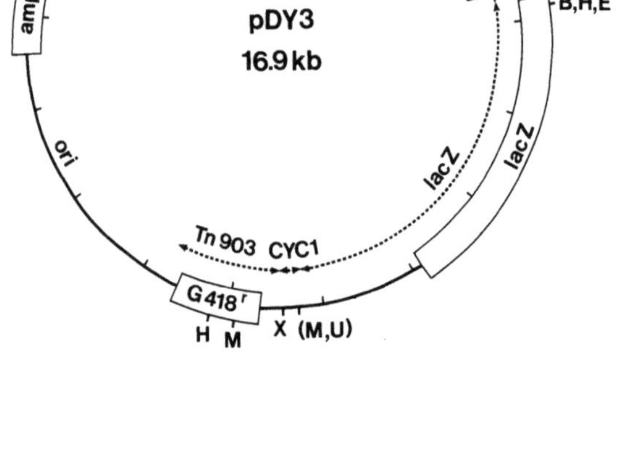

Fig. 5. Partial restriction endonuclease site map of the plasmid pDY3 (Yocum 1984). The *HO* region, *GAL1* (the galactose 1 promoter) and *CYC1* (iso-1-cytochrome c promoter) are all from *S. cerevisiae*. The *lacZ* (*β*-galactosidase) gene is from E. coli, *Tn903* is the bacterial transposon 903 coding for resistance to G418. The rest of the plasmid is from pBR322. *A* XbaI; *B* BamHI; *E* EcoRI; *H* HindIII; *K* KpnI; *M* SmaI; *S* SalI; *TI* SstI; *TII* SstII; *U* StuI; *X* XhoI; () = former site, destroyed during construction. The marks on the plasmid indicate 1000 bp

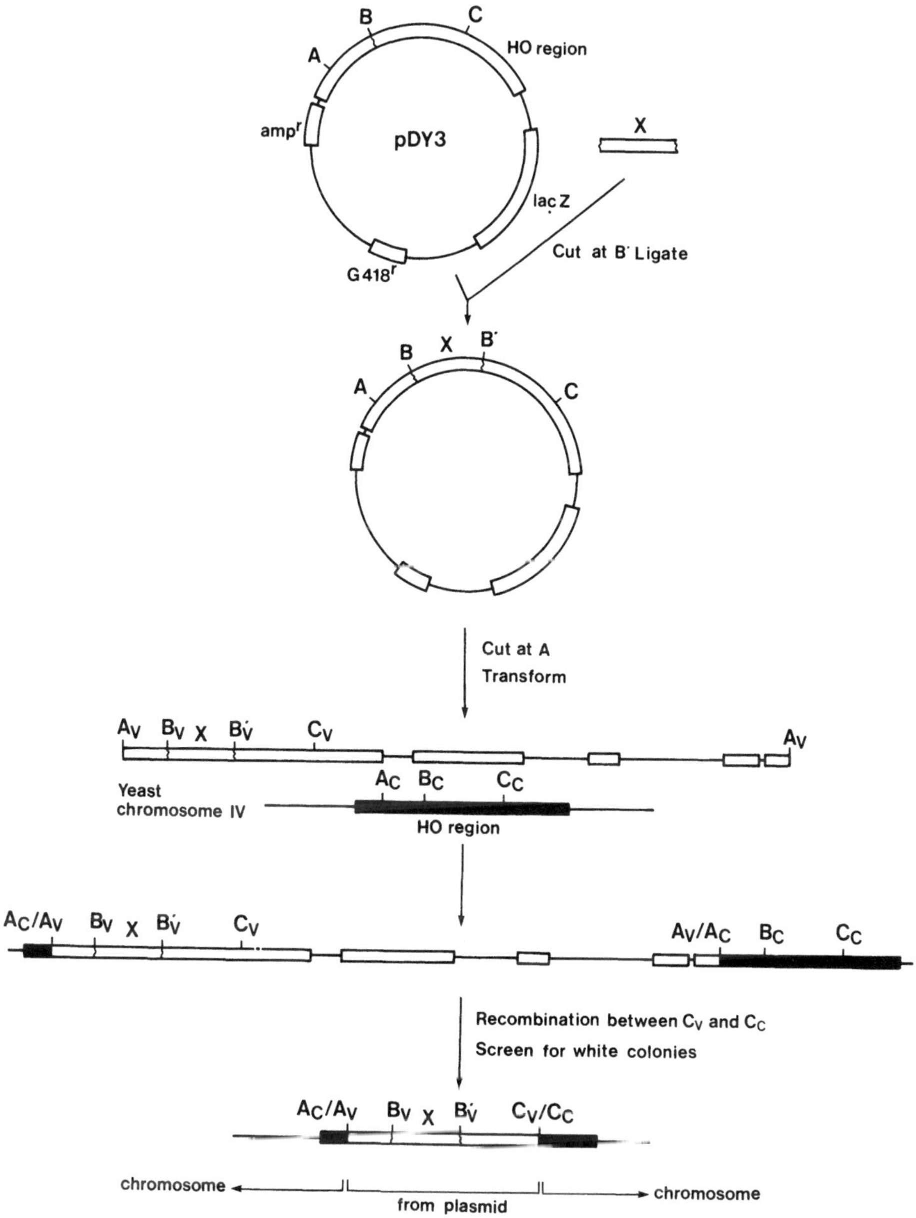

Fig. 6. Strategy for integration of a desired gene, X, in chromosome IV of *S. cerevisiae* by the vector pDY3

of chromosome IV of *S. cerevisiae*, namely the *HO* gene and adjacent regions. By insertion of the desired gene in the *HO* region, followed by linearization with KpnI and transformation, transformants appearing on G418 containing medium will contain the desired gene as well as the rest of the vector integrated in chromosome IV (see Fig. 6). In addition to endowing the transformants with resistance to G418, introduction of the vector also results in production of β-galactosidase, when the yeast is grown with galactose as carbon source. On medium containing X-gal this results in blue colonies, see Sect. 4.2.2. With a certain frequency, crossing over between homologous sequences on the modified chromosome IV results in looping out of vector sequences, leaving behind only the desired gene. By growing transformants for 20–60 generations in unselective medium and subsequent plating on X-gal medium, several crossing-over events will reveal themselves as white colonies. Screening of these for the presence of the desired gene and the absence of vector sequences will reveal the clones wanted.

4.2.1 Transformation Procedure Using a Non-Replicating Plasmid

The procedure is a modification of that described in Sect. 4.1.1: in order to determine the transformability of several strains pDY3 is linearized with KpnI prior to the transformation to facilitate integration into the HO gene of the recipient strain. After transformation, the cells are suspended in 400 µl YPD and shaken at 30 °C for 3 h before plating on the selective medium (YPD containing G418). During incubation in YPD, the plasmid could be integrated and *Tn903* could be expressed. The transformation frequencies are shown in Table 3. It should be mentioned that addition of 1% $(NH_4)_2SO_4$, 0.2% $CaCl_2$ or 0.2% $MgSO_4$ to the medium results in resistance of yeast to G418 up to 2 mg/ml (unpublished results; Webster and Dickson 1983). Thus, if the use of synthetic media is needed, the nitrogen source should be one or several amino acids.

4.2.2 Screening for β-Galactosidase Activity

Colonies with expression and secretion of β-galactosidase will appear as blue colonies on X-gal medium due to the conversion of 5-bromo-4-chloro-3-indoxyl-β-D-galactoside (X-gal, colourless, Sigma No. B4252) to D-galactose and 5-bromo-4-chloro-indole-3-ol. The latter is spontaneously oxidized to a blue substituted indigo. X-gal medium contains per liter, 13.6 g KH_2PO_4, 2 g $(NH_4)_2SO_4$, 0.5 mg

Table 3. Transformation frequency upon transformation with pDY3

Strain transformed	Concentration of G418 for selection of transformants mg/ml	Frequency of transformants no./µg DNA
A	0.5	200
A_{91}[a]	1.0	10
KK7	0.5	120
A_6[a]	1.0	50
KK194[b]	1.0	150

[a] Meiotic progeny of strain A.
[b] Diploid laboratory strain.

$FeSO_4 \cdot H_2O$, 4 mg thiamin, 200 mg biotin, 4 mg pantothenic acid, 20 mg inositol, 4 mg pyridoxine, appropriate amounts of required amino acids, 20 g galactose and 20 g agar for solidification. 40 mg X-gal and 1 ml 1 M $MgSO_4$ per liter medium are added after the heat sterilization. pH is adjusted to 7.0 with KOH.

Transformants were tested for β-galactosidase activity on X-gal medium. Differences in the behaviour are found for transformants of different strains as well as transformants of the same strain in the intensity of the blue colour. This could be due to differences in the number of β-galactosidase genes integrated.

References

Anderson E, Martin PA (1975) The sporulation and mating of brewing yeasts. J Inst Brew 81:242–247

Beggs JD (1978) Transformation of yeast by a replicating hybrid plasmid. Nature 275:104–109

Benítez T, Castello LD, Aguilera A, Conde J, Cardáolmedo E (1983) Selection of wine yeasts for growth and fermentation in the presence of ethanol and sucrose. Appl Environ Microbiol 45:1429–1436

Borstel RC, Mehta RD (1976) Mutation and selection systems for yeasts. In: Schlessinger D (ed) Microbiology. American Society for Microbiology, Washington DC, pp 507–509

Cameron JR, Phillippsen P, Davis RW (1977) Analysis of chromosomal integration and deletions of yeast plasmids. Nucl Acids Res 4:1429–1448

Chinault AC, Carbon J (1979) Overlap hybridization screening: Isolation and chracterization of overlapping DNA fragments surrounding the *LEU2* gene on yeast chromosome III. Gene 5:111–126

Dobson MJ, Kingsman SM, Kingsman AJ (1981) Sequence variation in the *LEU2* region of the *Saccharomyces cerevisiae* genome. Gene 16:133–139

Falco SC, Li Y, Broach JR, Botstein D (1982) Genetic properties of chromosomally integrated 2 μ plasmid DNA in yeast. Cell 29:573–584

Fincham JRS (1983) Genetics. Wright, Bristol

Fowell RR (1969) Sporulation and hybridization of yeasts. In: Rose AH, Harrison J (eds) The yeasts, vol I. Academic, New York, pp 303–383

Gjermansen C (1983) Mutagenesis and genetic transformation of meiotic segregants of lager yeast. Carlsberg Res Commun 48:557–565

Gjermansen C, Sigsgaard P (1981) Construction of a hybrid brewing strain of *Saccharomyces carlsbergensis* by mating of meiotic segregants. Carlsberg Res Commun 46:1–11

Gunge N (1966) Breeding of bakers' yeast–determination of the ploidy and an attempt to improve practical properties. Jpn J Genet 41:203–214

Hawthorne DC, Mortimer RK (1960) Chromosome mapping in *Saccharomyces:* centromere-linked genes. Genetics 45:1085–1100

Haynes RH, Kunz BA (1981) DNA repair and mutagenesis in yeast. In: Strathern JN, Jones EW, Broach JR (eds) The molecular biology of the yeast Saccharomyces. Life cycle and inheritance. Cold Spring Harbor Lab, New York, pp 371–414

Hinchliffe E, Daubney CJ (1986) The genetic modification of brewing yeast with recombinant DNA. J Am Soc Brew Chem (in press)

Hinnen A, Hicks JB, Fink Gr (1978) Transformation of yeast. Proc Natl Acad Sci USA 75:1929–1933

Holmberg S (1982) Genetic differences between *Saccharomyces Carlsbergensis* and *S. cerevisiae*. II. Restriction endonuclease analysis of genes in chromosome. III. Carlsberg Res Commun 47:233–244

Holmberg S, Petersen JGL, Nilsson-Tillgren T, Kielland-Brandt MC (1979) Molecular characterization of a *Saccharomyces* plasmid containing the *HIS4* gene. Carlsberg Res Commun 44:269–282

Holmberg S, Nilsson-Tillgren T, Kielland-Brandt MC, Petersen JGL (1982) Chromosome breakage in yeast caused by sister-chromatid recombination in an integrated 2-micron plasmid. Carlsberg Res Commun 47:355–369

Ismail AA, Ali AMM (1971) Selection of high ethanol-yielding *Saccharomyces*. II. Genetics of ethanol tolerance. Folia Microbiol 16:350–354

Ito H, Fukuda Y, Murata K, Kimura A (1983) Transformation of intact yeast cells treated with alkali cations. J Bacteriol 153:163–168

Jansen GP, Barney MC, Helbert JR, Esposito MS (1979) Nonvectorassociated transformation in *Saccharomyces*. J Appl Biochem 1:369–376

James AP, Inhaber ER (1975) Evidence of preferential pairing of chromosomes at meiosis in aneuploid yeast. Genetics 79:561–571

Jimenez A, Davies J (1980) Expression of a transposable antibiotic resistance element in *Saccharomyces*. Nature 287:869–871

Keiding AK (1985) Genetic and molecular characterization of a distiller's yeast. Carlsberg Res Commun 50:95–125

Kielland-Brandt MC, Nilsson-Tillgren T, Holmberg S, Petersen JGL, Svenningsen BA (1979) Transformation of yeast without the use of foreign DNA. Carlsberg Res Commun 44:77–87

Kielland-Brandt MC, Gjermansen C, Nilsson-Tillgren T, Petersen JGL, Holmberg S, Sigsgaard P (1981) Genetics of a lager production yeast. Tech Q Master Brew Assoc Am 18:185–187

Klapholz S, Esposito RE (1982) A new mapping method employing a meiotic Rec$^-$ mutant of yeast. Genetics 100:387–412

Maniatis T, Fritsch EF, Sambrook J (1982) Molecular cloning. A laboratory manual. Cold Spring Harbor Lab, New York

Mortimer RK, Hawthorne DC (1973) Genetic mapping in *Saccharomyces*. IV. Mapping of temperture-sensitive genes and use of disomic strains in localizing genes. Genetics 74:33–54

Mortimer RK, Schild D (1982) Genetic map of *Saccharomyces cerevisiae*. Microbiol Rev 44:519–561

Nasmyth KA, Tatchell K (1980) The structure of transposable yeast mating type loci. Cell 19:753–764

Nilsson-Tillgren T, Gjermansen C, Kielland-Brandt MC, Petersen JGL, Holmberg S (1981) Genetic differences between *Saccharomyces carlsbergensis* and *S. cerevisiae*. Analysis of chromosome III by single chromosome transfer. Carlsberg Res Commun 46:65–76

Orr-Weaver TL, Szostak JW, Rothstein RJ (1983) Genetic applications of yeast transformation with linear and gapped plasmids. In: Wu R, Grossman L, Moldave K (eds) Methods in enzymology, vol 101. Academic, New York, pp 228–245

Parry EM, Cox BS (1972) The general properties of a strain of yeast with less DNA than normal. J Gen Microbiol 70:129–149

Rigby PWJ, Doeckmann M, Rhodes C, Berg P (1977) Labelling deoxyribonucleic acids to high specific activity in vitro by nick translation with DNA polymerace. I. J Mol Biol 113:237–251

Riley MI, Manney TR (1978) Tetraploid strains of *Saccharomyces cerevisiae* that are trisomic for chromosome III. Genetics 89:667–684

Roman H, Phillips MM, Sands SM (1955) Studies of polyploid *Saccharomyces*. I. Tetraploid segregation. Genetics 40:546–561

Rothstein RJ (1983) One-step gene disruption in yeast. In: Wu R, Grossman L, Moldave K (eds) Methods in enzymology. Recombinant DNA, vol 101. Academic, New York, pp 202–211

Schaeffer AB, Fulton M (1933) A simplified method of staining endospores. Science 77:194

Scherer S, Davis RW (1979) Replacement of chromosome segments with altered DNA sequences constructed in vitro. Proc Natl Acad Sci USA 76:4951–4955
Shaffer B, Brearley I, Littlewood R, Fink GR (1971) A stable aneuploid of *Saccharomyces cerevisiae*. Genetics 67:483–495
Sherman F (1975) Use of micromanipulators in yeast studies. In: Prescott DM (ed) Methods in cell biology, vol 11. Academic, New York, pp 189–199
Slonimski PP, Perrodin G, Croft JH (1968) Ethidium bromide induced mutation of yeast mitochondria: complete transformation of cells into respiratory deficient non-chromosomal "petites". Biochem Biophys Res Commun 30:232–239
Southern EM (1975) Detection of specific sequences among DNA fragments separated by gel electrophoresis. J Mol Biol 98:503–517
Storms RK, McNeil JB, Khandekar PS, An G, Parker J, Friesen DJ (1979) Chimeric plamids for cloning of deoxyribonucleic acid sequences in *Saccharomyces cerevisiae*. J Bacteriol 140:73–82
Strathern JN, Spatola E, McGill C, Hicks JB (1980) Structure and organization of transposable mating type cassettes in *Saccharomyces* yeasts. Proc Natl Acad Sci USA 77:2839–2843
Struhl K, Stinchcomb DT, Scherer S, Davis RW (1979) High-frequency transformation of yeast: Autonomous replication of hybrid DNA molecules. Proc Natl Acad Sci USA 76:1035–1039
Thomsen KK (1983) Mouse α-amylase synthesizes by *Saccharomyces cerevisiae* is released into the culture medium. Carlsberg Res Commun 48:545–555
Walton EF, Carter BLA, Pringle JR (1979) An enrichment method for temperature-sensitive and auxotrophic mutants of yeast. Mol Gen Genet 171:111–114
Webster TD, Dickson RC (1983) Direct selection of *Saccharomyces cerevisiae* resistant to the antibiotic G418 following transformation with a DNA vector carrying the kanamycin-resistence gene of *Tn903*. Gene 26:243–252
Williamson VM (1983) Transposable elements in yeast. Int Rev Cytol 83:1–25
Winge Ø (1944) On segregation and mutation in yeast. C R Lab Carlsberg, Ser Physiol 22:235–247
Yocum R (1984) Yeast vector. US Pat Appl 612796, Eur Pat Appl 85303625.9

Automated Testing of Brewer's Yeast Strains for Fermentation Characteristics (Multiferm, System Carlsberg)

P. SIGSGAARD

1 Introduction

The production of beer is based on a thousand years of tradition. Being an unstirred process, the fermentation is highly dependent on the technological behaviour of the yeast strains. The yeast is added to the aerated wort at the start of fermentation and must remain in suspension to be able to convert the majority (approx. 95%) of the fermentable carbohydrates during the main fermentation. At the end of this process the yeast must flocculate and sediment to the bottom of the fermentor to enable separation of the yeast from the beer and re-use for the following fermentation. However, sufficient yeast must remain in suspension to be active in the maturation processes during the secondary fermentation of the green beer. In beer more than 800 chemical components are identified, many of which are produced during the fermentation and contribute to the overall flavour of the final beer (Meilgaard 1981, 1982).

Significant changes have occurred during the last 30 years in the methods of beer production, including changes in fermentation temperatures, wort composition, tank design and lay-out. An increasing demand for more economical and flexible production methods have resulted in shortening of the overall production time, especially in the fermentation and maturation processes (Annemüller 1981; Hackstaff 1978; Knudsen 1985; Maule 1986; Miedaner 1978; Narziss 1981). A yeast strain used in traditional brewing processes does not necessarily perform well under modern conditions. Therefore breeding of yeast strains adapted to new technologies has become an increasingly important task. Only recently have the genetic methods become available to carry out systematic breeding work with brewer's yeast. These include selection of spore clones expressin mating type, hybridization, mutagenesis and transformation (Gjermansen and Sigsgaard 1981; Ishimaru et al. 1967; Kielland-Brandt et al. 1981; Molzahn 1977; Spencer and Spencer 1983; Stewart et al. 1983; von Wettstein et al. 1984). Brewer's yeast with completely new properties, i.e. no diacetyl production (Ramos-Jeunehomme and Masschelein 1977, Petersen et al. 1983), amylase activity (Innis et al. 1985; Meaden et al. 1985; Rothstein 1984; Thomsen 1983; Tubb 1986), β-glucanase activity (Jackson et al. 1986) or anti-microbial activity (Hammond and Eckersley 1984; Sasaki et al. 1984) might be used either to produce special beer varieties or to solve specific production problems. A breeding programme for yeast strains with improved properties demands adequate facilities for testing the brewing performance of a large number of genotypes. The yeast must be tested under conditions equivalent to that in the brewery with consideration to environmental factors such as wort composition, temperature, and tank geometry. In this article suitable test procedures are described.

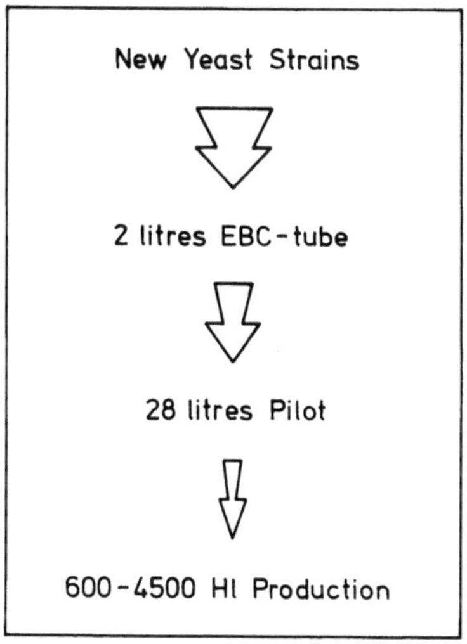

Fig. 1. Yeast evaluation programme of the Carlsberg Research Center

2 Testing Techniques

In the yeast breeding programme at the Carlsberg Research Center the brewing performance is being evaluated sequentially in three steps (Fig. 1). The first step is performed on the MULTIFERM system in 2-l scale. Promising strains are tested in trials comprising 28-l volumes (Ahrenst-Larsen 1967). The final evaluation is in production scale fermentations.

2.1 Small-Scale Fermentations

Small-scale fermentations have been used by many workers in the evaluation of brewer's yeast strains, wort composition, fermentation conditions etc. Fermentors consisting of tall narrow tubes will generally give conditions to the yeast related to the conditions experienced in production scale fermentations in the brewery (Cook 1963; Dixon 1967, Ishimaru et al. 1967; Kruger et al. 1982; Weinfurtner et al. 1961). Even if the small-scale laboratory tests not necessarily reproduce numerically the behaviour in large scale, the relative differences between different yeast strains correspond to the differences found in the brewery. Based upon work carried out within the European Brewery Convention-yeast group a standard technique was adapted in 1977 by the EBC-Analytica Microbiologica as Method 2.5.4. Tubes EBC (EBC Analytica Microbiologica 1977).

2.2 MULTIFERM Fermentations

Based on the principle of the EBC Tubes, an automated equipment with 60 tubes holding 2 l each has been developed at the Carlsberg Research Center. It allows routine screeing for improved brewing yeast strains with 60 fermentations per week under strictly standardized conditions (Sigsgaard and Nørager 1985).

2.2.1 Description of the MULTIFERM System

In the automated MULTIFERM the fermentations and operations such as wort filling, temperature regulation, sampling and cleaning are all controlled by a micro-computer (Figs. 2 and 3). The MULTIFERM is placed in a cold room and has two identical carousels each holding 30 tubes. The arrangement of the tubes on turntables saves space and allows them to be positioned at a station for wort filling, sampling and cleaning in place. Use of two carousels permits two independent sets of fermentations and alternative maintenance repair without the need to shut down the entire plant. The EBC tubes are made from stainless steel and equipped with cooling jackets for temperature control (Fig. 4). Each tube is supplied with cooling water from the center of the carousel. The temperature in the tubes can be varied from 5 ° to 18 °C with an accuracy of ±0.1 °C. Each EBC tube is fitted with a rubber membrane located 75 cm above the bottom of the tube. During automatic sampling this membrane is penetrated with an injection needle (see later). Each carousel has three aluminium platforms (90 cm in diameter), the tubes being clamped between the two upper ones. Stainless steel centrifuge beakers function as bottoms of the tubes and are placed in holes in the lower-

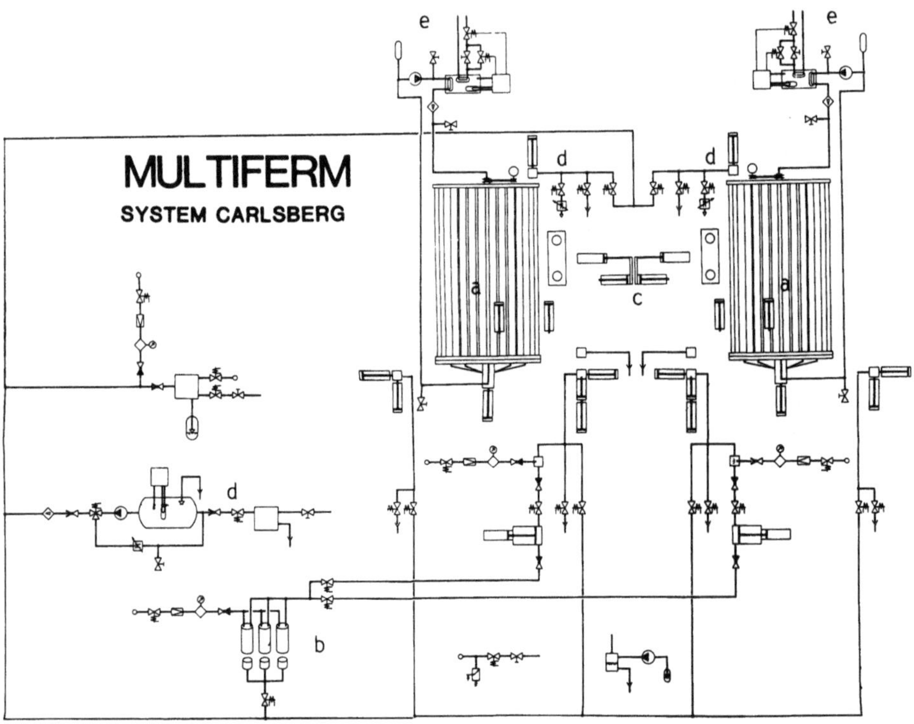

Fig. 2. Schematic set-up of the MULTIFERM. *a* 2 × 30 EBC-tubes; *b* wort filling; *c* automated sampling equipment; *d* CIP; *e* cooling system

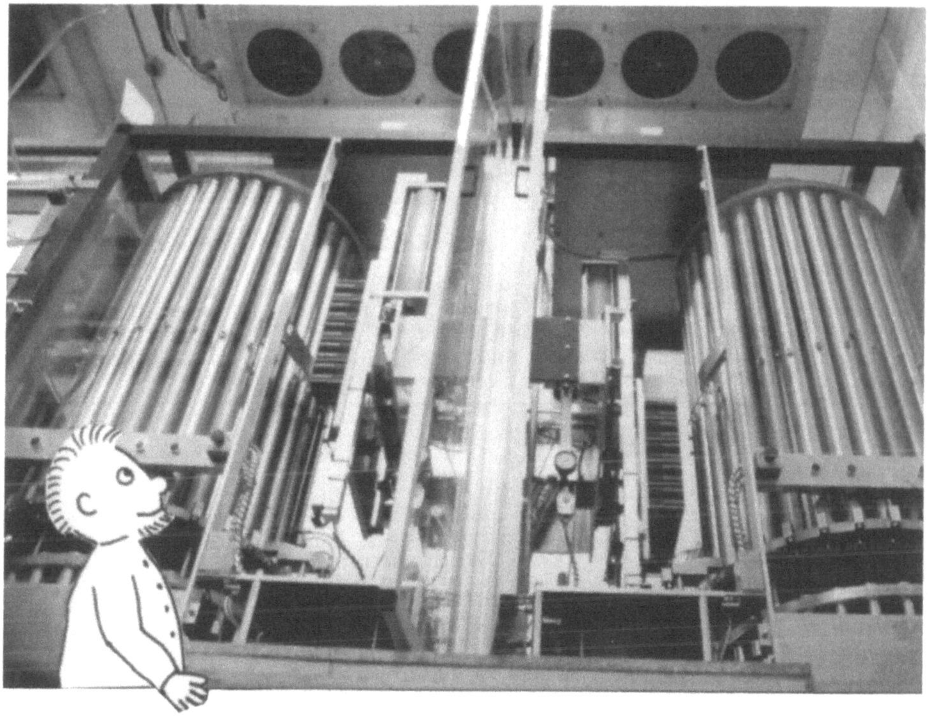

Fig. 3. MULTIFERM. Two independent carousels, each holding 30 EBC tubes. Lowermost platforms are shown in the lower position. In *center* filling and sampling equipment

most platform. This platform can be lowered, whereby the beakers move away from the EBC tubes to gain access for filling them with yeast or for centrifuging the yeast after completed fermentations. (Fig. 5). A valve at the top of the EBC tube allows the CO_2 to escape during fermentation and when open serves as a conus to spread the detergent and rinsing water during the cleaning of the tube (Fig. 6). Step motors controlled by the computer turn the carousel stepwise, allowing each EBC tube to be moved into position for filling with wort, or sampling or cleaning in place. Piston pumps measure 1.9 l of wort into each EBC tube. The wort is aerated on line by sterile filtered compressed air to obtain an oxygen content in the wort of approximately 8.5 ppm. Owing to the downward-pointing filling pipe stub, the yeast in the bottom of the closed EBC tube is suspended effectively in the wort.

Fermentations are performed at a preset temperature, the standard being 12 °C. After 6 days' fermentation, the temperature is lowered to 7 °C. The temperature is controlled by circulating the water of the cooling jackets of the EBC tubes through a heat exchanger (−4 °C glycol + electric heating element). The temperature in the EBC tubes can be altered at a rate of at least 0.1 °C/10 min.

Samples are taken daily during fermentation by an automated sampler containing 3 × 30 collectors, each holding one evacuated test tube of the type used for blood collection. A double-headed injection needle penetrates with one tip the

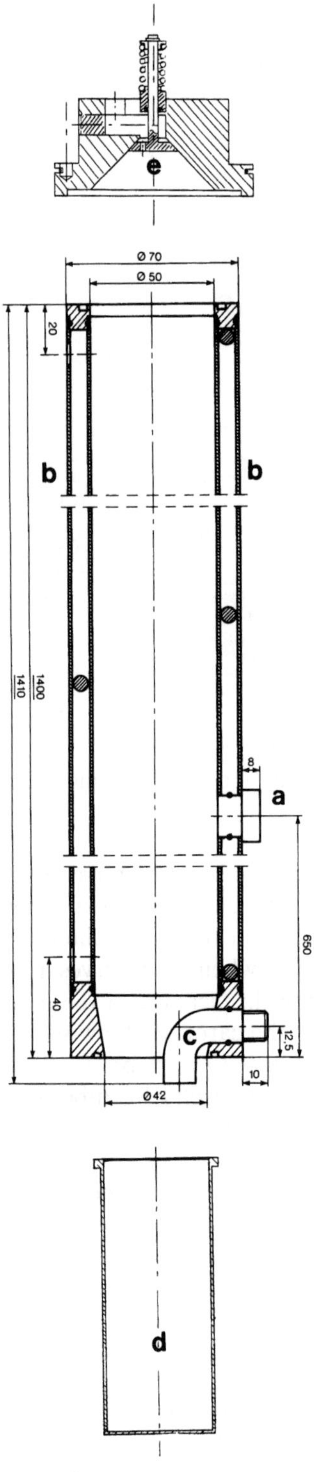

Fig. 4. Drawing of an EBC-tube. *a* Sampling device. *b* Water jacket for temperature control. *c* Wort filling tube. *d* Centrifuge beaker. *e* Top valve

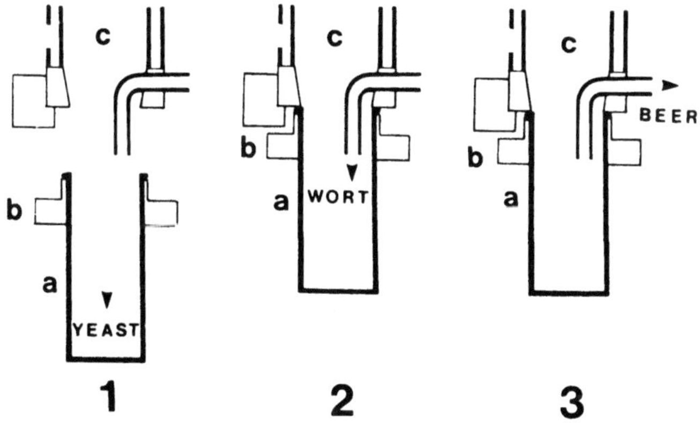

Fig. 5. Bottom of EBC tube. *a* Centrifuge beaker; *b* lowermost aluminium platform; *c* EBC tube. **1** Centrifuge beaker lowered during yeast filling. **2** Centrifuge beaker elevated, wort introduced. **3** Beer drained off

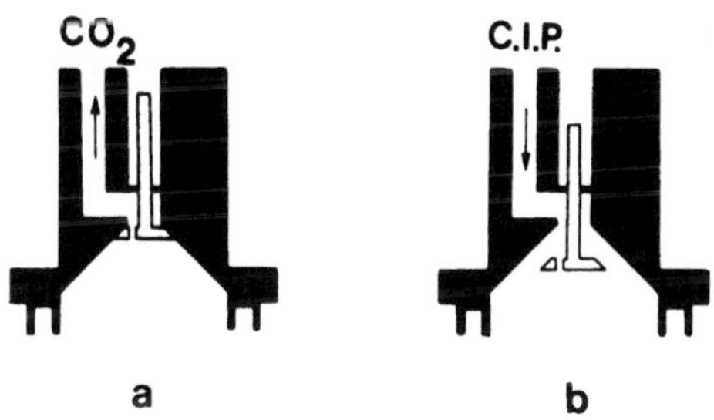

Fig. 6 a, b. Top of EBC tube. **a** Valve closed. **b** Valve opened

rubber membrane of the EBC tube and with the opposite tip the seal of the evacuated test tube. After approx. 20 s, the needle is retracted and the sample has been collected. A photocell ensures that the needle has actually been removed from the tube before the next EBC tube is moved in position for sampling. Individual sterile needles are used for each EBC tube (see also Sect. 2.2.2.4).

At the end of fermentation, the beer in the EBC tubes is automatically emptied and the sedimented yeast collected by centrifugation for analysis and possible reuse. Often the supernatant in the centrifuge tubes provides a sufficient test sample of the green beer. The EBC tubes are automatically cleaned (see Sect. 2.2.2.6).

2.2.2 Working Procedures

The procedures used with the multiple fermentations in the MULTIFERM are described in the following.

2.2.2.1 Propagation of Yeast

Yeast from an agar slant culture or a frozen culture is spread and grown on solid yeast medium (e.g. wort agar or yeast extract dextrose agar). A loopful of yeast

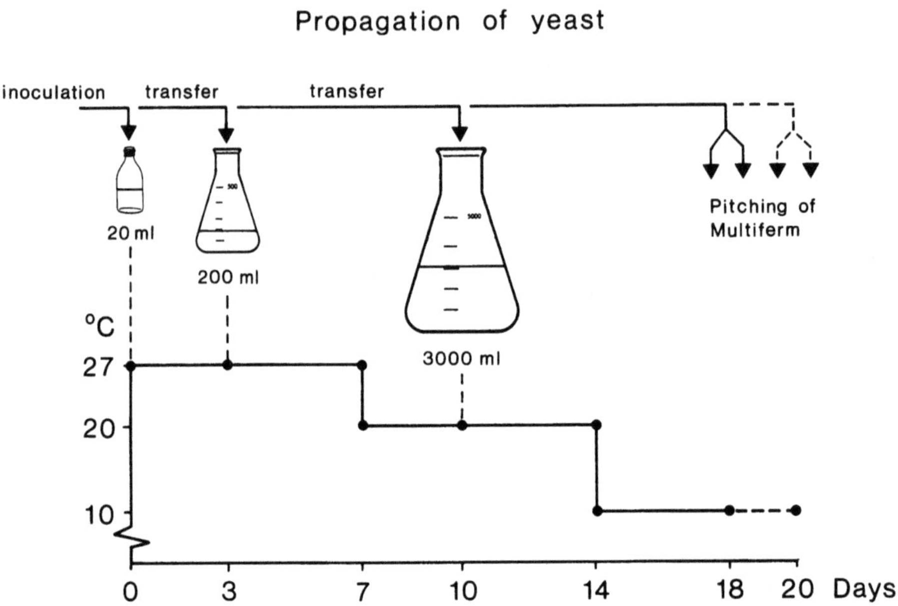

Fig. 7. Yeast propagation system

is transferred to 20 ml sterile wort in a 50-ml medical bottle (Fig. 7). After 3 days at 27 °C the content is poured into 200 ml wort in 500-ml Erlenmeyer (EM) flasks. After 4 days at 27 °C followed by 3 days at 20 °C (shaken daily by hand), the content is transferred to 3000 ml sterile aerated wort in 5000 ml EM flasks. After 4 days at 20 °C (shaken daily by hand), the temperature is lowered to 10 °C to allow the yeast to settle. After 4–6 days the supernatant is poured off and the sediment is centrifuged (1200 *g* in 25 min). Ten g centrifuged yeast is used for pitching one EBC tube.

2.2.2.2 Wort Collection and Handling

Twentyfive litres kegs normally used for kegged beer are cleaned and sterilized by steam. After being filled in the plant brewhouse with 80 °C hot filtered wort, the kegs are closed and rolled to ensure heating of all inner surfaces. The kegs are air-cooled and the wort can be kept at 10 °C for several months without microbiological problems.

Before use the kegs are rolled again to suspend trub-particles sedimented during the storage.

2.2.2.3 Start of Fermentations

Ten g centrifuged yeast, either newly propagated or remaining in the centrifuge beaker from a previous EBC tube fermentation is suspended in 90 ml sterile wort, whereafter the beakers are placed in the holes on the lowermost platform of the MULTIFERM.

The top of three kegs with cooled wort is disinfected with 80% alcohol. The wort-filling system is automatically cleaned, whereafter the kegs are connected to the MULTIFERM. One by one the EBC tubes are then automatically filled with aerated wort and the fermentations will start at the preset temperature. Finally, the wort-filling system is again automatically cleaned and sterilized.

2.2.2.4 Sampling During Fermentations

Samples are taken at desired intervals with the above mentioned sampler. Evacuated 10-ml test tubes are prefilled: a specially constructed dosage pump fills 0.5 ml of a 2.5% w/w solution of $CuSO_4$ into the tubes without breaking the vacuum. The final concentration in the sample (approx. 1000 ppm Cu) will almost instantly stop the yeast activity and keep the sample unchanged until further analysis for apparent extract and yeast cell number in suspension (Fig. 8).

Prior to these analyses, the gaseous carbon dioxide must be removed. This can conveniently be done by placing the opened test tube on a shaking water bath at 60 °C for 10 min. The changes in the sample due to the addition of $CuSO_4$-solution and evaporation during decarbonation are corrected for according to empirical functions found when measuring parallel preserved, decarbonated samples and unpreserved samples (Fig. 9).

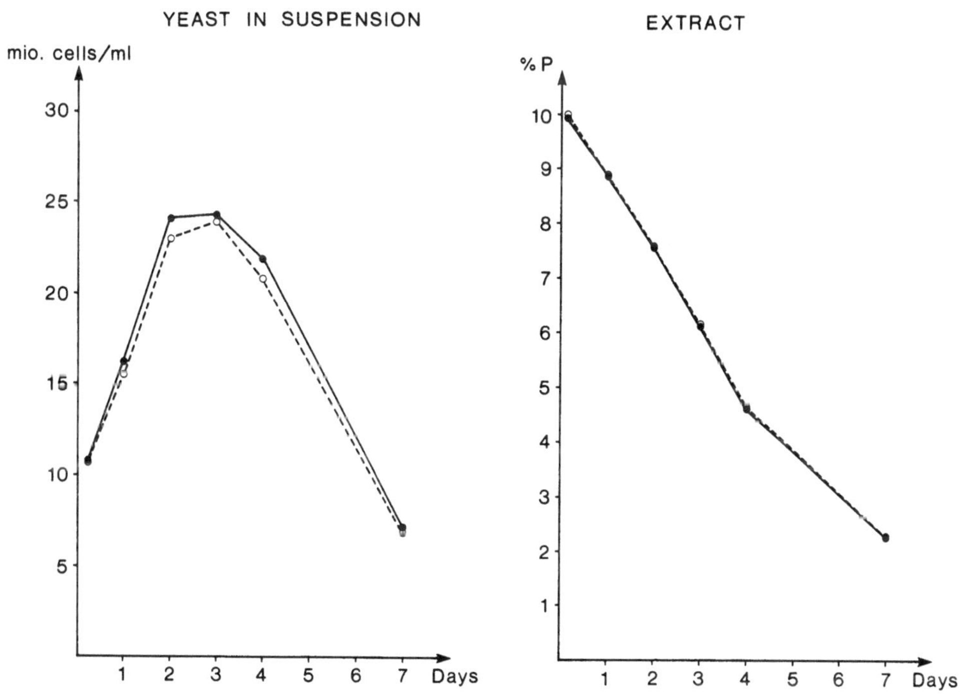

Fig. 8. Yeast cell number and apparent extract in unpreserved ●—● and preserved ○- - -○ samples during fermentation. The unpreserved samples are tested immediately after sampling, whereas the preserved samples are kept for 4 days at 10 °C

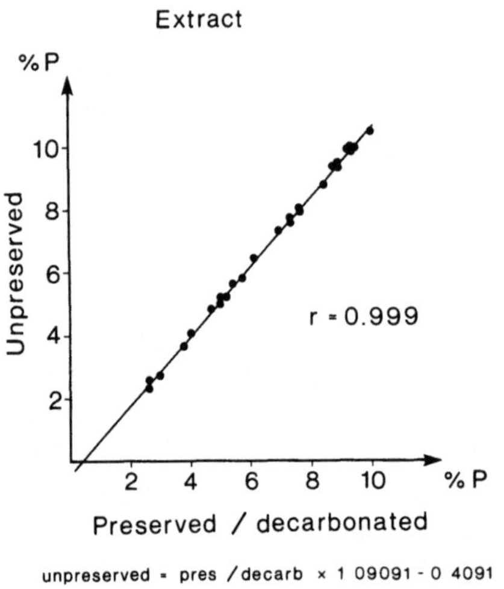

Fig. 9. Correction of apparent extract due to $CuSO_4$ addition and decarbonation on water bath

2.2.2.5 End of Fermentations

At the end of the main fermentation the temperature in the EBC tubes is lowered and most of the yeast will settle to the bottom of the centrifuge beaker.

Unpreserved samples are taken either by the automatic sampling system or by manual valves (greater volumes of samples), whereafter the tubes are emptied to a level of approx. 2 cm below the upper part of the centrifuge beaker. The lowermost platform is lowered and the centrifuge beakers removed from the MULTIFERM.

2.2.2.6 Cleaning in Place of the MULTIFERM

After each fermentation the EBC tubes are cleaned. During cleaning a stainless steel cup with a flow-through opening replaces the beaker at the bottom of the tubes. The detergent used is a basic foam cleaner (pH 10.7). After cleaning, it is removed by two rinsings with 80 °C demineralized water introduced alternatively from both ends of the EBC tube (Fig. 10).

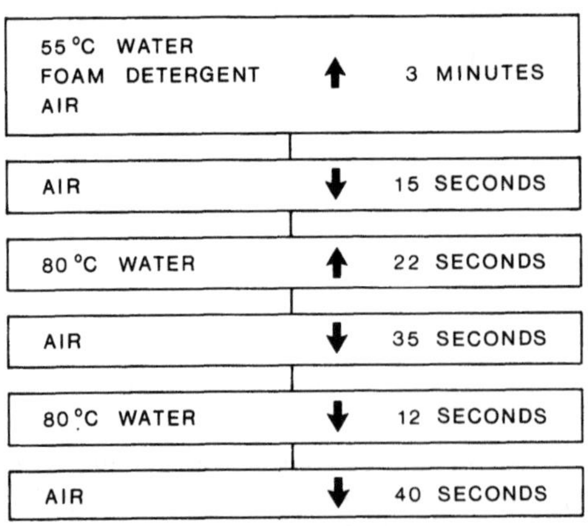

Fig. 10. Cleaning in place procedure

2.2.2.7 *Re-Use of Yeast*

To test for genetic stability of the new yeast strains the yeast is reused for at least five consecutive fermentations (generation 1 to 5). The harvesting of yeast and the pitching of the following 30 EBC tubes can be performed in 1 day due to the automatic cleaning and filling possibilities.

2.2.3 Analyses Performed

The standard program of analyses on the MULTIFERM is as follows:

At start:
- yeast cell size
- pitching rate
- extract of wort

Daily during fermentation (on preserved samples):
- extract
- yeast cell number in suspension

At end:
- yeast yield
- aroma profile of beer
- diacetyl/pentandione content of beer

2.2.3.1 *Yeast Cell Size and Yeast Cell Number*

Yeast cell number is measured on $CuSO_4$-preserved samples. The yeast cell size, however, is significantly smaller in the preserved samples due to the killing of the yeast cells. The cell size can therefore only be determined on unpreserved samples.

The principle for both analyses is an electronic counting (EBC Analytica Microbiologica 1981). A sample of 0.025 ml is diluted in 10 ml electrolyte (Diluid) and counted on a Coulter Counter ZB (Coulter Electronics Ltd., Coldharbour Lane, Harpenden, Hertfordshire, England). The cell number is calculated on a HP 85 computer (Hewlett Packard, Corvallis, Oregon, USA). A programme takes into account the dilution factor as well as the addition of $CuSO_4$ solution. The cell size distribution is determined on a VDA 740 Volume Distribution Analyzer (Analysis Instrument AB, Stockholm, Sweden), which sorts the signals from the Coulter Counter according to cell size.

2.2.3.2 *Pitching Rate*

As a standard, 10 g centrifuged yeast (1200 *g* in 25 min) is used for pitching 2 l of wort. Due to the difference in yeast cell size this amount of yeast will correspond to yeast cell number varying from 10 to 25×10^6/ml.

2.2.3.3 Extract Determination

The amount of extract during fermentation is measured on $CuSO_4$-preserved samples. After decarbonation on a shaking water bath, the specific density of the samples are measured on an Anton Paar DMA 46 density meter (Anton Paar K.G., Graz, Austria). A thermostated u-shaped hollow oscillator is filled with sample. The mass of the liquid changes the natural frequency of the oscillator and, as the volume is constant, the specific gravity (SG_E) can be determined. The extract (E) is calculated according to the following polynomial approximation from table values (Goldiner et al. 1966):

$$E = -468.08 + 677.859\ SG_E - 209.782\ SG_E^2.$$

The extract value (preserved/decarbonated) is corrected for changes due to the $CuSO_4$ addition and decarbonation (Fig. 9) and the "true" extract value (unpreserved) is found.

2.2.3.4 Yeast Yield

After centrifugation of the settled yeast at the end of the fermentation, the yeast is weighed and the yeast yield expressed as gram harvested yeast per gram pitched yeast.

2.2.3.5 Vicinal Diketones and Ester/Alcohol (Aroma Profile) Analyses

At the end of fermentation unpreserved samples are taken. After centrifugation the supernatant is analyzed.

The determination of vicinal diketones (VDK) and ester/alcohols is performed using a headspace sampling technique and analyzing the gas phase above the sample by gas chromatography using capillary columns with electron capture or flame ionization detectors. Headspace gas chromatography makes use of the equilibrium between the volatile components of the sample and gas phase in a sealed vial.

In the sample preparation, 5 ml of sample is pipetted into a 20-ml vial to which 1.5 g NaCl and a stirring bar have been added. An internal standard is added to each sample – 2,3-hexadione to the VDK's and N-pentylacetate to the aroma profile samples. The vials are then purged with N_2 and sealed with Teflon-lined crimped caps. The samples are heated with stirring at 60 °C for 30 min and then transferred to a thermostated automatic headspace sampler. The VDK's are analyzed by electron capture and the ester/alcohols by flame ionization detectors. Quantitation of the components of interest is performed on line by a dedicated data system using the internal standard method based on peak area measurements.

The anlytical conditions are as follows:

Instrument: VDK: DANI HSS 3950 (DANI SPA, Monza, Italy) with Varian 6500 6C (Varian Instruments, Walnut Creek, California). Aroma profile (ester/alcohols): Perkin Elmer Multifract F 45 (Perkin Elmer, Norwalk, Conneticut).

Detectors: Flame Ionization Detector and Ni^{63} Electron Capture Detector.

Column: VDK: Fused Silica BP-5, 25 m × 0.32 mm ID (SGE, Victoria, Australia). Aroma profile (ester/alcohols): Fused Silica Carbowax 20 M, 50 m × 0.32 mm ID (SGE, Victoria, Australia).

Carrier gas: N_2, prepurified.

Temperatures: VDK:	– inlet	150 °C		
	– detector	175 °C		
	– oven	60 °C		
Aroma profile (ester/alcohols):				
	– inlet	100 °C		
	– detector	200 °C		
	– oven	initial	60 °C	
		hold	10 min	
		ramp	30 °C/min	
		final	130 °C	
		hold	3 min	

2.2.4 Data

2.2.4.1 Data Handling and Storage

A MULTIFERM "run" consisting of 30 fermentations in five consecutive generations will produce more than 6000 single data. All data are entered in the central data handling system ANASYS of the United Breweries (Rosendal et al. 1983). One of the 15 pages containing raw data from the five consecutive fermentations is shown in Fig. 11. After completion of a "run" the data are transferred to the Vax-computer system of the Carlsberg Research Center for further evaluation.

2.2.4.2 Data Evaluation

In all "runs" a presently used production strain of brewer's yeast is used as an internal reference and the other strains are compared to this strain. Fermentations are often performed in duplicate using the same yeast strain in two individual EBC tubes. The t-test is used to determine whether significant differences exist between a test strain and the reference strain. Generally a difference of the order of 10% in a parameter is sufficient to give a significance level of 99.5%. In the initial sorting of yeast for a desired changed characteristic we consider a 20% difference as a minimum requirement. Strains showing greater than 20% difference in a parameter for at least four of the five consecutive fermentations are considered to be of interest for further evaluation. In Fig. 12 are shown the evaluated data of one yeast strain obtained by mating of meiotic segregants from the reference yeast strain 374.02 (Tuborg production strain) and selected for rapid growth at 10 °C. The MULTIFERM revealed the strain to have an increased fermentation rate, increased yeast yield and an increase in the aroma components propa-

UNITED BREWERIES
QUALITY SERVICES
CENTRAL LABORATORY

	REPORT NUMBERS									
	95825	95826	95827	95828	95829	95830	95831	95832	95833	95834
QUALITY PARAMETER										
START OF FERM., DATE.H	50815.15	50815.15	50815.15	50815.15	50815.15	50815.15	50815.15	50815.15	50815.15	50815.15
YEAST STRAIN NO.	324.02	324.02	325.02	325.02	215.02	215.02	326.02	326.02	327.02	327.02
GENERATION	2	2	2	2	2	2	2	2	2	2
EBC-TUBE NO.	1	2	3	4	5	6	7	8	9	10
APP.EXTR.TIME 0, %P	10.17	10.06	10.30	10.27	10.09	9.67	10.17	10.14	10.17	9.98
APP.EXTR.TIME 1, %P	8.71	9.24	8.52	9.61	9.21	7.37	9.13	9.06	9.27	9.32
APP.EXTR.TIME 2, %P	7.07	7.13	8.04	8.12	6.64	6.56	6.64	6.70	6.94	6.94
APP.EXTR.TIME 3, %P	4.80	4.80	6.02	6.26	3.92	4.00	4.74	4.80	4.55	4.63
APP.EXTR.TIME 4, %P	3.12	3.07	4.28	4.63	2.24	2.24	3.51	3.53	2.68	2.73
APP.EXTR.TIME 5, %P	1.99	2.07	3.01	3.37	1.62	1.60	2.82	2.96	1.65	1.62
APP.EXTR.TIME 6, %P	1.71	1.65	1.93	2.32	1.51	1.48	2.54	2.65	1.43	1.48
APP.EXTR.TIME 7, %P	1.51	1.62	1.79	1.85	1.45	1.60	2.49	2.57	1.46	1.51
CELL COUNT TIME 0, MIO./ML.	13.0	12.6	8.5	9.4	16.5	15.1	18.3	17.6	14.1	14.8
CELL COUNT TIME 1, MIO./ML.	13.7	13.6	11.5	11.3	22.2	22.2	18.1	16.9	19.6	20.0
CELL COUNT TIME 2, MIO./ML.	19.4	20.1	23.2	22.6	39.8	41.5	30.6	28.9	36.5	36.8
CELL COUNT TIME 3, MIO./ML.	23.5	22.6	34.3	31.3	46.2	46.2	30.5	29.0	43.8	40.7
CELL COUNT TIME 4, MIO./ML.	22.5	22.2	35.9	33.9	38.9	38.5	17.6	15.3	40.8	38.6
CELL COUNT TIME 5, MIO./ML.	14.8	14.9	32.2	30.0	15.5	16.2	8.2	7.6	31.0	31.9
CELL COUNT TIME 6, MIO./ML.	5.3	5.7	31.7	29.8	8.0	7.7	4.3	4.1	22.8	23.3
CELL COUNT TIME 7, MIO./ML.	3.1	3.6	30.4	30.1	5.1	4.5	4.0	4.3	17.7	16.9
CELL DIAMETER TIME 0, MY M	6.51	6.49	6.79	6.74	6.54	6.45	6.07	6.15	6.55	6.72
YEAST HARVEST, G	30.7	30.0	10.2	11.7	31.9	31.7	29.3	28.8	20.0	19.8
YEAST YIELD	3.1	3.0	1.3	1.3	3.2	3.2	2.9	2.9	2.0	2.0
TOTAL DIACETYL TIME 7, PPM	0.89	0.87	1.42	1.23	1.11	1.21	0.96	1.01	0.83	0.88
T. PENTANEDIONE TIME 7,PPM	0.64	0.57	0.76	0.69	0.57	0.59	0.50	0.52	0.46	0.48
SAMPLE TIME 0, DATE.H	50815.18	50815.18	50815.18	50815.18	50815.18	50815.18	50815.18	50815.18	50815.18	50815.18
SAMPLE TIME 1, DATE.H	50816.07	50816.07	50816.07	50816.07	50816.07	50816.07	50816.07	50816.07	50816.07	50816.07
SAMPLE TIME 2, DATE.H	50817.07	50817.07	50817.07	50817.07	50817.07	50817.07	50817.07	50817.07	50817.07	50817.07
SAMPLE TIME 3, DATE.H	50818.07	50818.07	50818.07	50818.07	50818.07	50818.07	50818.07	50818.07	50818.07	50818.07
SAMPLE TIME 4, DATE.H	50819.07	50819.07	50819.07	50819.07	50819.07	50819.07	50819.07	50819.07	50819.07	50819.07
SAMPLE TIME 5, DATE.H	50820.06	50820.06	50820.06	50820.06	50820.06	50820.06	50820.06	50820.06	50820.06	50820.06
SAMPLE TIME 6, DATE.H	50821.07	50821.07	50821.07	50821.07	50821.07	50821.07	50821.07	50821.07	50821.07	50821.07
SAMPLE TIME 7, DATE.H	50822.06	50822.06	50822.06	50822.06	50822.06	50822.06	50822.06	50822.06	50822.06	50822.06
GASCROMATOGRAPHY ,PPM										
HYDROCARBON 1	0.280	0.340	0.180	0.200	0.280	0.800	0.300	0.220	0.040	0.230
ACETALDEHYDE	26.7	23.1	20.2	21.4	18.1	22.1	32.5	26.4	16.9	13.8
DIMETHYL SULPHIDE	0.081	0.066	0.078	0.071	0.050	0.072	0.078	0.070	0.160	0.048
ETHYL FORMATE	0.90	0.89	0.90	0.98	1.32	1.09	1.02	1.04	1.08	0.90
ETHYL ACETATE	23.8	22.9	25.7	23.3	26.6	27.4	20.5	20.0	33.9	30.4
ISOBUTYL ACETATE	0.610	0.580	0.590	0.540	0.580	0.590	0.490	0.490	0.640	0.640
1-PROPANOL	23.9	21.6	23.5	23.7	23.4	24.7	20.3	19.7	24.1	23.5
2-METHYL-1-PROPANOL	13.1	11.6	16.4	16.1	11.1	11.5	8.2	7.9	14.2	13.6
3-METHYLBUTYL ACETATE	3.1	2.8	3.3	2.4	4.1	4.5	1.8	1.8	4.6	4.6
1-BUTANOL	0.740	0.640	0.540	0.480	0.710	0.850	0.230	0.190	0.820	0.720
STANDARD	5.20	5.71	5.66	5.35	5.68	5.49	5.92	5.70	5.80	5.73
3-METHYL-1-BUTANOL	73.	69.	73.	71.	73.	78.	55.	53.	73.	72.
ETHYL HEXANOATE	0.24	0.27	0.21	0.18	0.34	0.38	0.39	0.30	0.31	0.32
ETHYL OCTANOATE	0.15	0.18	0.12	0.11	0.24	0.25	0.19	0.16	0.32	0.33

Fig. 11. Raw data from one MULTIFERM run (one page out of 15)

% DIFFERENCE FROM REFERENCE YEAST 374.02

GEN	YEAST	DEA0	DEA1	DEA2	DEA3	DEA4	DEA5	DEA6	DEA7	DCE0	DCE1	DCE2	DCE3	DCE4	DCE5	DCE6
1	2625.02	.	-3	-3	-14	-23	-54	-32	-35	.	7	7	.	36	-20	-13
2	2625.02	-0	-2	-7	-16	-24	-33	-39	-44	-7	3	6	7	1	-8	-1
3	2625.02	-0	-2	-8	-14	.	-28	-32	-37	-14	-17	-2	4	-31	-11	2
4	2625.02	0	-2	-11	-21	-30	-40	-49	-53	-6	-9	21	24	12	14	-5
5	2625.02	-0	-3	-10	-16	-27	-36	-41	-49	-11	19	13	20	17	-2	-5

GEN	DCE7	DDM0	DDM7	DYIELD	DDC7	DPT7	DHYDC1	DACETA	DDMS	DETFO	DETAC	DIBAC	DPROP	DM1P	DMBA	DBUT	DSTAND	DM1B	DEHEX	DEOCT
1	6	.	3	45	11	.	11	-2	.	3	15	.	40	37	46	.	-2	8	.	54
2	-6	.	7	31	19	.	23	28	.	7	19	.	38	23	55	.	-5	17	.	35
3	-14	.	4	26	46	.	9	19	.	25	8	.	27	16	28	.	-4	8	91	21
4	-5	.	2	36	64	.	-33	-42	.	1	219	.	42	28	251	.	-1	38		8
5	-10	.	5	32	-36	.	0	16	.	-12	2	.	39	30	31	.	2	28	.	53

Fig. 12. Evaluation sheet for one yeast strain with improved fermentation rate, increased yeast yield and increased amount of certain aroma components.
GEN Generation. *DEAO* to *DEA7* differences in apparent extract on fermentation day 0 to 7. *DCEO* to *DCE7* Differences in cell number in suspension on fermentation day 0 to 7. *DDMO* and *DDM7* differences in cell diameter on fermentation day 0 and 7. *DYIELD* difference in yeast yield. *DDC7* difference in diacetyl; *DPT7* difference in pentandione; *DHYDC1* difference in hydrocarbon 1 (2-methyl-2-butene + isopren); *DACETA* difference in acetaldehyde; *DDMS* difference in dimethylsulphide; *DETFO* difference in ethylformate; *DETAC* difference in ethylacetate; *DIBAC* difference in isobutylacetate; *DPROP* difference in 1-propanol; *DM1P* difference in 2-methyl-1-propanol; *DMBA* difference in 3-methylbutylacetate; *DBUT* difference in 1-butanol; *DSTAND* difference in internal standard (5.2 ppm 1-pentyl acetate); *DM1B* difference in 3-methyl-1-butanol; *DEHEX* difference in ethylhexanoate; *DEOCT* difference in ethyloctanoate all at end of fermentation

nol, 2-*m*ethyl-1-propanol, 3-*m*ethyl-butyl-acetate, and ethyl octanoate, whereas the other parameters tested did not differ significantly from those of the reference strain.

3 Conclusion

Repeated testings of various production strains and new yeast strains in the MULTIFERM system have revealed it to be well suited to detect significant differences in parameters of relevance in beer production.

The characteristics of yeast strains selected in the MULTIFERM have been confirmed in pilot scale and production scale fermentations. The capacity of the MULTIFERM makes it an indispensible tool in the search for yeast strains with improved fermentation properties.

The MULTIFERM is to a certain degree also suited for experiments aiming at the evaluation of the effect of different fermentation conditions on the characteristics of the product. The two independent carrousels allow to assess for example, the influence of fermentation temperature on several yeast strains at a time as well as the influence of certain changes in wort composition.

References

Ahrenst-Larsen B (1967) Nyt forsøgsbryggeri på Tuborg. Brygmesteren 24:109–128

Annemüller G (1981) Probleme der Hefetechnologie bei der Gärung und Reifung von Bier in zylindrokonischen Großtanks. Lebensmittelindustrie 28:549–555

Cook AH (1963) Yeast and the work of the yeast group. Proc Eur Brew Con Congr, Brussels, pp 477–486

Dixon IJ (1967) Hop substances and yeast behaviour. J Inst Brew 73:488–493

EBC Analytica Microbiologica (1977) Method 2.5.4 Tubes E.B.C. J Inst Brew 83:117–118

EBC Analytica Microbiologica (1981) Method 2.2.2.2. Electronic counter. J Inst Brew 87:303

Gjermansen C, Sigsgaard P (1981) Construction of a hybrid brewing strain of *Saccharomyces carlsbergensis* by mating of meiotic segregants. Carlsberg Res Commun 46:1–11

Goldiner F, Klemann H, Kämpf W (1966) Tafel zur Ermittlung der Gewichtsprozente und Gemischtprozente Rohrzucker aus den Gewichtsverhältnissen s_L 20/20 °C und Dichten 20/4 °C von Rohrzuckerlösungen. Inst Gärungsgewerbe, Berlin, pp 11–41

Hackstaff BW (1978) Various aspects of high gravity brewing. MBAA Tech Q 15:1–7

Hammond JRM, Eckersley KW (1984) Fermentation properties of brewing yeast with killer character. J Inst Brew 90:167–177

Innis MA, Holland MJ, McCabe PC, Cole GE, Wittman VP, Tal R, Watt KWK, Gelfand DH, Holland JP, Meade JH (1985) Expression, glycosylation, and secretion of an *aspergillus* glucoamylase by *Saccharomyces cerevisiae*. Science 228:21–26

Ishimaru S, Kudo S, Hattan M, Yoshida T, Kataoka J (1967) Selection of small vessels for fermentation tests in the laboratory. Rep Res Lab Kirin Brew Co. Ltd. 10:61–65

Jackson EA, Ballance GM, Thomson KK (1986) Construction of a yeast vector directing the synthesis and release of barley 1→3, 1→4)-β-glucanase. Carlsberg Res Commun 51:445–458

Kielland-Brandt MC, Gjermansen C, Nilsson-Tillgren T, Petersen JGL, Holmberg S, Sigsgaard P (1981) Genetics of a lager production yeast. MBAA Tech Q 18:185–187

Knudsen FB (1985) Fermentation variables and their control. ASBC J 43:91–95

Kruger L, Ryder DS, Alcock C, Murray JP (1982) Malt quality. I. Prediction of malt fermentability. MBAA Techn Q 19:45–51

Maule DR (1986) A century of fermenter design. J Inst Brew 92:137–145

Meaden P, Ogden K, Bussey H, Tubb RS (1985) A DEX gene conferring production of extracellular amyloglucosidase on yeast. Gene 34:325–334

Meilgaard MC (1981) Beer flavor. Thesis 81 26 165. Univ Microfilms Int, 300 N. Zeeb Road, Ann Arbor M 1 48106

Meilgaard MC (1982) Prediction of flavor differences between beers from their chemical composition. J Agric Food Chem 30:1009–1017

Miedaner H (1978) Optimisation of fermentation and conditioning during the production of lager. The Brewer, pp 33–39

Molzahn SW (1977) A new approach to the application of genetics to brewing yeast. ASBC J 35:54–59

Narziss L (1981) Moderne Gärung und Reifung. Der Weihenstephaner 2:70–77

Petersen JGL, Kielland-Brandt MC, Holmberg S, Nilsson-Tillgren T (1983) Mutational analysis of isoleucine-valine biosynthesis in *Saccharomyces cerevisiae*. Mapping of ilv2 and ilv5. Carlsberg Res Commun 48:21–34

Ramos-Jeunehomme C, Masschelein CA (1977) Contrôle génétique de la formation des dicétones vicinales chez *Saccharomyces cerevisiae*. Proc Eur Brew Conv Congr, Amsterdam, pp 267–283
Rosendal I, Nørgaard O, Schmidt F (1983) ANASYS, an EDP-based analysis administrative system for quality control. Proc Eur Brew Conv Congr London, pp 611–618
Rothstein SJ, Lazarus CM, Smith WE, Baulcombe DC, Gatenby AA (1984) Secretion of a wheat α-amylase expressed in yeast Nature 308:662–665
Sasaki T, Watari J, Kohgo M, Nishikawa N, Matsui Y (1984) Breeding of a brewer's yeast possessing anticontaminant properties. ASBC J 42:164–166
Sigsgaard P, Nørager JR (1985) Screening of the brewing performance of new yeast strains. ASBC J 43:104–108
Spencer JFT, Spencer DM (1983) Genetic improvements of industrial yeasts. Annu Rev Microbiol 37:121–142
Stewart GG, Panchal CJ, Russel I (1983) Current development in the genetic manipulation of brewing yeast strains–a review. J Inst Brew 89:170–188
Thomsen KK (1983) Mouse α-amylase synthesized by *Saccharomyces cerevisiae* is released into the culture medium. Carlsberg Res Commun 48:545–555
Tubb RS (1986) Amylolytic yeasts for commercial applications. TIBTECH, pp 98–104
Von Wettstein D, Gjermansen C, Holmberg S, Kielland-Brandt MC, Nilsson-Tillgren T, Pedersen MB, Petersen JGL, Sigsgaard P (1984) Genetic engineering in the improvement of brewer's yeast. MBAA Tech Q 21:45–61
Weinfurtner F, Wullinger D, Piendl A (1961) Eine Kleingärapparatur zur Ermittlung der Gäreigenschaften von Brauereihefen. Brauwissenschaft 3:109–119

The Use of Nucleotide Sequence Polymorphisms and DNA Karyotyping in the Identification of Brewer's Yeast Strains and in Microbiological Control

M. BOHL PEDERSEN

1 Introduction

Several methods are employed for differentiation of yeast strains. These comprise analyses of microbiological and physiological characters such as the carbohydrate fermentation patterns as used in *The Yeasts* (ed. Lodder 1970), capacity of spore formation and ability to produce fertile offspring by hybridization as described by Winge and Laustsen (1939), or by analyses of isozymes and serologically distinguishable components as described by Donhauser et al. (1981) and by Yarrow and Nakase (1975).

Analyses of the genetic material, the nucleic acids, have been performed by determination of the average base composition of G-C and A-T pairs in the genome and expressed as % G+C (Yarrow and Nakase 1975), DNA-RNA homologies and DNA-DNA reassociation kinetics (Martini and Kurtzman 1985). By these methods several yeast genera and groups of species can be recognized, but they are not suitable to differentiate individual strains unequivocally. Identification of yeast strains by finger printing their DNA involves variations of three techniques: separation of restriction endonucelase fragments by electrophoresis, recognition of fragment polymorphisms by staining or hybridization to radioactively labelled cloned probes and karyotyping of chromosome sized DNA molecules (Pedersen 1983, 1986a).

The cutting of isolated DNA by restriction endonucleases, which recognize palindromic sequences of 4, 5, 6 or even 8 base pairs, generates a population of linear DNA molecules in a wide range of sizes. By electrophoresis in a flatbed agarose gel the molecules are separated according to size. When staining the separated DNA molecules in the agarose gel by ethidium bromide and illuminating the gel in UV, the DNA molecules are visualized. This procedure is only suitable for repetitive DNA molecules giving discrete bands. For genes and genetic elements present in single or a few copies, specific hybridization probes applied to filter blots of the separated molecules have to be used to visualize by autoradiography the polymorphisms, e.g. restriction fragment length polymorphisms are detectable even in closely related strains because of the large number of different sites, recognizable by the restriction endonucleases available. The base sequence differences distinguishing two strains may be silent, i.e. occur within the coding region of a gene due to an alternative triplet specifying the same amino acid or outside the coding region without affecting the expression of the gene. The restriction site polymorphisms may signify, however, also differences in coding and promoter regions of genes giving a phenotype not – or not yet – recognizable by the fermentation or morphological characteristics of the strain. Only rarely will it be possible to relate a nucleotide sequence polymorphism used in strain characterization to a specific property of the line under investigation.

1.1 Restriction Fragment Length Polymorphisms

In Fig. 1 are summarized the RFLP's (*R*estriction *F*ragment *L*ength *P*olymorphisms) of four types of gene used to identify taxons of the genus *Saccharomyces*

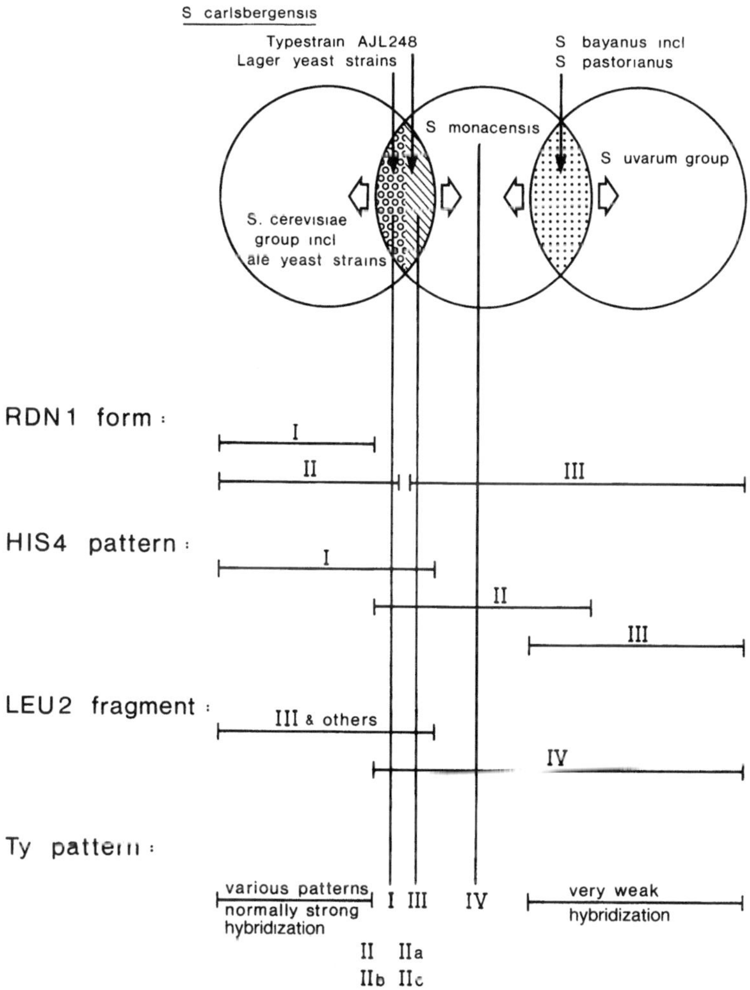

Fig. 1. Composite diagram of the restriction endonuclease fragment length polymorphisms (RFLP) of four types of genes analyzed. *Roman numerals* above the horizontal bars correspond to the allele of the particular gene, gene cluster or genetic element. *RDN1:* gene cluster encoding the ribosomal RNA molecules 5S, 5.8S, 18S, and 25S; *HIS4* one region covering the HIS4 gene located on chromosome III; *LEU2* SalI fragment pattern of the LEU2 gene located on chromosome III; *Ty* the pattern of fragments hybridizing to the Tyl p14 probe from *S. cerevisiae*.

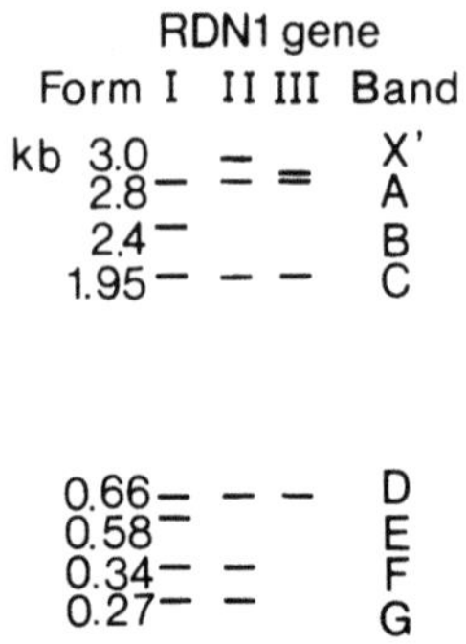

Fig. 2. Diagnostic key to the three EcoRI patterns of the RDN1 gene cluster as revealed by the hybridization probe covering an entire RDN1 cassette. Fragment sizes are in kb (thousands of base pairs). Note: (The two fragments F and G are presumably present in form III as well, but they have not been detected at the chosen stringency in 0.7%–1% agarose Southern blots)

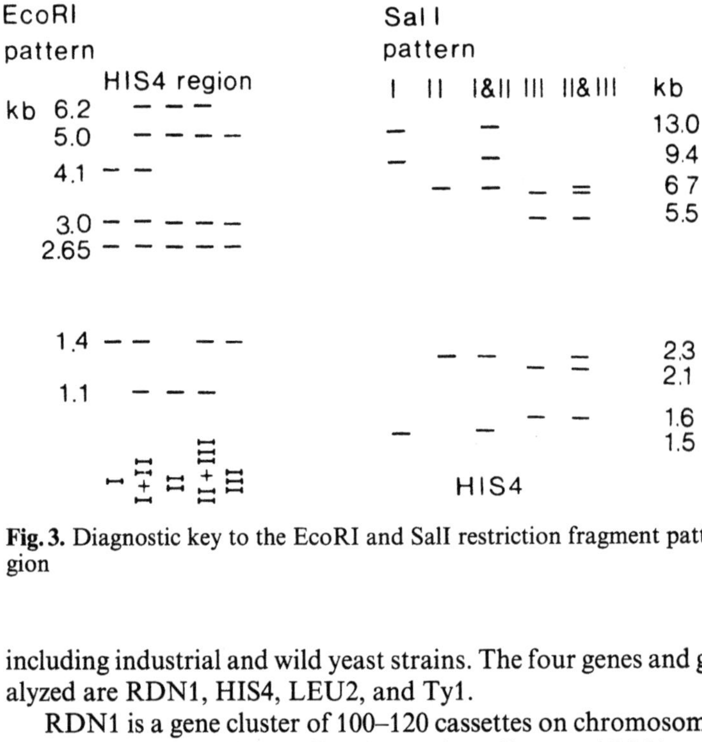

Fig. 3. Diagnostic key to the EcoRI and SalI restriction fragment patterns of the HIS4 region

including industrial and wild yeast strains. The four genes and gentic elements analyzed are RDN1, HIS4, LEU2, and Ty1.

RDN1 is a gene cluster of 100–120 cassettes on chromosome XII (Petes 1979) and Petes et al. (1978) encoding the nuclear ribosomal RNA molecules with a size of 5S, 5.8S, 18S, and 25S.

Three forms have been detected in the analyzed strains of which forms I and II (Petes et al. 1978), are found in the *S. cerevisiae* group while in lager strains only form II is present. Form III characterizes *S. carlsbergensis* type strains, *S. monacensis, S. bayanus/S. pastorianus* and *S. uvarum*. The diagnostic patterns for the three forms are given in Fig. 2, (Pedersen 1983, 1985a, b, 1986a).

HIS4 is a multifunctional gene encoding three enzyme functions catalyzing steps in the histidine de novo biosynthesis and has been detected in three allelic patterns among industrial strains, namely patterns I and II (Nilsson-Tillgren et al. 1981), and pattern III (Pedersen 1983) (Fig. 3).

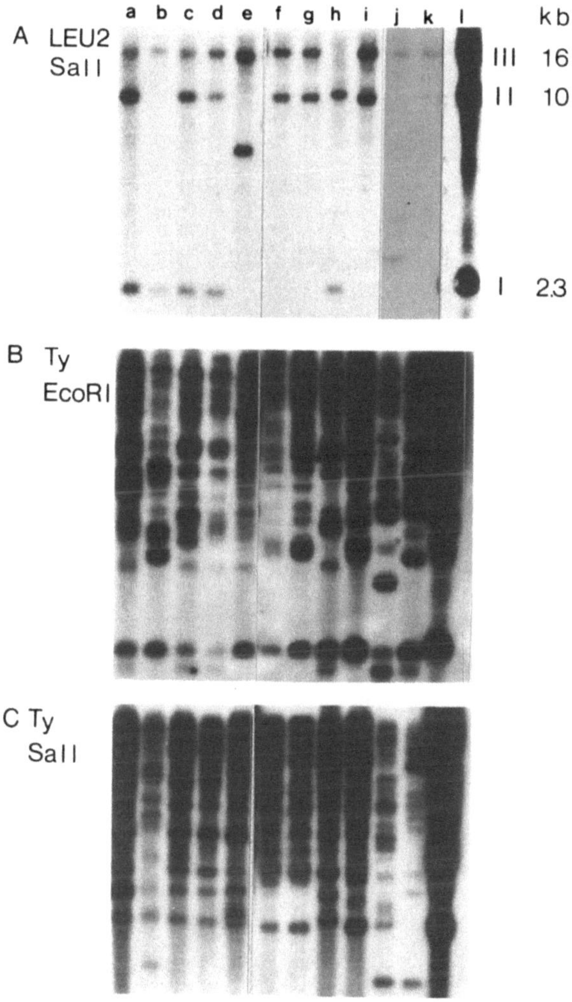

Fig. 4 A–C. Analyses of 12 *S. cerevisiae* top fermenting strains. **A** Southern blots of the SalI restriction fragment patterns of the 12 strains probed with the LEU2 fragment. **B** and **C** Southern blots probed with the Ty element where **B** contains EcoRI digested DNA's and **C** contains SalI digested DNA's. Strain *a* C83–FP1 (Malta); strain *b* C82–FP10 (Canada); strain *c* C82–FP15 (Great Britain); strain *d* C82–FP16 (Great Britain); strain *e* C82–FP17 (Great Britain); strain *f* Danish top fermenting strain BK3300; strain *g* Danish top fermenting strain BK3302; strain *h* British top fermenting strain BK3304; strain *i* Danish top fermenting strain BK3305; strain *j* British top fermenting strain BK3331; strain *k* Swedish top fermenting strain BK3313; strain *l* *S. cerevisiae* type strain CBS1171

HIS4 pattern I is found in homozygous condition in the *S. cerevisiae* group including top fermenting strains and some wild yeast strains. Pattern II is found in heterozygous condition with either pattern I (nearly all *S. carlsbergensis* strains) or with pattern III as found in the *S. bayanus* strains. Pattern II has been seen in homozygous condition in two industrial strains namely the Pschorr lager strain (Pedersen 1983), and another old bottom fermenting strain *S. monacensis* CBS 1503. No wild yeasts have been found to contain HIS4 patterns I and II or II alone. Until now this seems to be the case for the combination of HIS4 patterns II and III as well.

S. uvarum, including *S. validus*, contain HIS4 pattern III. A couple of wild yeast strains with HIS4 pattern III have been detected.

LEU2 (encoding β-isopropyl malate dehydrogenase) occurs in several alleles (here represented by their SalI fragment patterns).

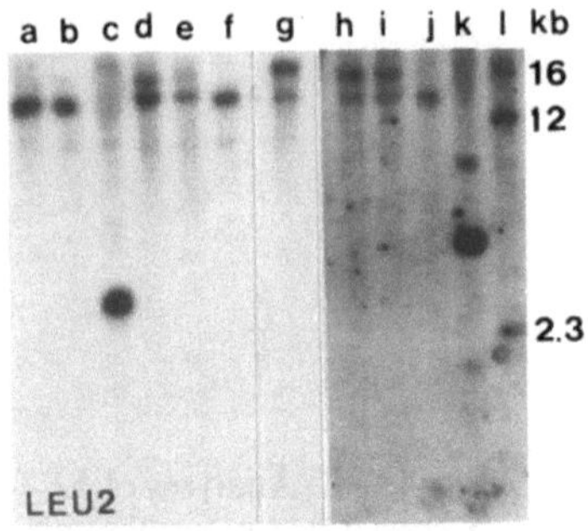

Fig. 5. SalI restriction fragment patterns hybridizing to LEU2 in 12 strains. The Southern blots of the digested DNA preparations have been probed with the EcoRI-SalI 0.9 kb LEU2 fragment from *S. cerevisiae*. Lane *a S. monacensis* CBS1503; lane *b S. bayanus* NCYC374; lane *c* genetic standard strain S288C; lane *d* Carlsberg lager strain BK2246; lane *e* Pschorr lager strain BK1101; lane *f S. uvarum* C81–1511; lane *g S. carlsbergensis* C84–AJL248; lane *h* lager strain BK2229; lane *i* Danish lager strain BK2236; lane *j S. validus* BK4404; lane *k S. odessa* BK4410; lane *l* ale strain C82–FP1. Approximate fragment sizes are in thousand of base pairs (kb)

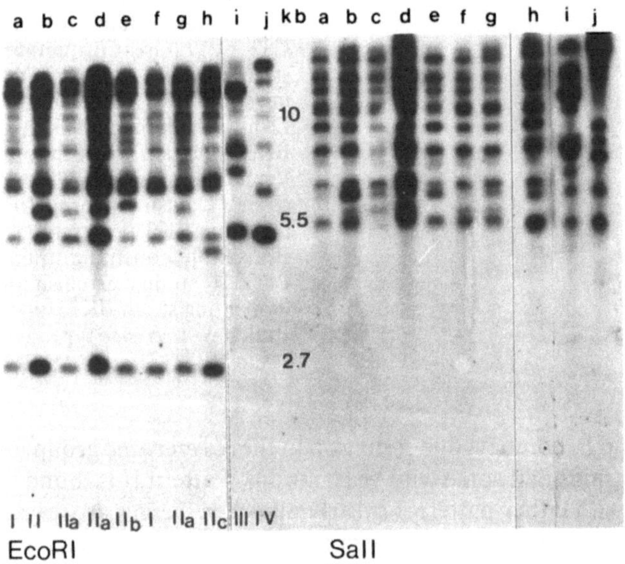

Fig. 6. EcoRI and SalI restriction endonuclease fragment patterns (RFLP's) of ten strains which have all been used as bottom fermenting strains. Panel *A* Southern blots of EcoRI digestions; panel *B* Southern blots of SalI digestions. The DNA's have all been habridized with the Tyl p14 probe. Lane *a* Carlsberg lager strain BK2246; lane *b* Tuborg lager strain BK2224; lane *c* Pschorr lager strain BK1101; lane *d* Danish lager strain BK2208; *e, f, g* German lager strains C83–FP34, C83–FP35 and C83–FP36; lane *h* Danish lager strain BK2236; lane *i S. carlsbergensis* AJL248 bottom fermenting strain No. 1 (Hansen); lane *j S. monacensis* CBS1503 bottom fermenting strain No. II (Hansen). *Roman numbers* at the bottom of the lanes indicate the pattern designation

Four SalI fragments, each corresponding to an allele of LEU2 have been seen regularly in the genus and sometimes accompanied by SalI fragments of different size. SalI fragments I, II, and III are found in all possible combinations in the *S. cerevisiae* group, (I and II and III, I and II, I and III, II and III) (Fig. 4).

All analyzed *S. carlsbergensis* strains contain LEU2 SalI fragments III (16 kb) and IV (12 kb) together while *S. monacensis* CBS 1503, *S. bayanus* and *S. uvarum* strains contain LEU2 SalI fragment IV alone (Pedersen 1986 a) (Fig. 5.).

Recognition of individual brewing strains has been possible by restriction fragment length polymorphisms of DNA segments hybridizing to the internal Tyl p14 probe without delta sequences (Philippsen et al. 1983).

A wide range of variation has been found for top fermenting strains with several unique fragment patterns (Pedersen 1985, 1986 a) (cf. Fig. 4). Also the *S. carlsbergensis* lager and type strains, including the *S. monacensis* strain CBS 1503, contain different Ty RFLP's (Pedersen 1985 a, b, 1986 a). At least seven Ty hybridizing patterns are present among the bottom fermenting strains (Decock and Iserentant 1985; Martens et al. 1985; Pedersen 1985 a; (cf. Fig. 6).

1.2 DNA Karyotypes

The grouping of the strains in the genus *Saccharomyces* based on chromosome length polymorphisms (CLP), as revealed by separation of chromosome-sized DNA molecules, is consistent with that based on the restriction fragment length polymorphisms (RFLP).

In Fig. 7 are shown the karyotypes of ten bottom fermenting strains and in Fig. 8 are shown the karyotypes of strains carrying RDN1 form III as obtained by orthogonal field alternation gel electrophoresis (OFAGE) of chromosome-sized DNA molecules. Identification of chromosomes is done by molecular hybridization of specific probes to Southern transfers of the gel (Carle and Olson 1984, 1985). An example of such chromosome identification is shown in Fig. 8. The chromosomal DNA's from the nine strains have been hybridized with a *S. cerevisiae* HIS4 probe (located on chromosome III), Pedersen (1986 b). The probe used was from pC503 containing the 9.4 kb HIS4 from *S. cerevisiae* (Holmberg et al. 1979). The hybridization of the HIS4 probe revealed chromosome length polymorphisms of chromosome III.

Two isolates (Fig. 8, lanes d and c) contain three bands which hybridize to the HIS4 probe, and isolate, which seems to be a subclone of the strain in lane d, hybridizes to two bands. The three isolates show similar karyotypes in panel A, but a chromosome rearrangement seems to have occurred in the *S. carlsbergensis* strain NCYC 396. All other strains, namely *S. uvarum* (lanes a and b), *S. pastorianus* (lanes f and g) and *S. bayanus* (lanes h and i) contain chromosome III of equal lengths.

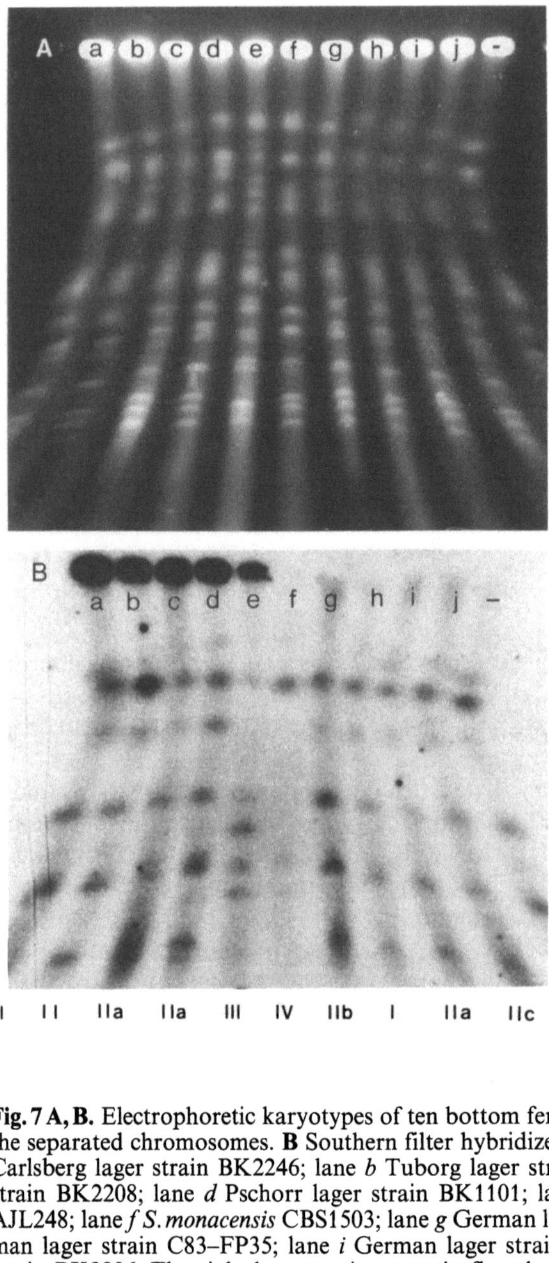

Fig. 7 A, B. Electrophoretic karyotypes of ten bottom fermenting strains. **A** Agarose gel of the separated chromosomes. **B** Southern filter hybridized with the probe Tyl p14. Lane *a* Carlsberg lager strain BK2246; lane *b* Tuborg lager strain BK2224; lane *c* Danish lager strain BK2208; lane *d* Pschorr lager strain BK1101; lane *e* *S. carlsbergensis* No. I C84–AJL248; lane *f* *S. monacensis* CBS1503; lane *g* German lager strain C83–FP34; lane *h* German lager strain C83–FP35; lane *i* German lager strain C83–FP36; lane *j* Danish lager strain BK2236. The eight lager strains contain five chromosomal bands which hybridize to the Tyl p14 probe, while *S. carlsbergensis* C84–AJL248 contains five and *S. monacensis* CBS1503 three bands. Two of the bands in the latter two strains are shared by the lager strains, while a third band is also found in the old Pschorr bottom fermenting strain, but not present in the modern lager brewing strains. *Roman numerals* below panel **B** correspond to the Ty restriction endonuclease patterns as found in the ten bottom fermenting strains. The chromosomal separations were carried out with a pulse time of 50 s, a constant current of 200 mA in a 1% agarose gel containing 0.5 × TBE buffer. The electrophoresis was run for 18 h at 8–12 °C

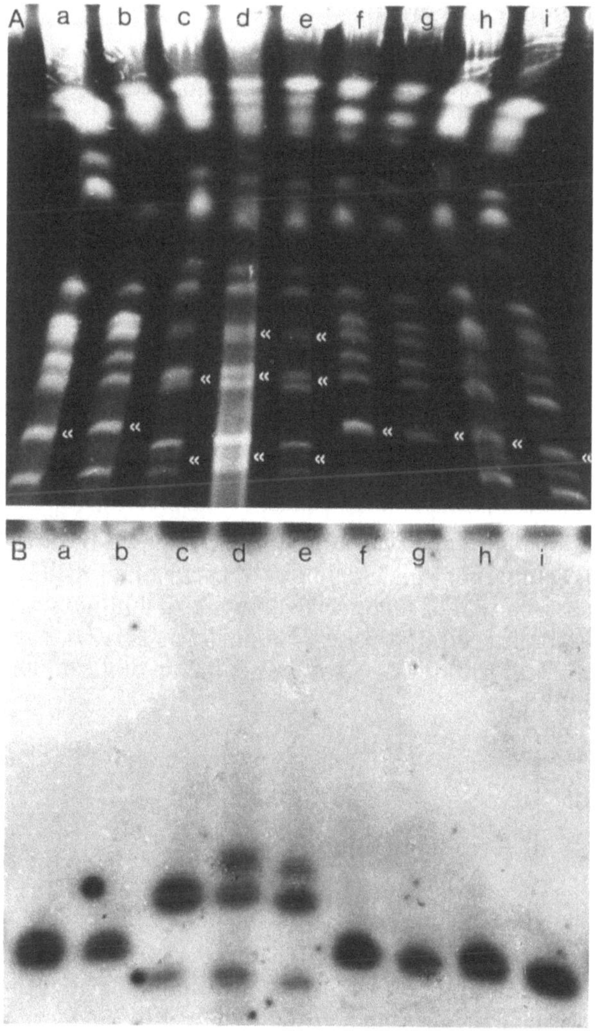

Fig. 8 A, B. A OFAGE separations of chromosomes from nine *Saccharomyces* species, which all contain RDN1 form III. Lane *a S. uvarum* CBS395; lane *b S. uvarum* C81–1511; lane *c S. carlsbergensis* NCYC396; lane *d S. carlsbergensis* C84–AJL248 (mother strain for NCYC396 and CBS1513); lane *e S. carlsbergensis* CBS1513; lane *f S. pastorianus* NCYC392; lane *g S. pastorianus* ATCC12752; lane *h S. bayanus* NCYC374; lane *i S. bayanus* CBS380. The karyotypes of *S. bayanus*, *S. pastorianus*, and *S. uvarum* seem to be closely related, since the distribution of chromosome bands is almost identical. The three clones of the *S. carlsbergensis* bottom fermenting strain are identical, and the chromosomal band patterns are significantly different from the other three species. *Arrowheads* point to the chromosomes which hybridize with the HIS4 gene probe. **B** Corresponding Southern filter of the gel in **A** hybridized with the pC503 HIS4 probe used to monitor the chromosome IIIs in the different species. Even though the electrophoretic karyotypes of the three *S. carlsbergensis* clones are identical, one difference between NCYC396 and the other clones C84–AJL248 and CBS1513 is seen. The uppermost of the three to HIS4 hybridizing bands in *S. carlsbergensis* C84–AJL248 and CBS1513 is missing in NCYC396

2 Storage and Propagation of Yeast Strains

1. A convenient way to store yeast strains is at −80 °C in fluid medium (YPD/Glycerol) 100:24 (v:v).

2. YPD/Glycerol stock[a]

A.	1% Bacto-Yeast extract	6 g
	2% Bacto-Yeast-Peptone	12 g
	Deionized water	500 ml
B.	2% Glucose	12 g
	Deionized water	100 ml
C.	80% Glycerol	144 ml

Autoclave A, B, and C by separately and mix B and C into A at room temperature.

[a] Adapted from Sherman et al. (1979).

3. When propagating a strain for DNA analysis, use single cell colonies as the source for the DNA isolation. The single cell colonies are prepared by either streaking out the original sample for single colonies on a maximal medium plate (YPD or wort agar) or isolation of single cell colonies by a dilution series where part of the original sample is diluted in sterile 10 ml 0.9% NaCl or H_2O and plated as (100, 10,000, 1×10^6) dilutions. A volume of 0.2 ml of the diluted samples per plate is appropriate.

3 Restriction Site Analysis of Yeast Genomic DNA

3.1 Isolation of Yeast Genomic DNA for Restriction Site Analysis

1. Scrape cells from a YPD plate (an amount of cells corresponding to 1 ml) which have been previously incubated for 48 h at room temperature.
2. Suspend the cells in 5 ml of 1 *M* sorbitol, spin cells down in a table top centrifuge (4000 *g*) and resuspend the cells in 5 ml of 1 *M* sorbitol containing 20 m*M* dithiothreitol (Sigma D-0632) and incubate the cell suspension at 30 °C for 30–45 min. Then add 1.4 mg of Novozyme 234 and incubate at 30 °C for 30 min.
3. Cell lysis is induced by the addition of 0.2 ml 10% sodium dodecyl sulfate (SDS), 0.5 ml of 0.5 *M* EDTA (pH 8.5) and 0.5 ml 2 *M* TRIS base [tris-(hydroxymethyl)-amino methane]).
4. After stirring, heat the solution to 65 °C for 20 min. Cool the tubes on ice and add 1 ml of 5 M potassium acetate. Incubate at 0 °C for 2 h and spin the precipitate down at full speed in the table top centrifuge for 10–15 min.
5. Decant the supernatant and mix it with 1 vol 2-propanol: water 5:1 (v/v) and mix thoroughly. After 1 h at −20 °C let the tubes equilibrate to room temperature, spin the precipitate down for 10 min, discard the supernatant and wipe

the inside of the tubes with a piece of paper towel to remove drops of propanol and water. Remember not touch to the pellet with the paper towel.

6. Suspend the pellet in 1 ml 10 m*M* TRIS 1 mM EDTA buffer (pH 7.5) and add 2.5 μg ribonuclease A (Sigma F-5125). Incubate the solution at room temperature for 30 min. Transfer the solution to two or more 1.5-ml Eppendorf tubes. Add 1 vol 2-propanol: water and 1/10 vol saturated sodium acetate and incubate overnight at −20 °C.
7. Spin down the DNA the next day and suspend it in TE buffer (100–200 μl, depending on the yield). To remove residual debris in the DNA solution, spin the Eppendorf tubes for 10–15 min and transfer the supernatant to new tubes. The DNA solution is ready for use. Always store the DNA solution at −20 °C when not in use, and keep buffers, enzymes and DNA on 0 °C ice when in use.

3.2 Analysis of Restriction Fragment Length Polymorphisms (RFLP)

Two levels of the RFLP analyses are possible: Level A: analysis of total genomic DNA (visual analysis of agarose gel), Level B: analysis of specific regions by molecular hybridization of radioactively labelled probes.

Before digesting the genomic DNA by various restriction endonucleases, make a visual estimate of the DNA concentration in the DNA preparations. The estimate is done by comparison of the yeast DNA preparation to a dilution series of an undigested DNA preparation, for instance a vector or lambda phage of known concentration (commercially available). The comparison is done in an ethidium bromide stained agarose gel (1 μg/ml). When running digested DNA preparations on an agarose gel always leave room for one to two lanes containing relevant DNA fragment length size markers.

If one suspects that the original yeast sample is a mixture of two or more strains, then pick 20–100 colonies which originate from a series of plates from the dilution series of the original sample and analyze each colony separately. Then compare the restriction endonuclease fragment patterns from the single colonies with the extracted DNA from the original sample.

3.2.1 RFLP Level A

Prepare a series of exposures of the UV-illuminated agarose gel on a 35-mm orthochromatic film and undertake the final RFLP analysis of the strains on an enlarged print of the film. It should be stressed and made very clear that the light from the UV illuminators are hazardous for the unprotected eye. Therefore always wear protective glasses when analyzing the stained gel.

Due to the small genomic size (in the order of 10^7 base pairs) and the presence of repeated DNA sequences it is possible to detect several discrete ethidium bromide stainable bands of electrophoresed digested yeast DNA.

3.2.2 RFLP Level B

To be able to detect single or few copy DNA regions one has to filter the information from the agarose gel in such a way that the only bands which are detected are the restriction endonuclease fragments one is seeking.

This is done by molecular hybridization of radioactively labelled single-stranded DNA from an already known DNA region in *S. cerevisiae* laboratory strains.

4 Southern Transfer and Molecular Hybridization of Radioactively Labelled Probes

The labelling of the DNA probe is done by nick translation, where the enzyme DNA polymerase I restores gaps in one of the DNA strands by cutting out nucleotides around the disrupted single strand and subsequently filling in the missing nucleotides. The order and kind of nucleotides filled in is governed by the complementary strand.

The nick translation is done with double-stranded DNA, while the probe has to be single-stranded when performing the molecular hybridization. Double-stranded DNA is made single-stranded by boiling the probe in TE buffer in a Pyrex tube for 5 min. The hydrogen bonds between the complementary nucleotides in the double strand will be broken and a population of single-stranded DNA obtained.

When the radioactively labelled DNA strands are added to the buffer in the hybridization vessel containing the nitrocellulose filters carrying the single strands of the DNA fragment that is to be compare with the probe, reassociation of single-stranded DNA in solution to the strands on the filter will occur.

Add an amount of nick translated probe to the hybridization container (tray, plastic bag, etc.) to give 2–10 ng/ml probe in the hybridization buffer and allow a hybridization period of 16–20 h (0.2–2 μg DNA to 200–400 ml of buffer).

The ratio between the added amount of nick translated probe and buffer is critical, since too high concentration of probe might give a high background on the final autoradiograph.

It is possible to hybridize between 6–10 filters (each 14 cm × 16 cm) in a volume of 400 ml buffer and 1–2.5 μg nick translated probe. For preparation of probes in *E. coli* consult Holmes and Quigley (1981) or Maniatis et al. (1982).

4.1 Transfer of DNA Molecules from Agarose Gels onto Nitrocellulose Membranes by Southern Transfer

Transfer of either restriction fragments or chromosomes from agarose gels onto nitrocellulose is done by the method known as Southern transfer (Maniatis et al. 1982). Before the actual transfer of the DNA molecules one has to pretreat the DNA for the transfer.

This is done in three steps:

A. HCl treatment (0.25 *M* HCl for 30 min).
B. NaOH treatment (0.4 *M* NaOH + 0.6 *M* NaCl for 1 h).
C. High salt treatment (1 *M* TRIS-HCl + 1.5 *M* NaCl for 2 h).

When preparing the nitrocellulose for DNA transfer always let the nitrocellulose float and later sink into distilled water before washing it in 3 × SSC or 6 × SSC. The SSC concentration depends on the hybridization conditions (stringency) which were preferred. After the overnight transfer, let the filter dry in a ventilated oven at 80 °C for 1 ½ hours. By this time the DNA strands are irreversibly linked to the nitrocellulose filter, and during the pretransfer bathes single strands have been produced.

4.2 Molecular Hybridization

4.2.1 Hybridization Buffers

A. Prewash: 3 × SSC[a], 30 min. ([a] SSC: 0.15 *M*-NaCl, 15 m*M* Na citrate.)
B. Wash: 3 × SSC + Denhardts solution, 2 h

Denhardts solution:

	2 g polyvinyl-pyrrolidone (SIGMA pvp 360)
In 1 l, 3 × SSC	2 g bovine serum albumin
	2 g Ficoll 400 (Pharmacia).

C. 1 h prior to the addition of the probe, add denatured salmon sperm DNA giving a final concentration of 50 μ/ml in the hybridization buffer.
D. Add the filtrated denatured probe and let it hybridize for 16–20 h.

4.2.2 Nick Translation

Nick translation mix:

n	μl	probe in TE buffer
5	μl	dTTP, dCTP and dGTP (each 1 m*M* in stock solution)
10	μl	nick translation buffer, 10 × (Maniatis et al. (1982)
10	μl	(α–^{32}P) dATP (100 μCi)
4	μl	DNA polymerase I (Boehringer Mannheim 104493)
X	μl	
100–X	μl	dist. sterile H_2O
100	μl	total volume

Perform the nick translation at 14 °–16 °C for 1–1 ¼ h. The process is stopped by the addition of 30 μl 0.5 *M* EDTA p 8.5 and the tube is put on ice in a lead container.

Filtrate the nick translated material behind a protective screen in a 5-ml column containing Sephadex SPC50 in 20 m*M* NaCl.

Elute three fractions from each sample. The first peak (relative counts measured by hand monitor) contains the probe while the broad second peak contains unincorporated material which should be stored in a tank for decay of ^{32}P.

After filtration, pool the probe containing samples and boil the eluate for 5 min at 100 °C and cool on ice for 3–5 min before adding the probe to the hybridization buffer.

4.2.3 Post Hybridization Wash

After the hybridization of the probe to the nitrocellulose filter, it is necessary to rinse off excess of radioactive material by rinsing the filters in three steps:

A. 1 × SSC, 0.2% SDS – 20 min, room temperature.
B. 1 × SSC, 0.2% SDS – ½–2 h, same temperature as the hybridization.
C. 1 × SSC, 0.2% SDS – 15 min, room temperature.
In all steps, gentle agitation of the trays is necessary.

The temperature and washing time in (B) is dependent on stringency of hybridization one wants to achieve.

High stringency: 68–73 °C for 1½–2 h.
Low stringency: 55–58 °C for ½–1 h.

It possible to vary the stringency as well by the salt concentration, using either 0.1 × SSC or 6 × SSC in the hybridization and in the wash. For more information consult Maniatis et al. (1982), pp. 387–389

Dry the filters at room temperature in a well-ventilated fume hood or wrap the filters in food-wrap before putting them, together with X-ray films (X-OMATS, XAR5, Kodak), in cassettes with screens (X-Omatic, Kodak). Expose the films at −80 °C for 4–24 h or longer if necessary.

5 Isolation of Yeast Genomic DNA for Chromosome Separation

5.1 Orthogonal Field Alternation Gel Electrophoresis

By the techniques described respectively by Schwartz and Cantor (1984) and Carle and Olson (1985), it is possible to isolate yeast chromosomal DNA as descrete large molecules and separate the chromosomal-sized DNA in an agarose gel in an electric alternating field (PFG or OFAGE) according to their size. This gives the opportunity to construct karyotypes for industrial and wild yeast strains, and together with the restriction fragment data valuable genetic information for each strain is obtained. The molecular size range for chromosomes of *Saccharomyces* strains are between 200 kb to approx. 1000 kb.

For the actual construction and wiring of an OFAGE apparatus consult Carle and Olson (1984, 1985).

5.2 Preparation of Chromosomal DNA for OFAGE Karyotypes

The DNA isolation method for yeast chromosomal DNA is a modification of the method introduced by Schwartz and Cantor (1984).

1. Scrape newly grown cells (24–48 h at room temperture of an YPD plate and transfer an amount of 100–150 µl cells to a 1.5 ml Eppendorf tube.
2. Suspend the cells in 1 ml of 50 m*M* EDTA + 10 m*M* Tris pH 7.5, spin cells down and resuspend in buffer twice.
3. After the second wash resuspend the cells in 150 µl of the TRIS-EDTA buffer and add 10 µl of Novozyme 234 (stock 20 mg/ml).
4. Add 250 µl of 1 M-dithiothreitol together with 400 µl of 1% low melting agarose (Seaplaque FMC) containing 0.125-*M* EDTA, pH 7.5 at 38 °C.
5. Mix carefully and put on ice immediately. When the agarose has hardened, put 500 µl of LET overlayer on top of the agarose plug and incubate (LET, 0.5 *M*-EDTA, 10 m*M*-TRIS, pH 7.5) overnight at 37 °C.
6. After the first overnight incubation remove the overlayer and replace with 500 µl of NDS (NDS = LET + 1% N-lauryl-sarcosine, pH 9.5 + 0.5 mg proteinase K) and incubate overnight at 50 °C.
7. The next day remove the NDS overlayer and rinse the agar plugs for 4 × 1 h in 1 ml LET buffer in each wash.
8. Store the chromosome preparations at 4 °C in a refrigerator with a 1-ml LET overlayer. Before the elctrohopresis appropriate samples (in the range of 20–40 µl) are cut out of the plugs and rinsed for 1 h in 50 m*M* EDTA + 10 m*M*-TRIS, pH 7.5.
9. After the samples are rinsed, remove the buffer and the tubes with the DNA containing agarose are transferred to heat block at 70 °C and incubated for 5 min. Immediately load the samples in the slot of the gel.
10. The agarose gels are made up with 0.5 × TBE[a] buffer. Electrophoresis is done in 3 l of 0.5 × TBE buffer at 200 mA constant current with a pulse period of 50 s. The electrophoresis is performed for 15–18 h at 8–12 °C. ([a]TBE: 0.089 *M* TRIS-Borate, 0.089 *M* Boric acid, 0.002 *M* EDTA.)
11. The gels are stained 0.5 × TBE containing 1 µg/ml ethidium bromide for at least 2 h and destained in 1 l of destilled water for at least 4 h prior to analysis.
12. Southern transfer of chromosomal DNA and molecular hybridization of nicktranslated probes are performed as described in Sect. 4.1–4.2.3.

References

Carle GF, Olson MV (1984) Separation of chromosomal DNA molecules from yeast by orthogonal field alternation gel electrophoresis. Nucl Acid Res 12:5647–5664

Carle GF, Olson MV (1985) An electrophoretic karyotype for yeast. Proc Natl Acad Sci USA 82:3756–3760

Decock JP, Iserentant D (1985) Ty hybridization patterns of yeast and their use for strain identification. Proc 20th Eur Brew Conv Congr Helsinki, pp 195–202

Donhauser S, Ritter H, Schmitt J (1981) Genetische Variabilität der Alkoholdehydrogenasen bei *Saccharomyces*. Proc 18th Eur Brew Conv Congr, Copenhagen

Holmberg S (1982) Genetic differences between *Saccharomyces carlsbergensis* and *S. cerevisiae*. II. Restriction endonuclease analysis of genes in chromosome III. Carlsberg Res Commun 47:233–244

Holmberg S, Petersen JGL, Nilsson-Tillgren T, Kielland-Brandt MC (1979) Molecular characterization of a *Saccharomyces* plasmid containing the HIS4 gene. Carlsberg Res Commun 44:269–282

Holmes DS, Quigley (1981) A rapid boiling method for the preparation of bacterial plasmid. Anal Biochem 114:193–197

Keiding AK (1985) Genetic and molecular characterization of a distillers yeast, Carlsberg Res Commun 50:95–125

Lodder (ed) (1970) The yeasts. North-Holland, Amsterdam

Maniatis T, Fritsch EF, Sambrook (1982) Molecular cloning. A laboratory manual. Cold Spring Harbor Lab, NY

Martens FB, van der Berg R, Harteveld PA (1985) DNA fingerprinting of yeast strains. Proc 20th Eur Brew Conv Congr Helsinki, pp 211–217

Martini AV, Kurtzman C (1985) Deoxyribonucleic acid relatedness among species of the genus *Saccharomyces* sensu stricto. Inst J Syst Bacteriol 35:508–511

Nilsson-Tillgreen T, Gjermansen C, Kielland-Brandt MC, Petersen JGL, Holmberg S (1981) Genetic differences between *Saccharomyces carlsbergensis* and *S. cerevisiae*. Analysis of chromosome III by single chromosome transfer. Carlsberg Res Commun 46:65–76

Pedersen MB (1983) DNA sequence polymorphisms in the genus *Saccharomyces*. I. Comparison of the HIS4 and ribosomal RNA genes in lager strains, Carlsberg Res Commun 48:485–503

Pedersen MB (1985a) Brewing yeast identification by DNA fingerprinting. Proc 20th Eur Brew Conv Congr Helsinki

Pedersen MB (1985b) DNA sequence polymorphisms in the genus *Saccharomyces*. II. Analysis of the genes RDN1, HIS4, LEU2 and Ty transposable elements in Carlsberg, Tuborg and 22 Bavarian brewing strains. Carlsberg Res Commun 50:263–272

Pedersen MB (1986a) DNA sequence polymorphisms in the genus *Saccharomyces*. III. Restriction endonuclease fragment patterns of chromosomal regions in brewing and other yeast strains. Carlsberg Res commun 51:163–183

Pedersen MB (1986b) DNA sequence polymorphisms in the genus *Saccharomyces*. IV. Homeologous chromosomes III of *Saccharomyces bayanus, S. carlsbergensis* and *S. uvarum*. Carlsberg Res Commun 51:185–202

Petes TD (1979) Meiotic mapping of yeast ribosomal deoxyribonucleic acid on chromosome XII. J Bacteriol 138:185–192

Petes TD, Hereford LM, Skryabin KG (1978) Characterization of two types of yeast ribosomal DNA genes. J Bacteriol 134:295–305

Philippsen P, Eibel H, Gafner J, Stotz A (1983) Ty elements and the stability of the *Saccharomyces cerevisiae* genome. In: Korhola M and Väisänen E (eds) Gene expression in yeast. Proc alko yeast symposium Helsinki. Found Biotech Ind Ferment Res 1:189–200.

Schwartz DC, Cantor CR (1984) Separation of yeast chromosome-sized DNAs by pulsed field gradient gel electrophoresis. Cell 37:67–75

Sherman F, Fink GR, Hicks JB (1979) Methods in yeast genetics. Cold Spring Harbor Lab, Cold Spring Harbor, NY 11724

Winge Ø, Laustsen O (1939) On 14 new yeast types produced by hybridization. CR Lab Carlsberg Ser Physiol 22:337–352

Yarrow D, Nakase T (1975) DNA base composition of the genus *Saccharomyces*. J Microbiol Antonie van Leeuvenhoek 41:81–88

Screening and Testing New Distillers' Yeasts for Their Potential in Molasses Ethanol Fermentations

B. E. CHRISTENSEN and H. ERIKSEN

1 Introduction

New strains and hybrids for ethanol production have to possess a range of desirable properties; apart from the ability to convert carbohydrates fast and with high efficiency into ethanol and carbon dioxide, tolerance to a prolonged contact with the formed alcohol is a prerequisite. In addition, the yeast must be able to withstand elevated temperatures and the presence of inhibitory substances in the medium without loss of viability and fermentation capacity. Furthermore, the yeast should not posses the ability to proliferate excessively during the fermentation. Ethanol tolerance is considered the main factor for fermentation rate and yeast performance, and some understanding of the factors involved has emerged in the last decade (Casey and Ingledew 1986).

The genetic background of a yeast determines to a large extent the amount of ehtanol tolerated or produced by the cells (Christensen 1987; Eschenbruch et al. 1982; Ismail and Ali 1971 a, b; Thornton 1982). Alikhanyan and Nalbandyan (1971) used mutagenesis for breeding ethanol-tolerant wine yeasts. Ethanol-tolerant variants have been recovered by selection in continuous cultures (Brown and Oliver 1982; Korhola 1983) and protoplast fusion (Seki et al. 1983) has been used to create strains with improved tolerance to high ethanol concentrations formed during the fermentation. Environmental factors play a prominent role in the performance of yeasts in ethanol fermentations. Temperature (Hacking et al. 1984; Nagodawithana et al. 1974; Nagodawithana and Steinkraus 1976), osmotic pressure (Panchall and Stewart 1980; Rose 1976), presence of unsaturated lipids (Thomas et al. 1978; Thomas and Rose 1979) and assimilable nitrogen (Thorne 1961; Bose and Ghose 1973; Irvali et al. 1986) influence the progress of the fermentation process.

Tolerance to substrates used in the distilling industry is a factor not much described. Beet and cane molasses are extensively used and are known to contain inhibitory substances which may influence the fermentation process (Burrows 1970; Tressl et al. 1976; Zauner et al. 1979).

Another important aspect in the evaluation of new distillers' yeasts is ease of handling: tedious propagation procedures and specific nutritional and physiological requirements tend to lower the usefulness of the strain.

In the distilling industry different fermentation processes are employed (Guidoboni 1984). The batch process is characterized as a fermentation where all carbohydrate to be converted is present at the onset of the fermentation; the yeast is consequently subjected to an initial osmotic stress followed by ethanol inhibition as the fermentation proceeds resulting in depressed fermentation performance. The yeast produced may be recycled totally or in part and after antibacterial treatment may provide a massive seed for the next fermentation, thus saving on substrate for yeast growth.

The fed-batch process is started as a batch fermentation into which substrate is subsequently pumped. The ethanol yield in this process is higher than in the batch process and fermentation time is generally shorter for the same amount of ethanol formed.

The basic continuous fermentation schemes are single stirred tank fermenters and the cascade system in which multiple tanks are serially connected. For both designs the yeast is recycled, thus limiting the amount of carbon source used for retaining viable yeast in the system.

The advantage of the continuous systems is a much improved productivity (Reed 1982), as substrate and end product inhibition do not act simultaneously.

The design and construction differences of the various fermentation systems renders comparison between results obtained with the systems difficult. The present chapter describes screening and testing methods for new distillers' yeast in both small-scale and full-scale fermentations using beet molasses as substrate and using a full-scale continuous cascade process.

2 Screening and Testing Techniques

In the breeding programme presently running at the Danish Distillers Ltd., new strains are being evaluated in three consecutive steps.

The first stage is small-scale (100 ml) batch fermentations monitoring the weight loss of the cultures. Promising strains are tested in continuous fermentations (4-l scale). The final evaluation is carried out as continuous production-scale fermentations.

2.1 Small-Scale Batch Fermentations

Even if results obtained in small-scale laboratory tests do not fully reproduce those from large-scale fermentations, the relative differences between various yeast strains reflect their behaviour in the production plant.

2.1.1 Description of Apparatus

Reliable indications on the fermentation performance of a novel yeast strain are obtained with batch fermentations (Christensen 1987). Erlenmeyer flasks containing 100 ml of medium are closed with a rubber stopper mounted with a fermentation lock with concentrated sulphuric acid. The flasks are agitated on a shaker during the fermentation, which takes place at 30 °C.

2.1.2 Propagation of Yeast

To insure reproducible results, yeast that is going to be evaluated in small-scale fermentations must be propagated under standard conditions. Cells from a discrete colony are inoculated into a 10 ml medium[1] in a test tube and allowed to grow overnight with shaking at 30 °C. Half a millilitre of this culture is inoculated into 250 ml propagation medium and allowed to multiply at 30 °C under aerobic conditions. The yeast is harvested in early stationary growth phase by centrifugation. Cells from 250 ml substrate are sufficient for two parallel fermentation cultures.

2.1.3 Fermentation and Analyses

Approximately 2×10^{10} newly propagated yeast cells, 1 g wet weight, are suspended in 5 ml sterile water and added to 100 ml fermentation medium[2] in fermentation flasks. The flasks are closed tightly and agitated at 30 °C for 44 h, during which time fermentation takes place and stops. Progress of the small-scale fermentations is monitored by recording the weight loss of the fermentation flasks at 0, 2, 4, 8, 20, 22, 24, 27, and 44 h. The weight of the carbon dioxide evolved is proportional to the ethanol formed during the fermentation.

Yeast density is determined prior to and at the end of the fermentation by measuring the optical density at 660 nm. Viability of the spent yeast is determined by plating appropriately diluted cell suspensions onto nutrient agar (Rose 1975) and/or by staining with methylene blue[3] (Pierce 1970) or Lissamine Green V[4] (Fowell 1954).

At the end of the fermentation, samples are analyzed for residual sugar and ethanol. Yeast in these samples is inactivated by addition of monoiodoacetic acid or monobromoacetic acid to a final concentration of 0.1% (w/v) (Inoue 1982).

Analysis for residual sugar and ethanol is performed by high pressure liquid chromatography.

2.2 Small-Scale Continous Fermentations

Yeast strains exhibiting higher weight losses than the reference strain are tried in small-scale continuous fermentations. The fermentation equipment is designed as a miniature version of the production plant of the Danish Distillers Ltd., see Sect. 2.3.2.

Three fermenters (Bioflo C-32, New Brunswick) with a working volume of 1.4 l each, are serially connected. Yeast recycling is accomplished by collecting the fermented wort at 4 °C followed by harvesting and washing the spent yeast by batch centrifugation. The yeast slurry obtained is kept under stirred conditions at 4 °C

[1] Propagation medium consists of 5% (w/v) beet molasses, 1‰ (w/v) $(NH_4)_2SO_4$, 0.1‰ (w/v) $NH_4)_2HPO_4$, 0.2‰ (w/v) $MgSO_4 \cdot 7H_2O$, pH 5.0 with sulphuric acid.

[2] Fermentation medium consists of 40% (w/v) beet molasses, adjusted to pH 5.0 with sulphuric acid.

[3] Methylene blue solution is 0.01% (w/v) methylene blue in 2% sodium citrate dihydrate. Live cells are colourless and non-viable cells stain blue when seen in the bright field microscope.

[4] 0.1% (w/v) Lissamine Green V in distilled water. Live cells are colourless and dead cells dark when seen through the microscope in orange light.

and is gradually pumped to the first fermenter in the system. This ensures a constant yeast concentration during the fermentation.

2.2.1 Propagation of Yeast

Pitching yeast for the continuous process is propagated essentially as described in Sect. 2.1.2. Yeast from 4 l of propagation medium is sufficient for a continuous fermentation experiment.

2.2.2 Fermentation and Analysis

Fermentation is started by adding approximately 6×10^{11} freshly harvested yeast cells, equal to 40–50 g yeast wet weight to aerated substrate[5] in the first fermenter. Substrate is added continuously at the rate of 180 ml/h. Aeration is stopped after 6 h. Sterile 80% (w/v) molasses is added to the fermenters depending on fermentation rate, yeast density and the concentration of residual sugar in the fermented substrate.

Samples collected twice daily from each fermenter are analyzed for yeast density and viability. This is done by staining with methylene blue (Pierce 1970) or Lissamine Green V (Fowell 1954; see Sect. 2.1.3), and counting in a haemocytometer. Lissamine Green V seems to give more reliable results than methylene blue when compared to plating techniques.

Residual sugar is determined with a urine sugar analysis kit (Clinitest, Ames Division, Miles Laboratories) and ethanol concentration according to the procedure described in Sect. 2.3.3.1.

2.3 Full-Scale-Fermentation

Fermentation in large scale gives a definite evaluation of the production potential of a new strain. Apart from ethanol production, ease of handling, yeast proliferation, viability and robustness in an industrial environment is assessed.

2.3.1 Propagation of Yeast

Yeast for ethanol production is essentially propagated in the same way as baker's yeast (Burrows 1970, 1979). Important parameters in pitching yeast are a high percentage of viable cells and a high total nitrogen content (Irvali et al. 1986).

Cells from a single colony are inoculated into 10 ml molasses medium[6] and allowed to ferment for 2 days at 27 °C.

---

[5] Fermentation substrate is 30% (w/v) beet-molasses, adjusted to pH 5.0 with sulphuric acid.

[6] Molasses media used in propagation consist of various dilutions of a standard 40 ° Balling molasses supplemented wih phosphoric acid, ammonium phosphate and magnesium sulphate. pH = 4.5.

Molasses used in the batch processes is 10 °–11 ° Balling and in the final fed-batch stage 40 ° Balling wort is fed exponentially to the culture.

The yeast is then transferred to 1 l molasses wort in a 2-l Erlenmeyer flask where fermentation for 24 h without shaking increases the biomass.

The supernatant is removed by decanting and 1 l new wort added to the yeast. This procedure is repeated three times. Yeast from six parallel flasks is added to a 500-l fermenter. Here propagation under aerobic conditions takes place as a batch fermentation. After 12 h the yeast is transferred to a 10,000-l aerobic fermenter, where the aerobic batch propagation is completed in 10 h. The resulting culture is pumped into the final stage, an aerated fed-batch fermentation.

The volume of yeast and substrate is at this stage brought up to 90,000 l. The final yeast density is approximately 165 g yeast per litre. The yeast is separated from the wort, washed with water in a separator and pumped to storage tanks as yeast cream.

In the two aerated propagation steps the air flow is limited so that half of the molasses is converted into ethanol and the other half to yeast biomass.

2.3.2 Fermentation Equipment

The main characteristic of the Danish Distillers Ltd. production plant is three serially connected fermentation tanks of each 170 m^3. Fermentation is continuous and yeast harvested from the fermented wort is recycled (Rosén 1978). Figure 1 shows the set-up of the production plant.

The fermentation is started by adding yeast cream equivalent to 3000 kg press yeast to 25 m^3 30% (w/v) molasses, pH adjusted to 5.0 with phosphoric acid and sulphuric acid in tank 1. The growth of the culture is boosted by aeration with sterile air and substrate is fed continuously. When the 80% level of filling in the tank is reached, the fluid overflows into tank 2. Aeration is now stopped. When the 80–90% level of filling is reached, the fermenting wort is allowed to finish the fermentation in tank 3. When the 80-90% level is reached in this tank the fermented wort is passed through a separator; the supernatant is led to the distilling plant and the yeast is after antibacterial acid treatment recycled to tank 1.

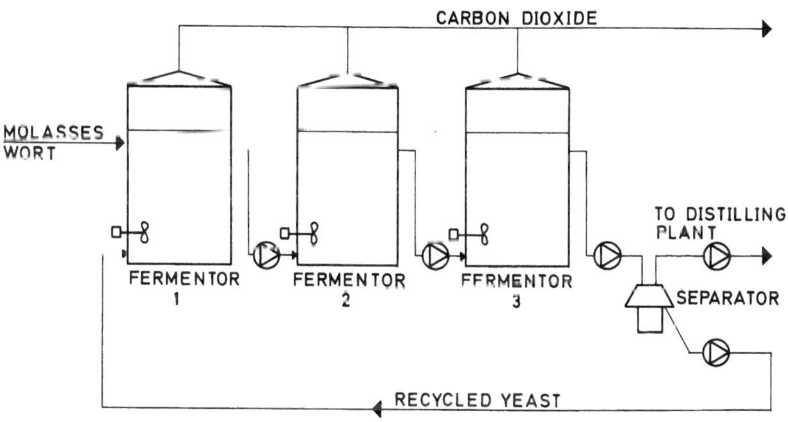

Fig. 1. Schematic representation of the ethanol production plant of the Danish Distillers Ltd

The residence time in each tank is approximately 7 h, which gives a yeast recycling time of 21 h. The fermentation temperature in all three tanks is 33 °C.

2.3.3 Analyses Performed

In samples drawn with 2-h intervals from each fermentation tank temperature and specific gravity of the fermenting wort is determined. Yeast density is monitored as g/l by centrifugation in graded Gerber tubes.

Ethanol concentration in each fermentation tank is determined every 8 h by the oxidative method (see Sect. 2.3.3.1), and the presence of residual sugar in the fermented wort from tank 3 is checked by a urine sugar analysis kit (Clinitest, Ames Division, Miles Laboratories).

Every 12 h the number of cells in suspension is determined by counting in a haemocytometer and the proportion of live and dead cells is determined by staining with methylene blue (Pierce 1970) or Lissamine Green V (Fowell 19854; see Sect. 2.1.3).

The total amount of molasses used is measured, as is the total amount of ethanol distilled. The concentration of sugar in the molasses is determined as described in Sect. 2.3.3.2 or after the method of Somogyi (1945).

2.3.3.1 Oxidation Method of Ethanol Determinatin

The method is modified from Amerine and Ough (1980). 10.0 ml wort with a maximum alcohol concentration of 1% (w/v) is poured, together with a little pumice, into a flask with a side tube. The flask is gently heated and ethanol vapours absorbed in a mixture of 10.0 ml potassium dichromate (42.59 g/l) and 10 ml 50% (v/v) H_2SO_4 contained in a "ball tube" (Fig. 2).

Empirically, distillation of ethanol has been found to be complete when the content of the ball tube is hot enough to allow white fumes to emerge. The content of the ball tube is transferred quantitatively to a 500-ml Erlenmeyer flask; 250 ml cold tap water is added together with 10 ml 10% (w/v) potassium iodide solution. The mixture is titrated with 1 N $Na_2S_2O_3$ (248.2 g/l). As indicator a solution of starch is used.

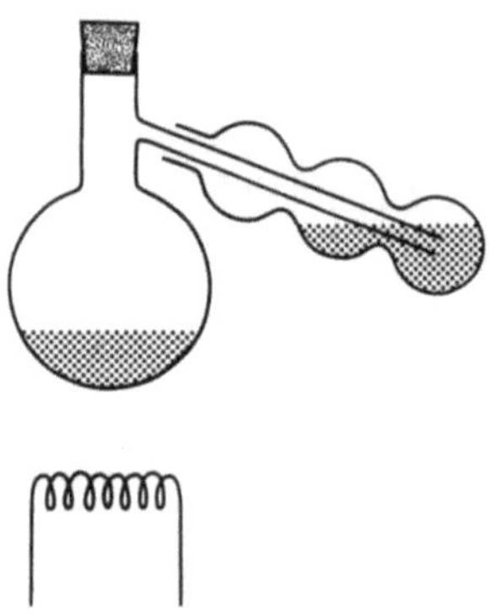

Fig. 2. Apparatus for ethanol determination

The thiosulphate solution is normalized against 10.00 ml of the dichromate solution; V_0 is the amount used in this procedure. The volume of thiosulphate used in the ethanol analysis = V. The ethanol concentration in weight % is calculated as

$$\frac{V_0 - V}{V_0}.$$

To convert the concentration to volume % multiply by 1.26.

2.3.3.2 Sugar Concentration Determined by Fermentation

A rapid and exact method of finding the amount of sugar available in a substrate is to determine the weight loss from a given volume by fermentation.

Carbohydrates are degraded stoichiometrically according to the formula

$$C_6H_{12}O_6 \rightarrow 2\,CH_3CH_2OH + 2\,CO_2.$$

If disaccharides or polysaccharides provide the fermentable carbohydrate, water is taken up by hydrolysis, thus:

$$C_{12}H_{22}O_{11} + H_2O \rightarrow 2\,C_6H_{12}O_6.$$

The theoretical weight loss is accordingly

$$\frac{\text{molecular weight of } CO_2 \times 2}{\text{molecular weight of carbohydrate}}$$

for each gram of carbohydrate fermented.

This gives a weight loss of

88/180 gram = 0.489 g per gram monosaccharide fermented

and

176/342 gram = 0.515 g per gram disaccharide fermented.

To 1000 ml substrate containing 200 g molasses, adjusted to pH 5.0 with sulphuric acid and 2 g diammonium phosphate, is added 10 g yeast (wet weight). Fermentation is allowed to take place in a container closed with a fermentation lock filled with concentrated sulphuric acid, at 30 °C with agitation. The flask is weighed at 12-h intervals until the weight is constant. A three-double determination is carried out.

The carbohydrate in beet molasses consists of approximately 98% sucrose; the rest is mainly invert sugar and raffinose. This implies that a weight loss of 1 g in a beet molasses medium corresponds to 1.94 g of fermentable carbohydrate present in the substrate.

The differences in sugar concentration found when comparing the fermentation method with the official analysis method is less than 2% (Offer and Haldenwanger 1984).

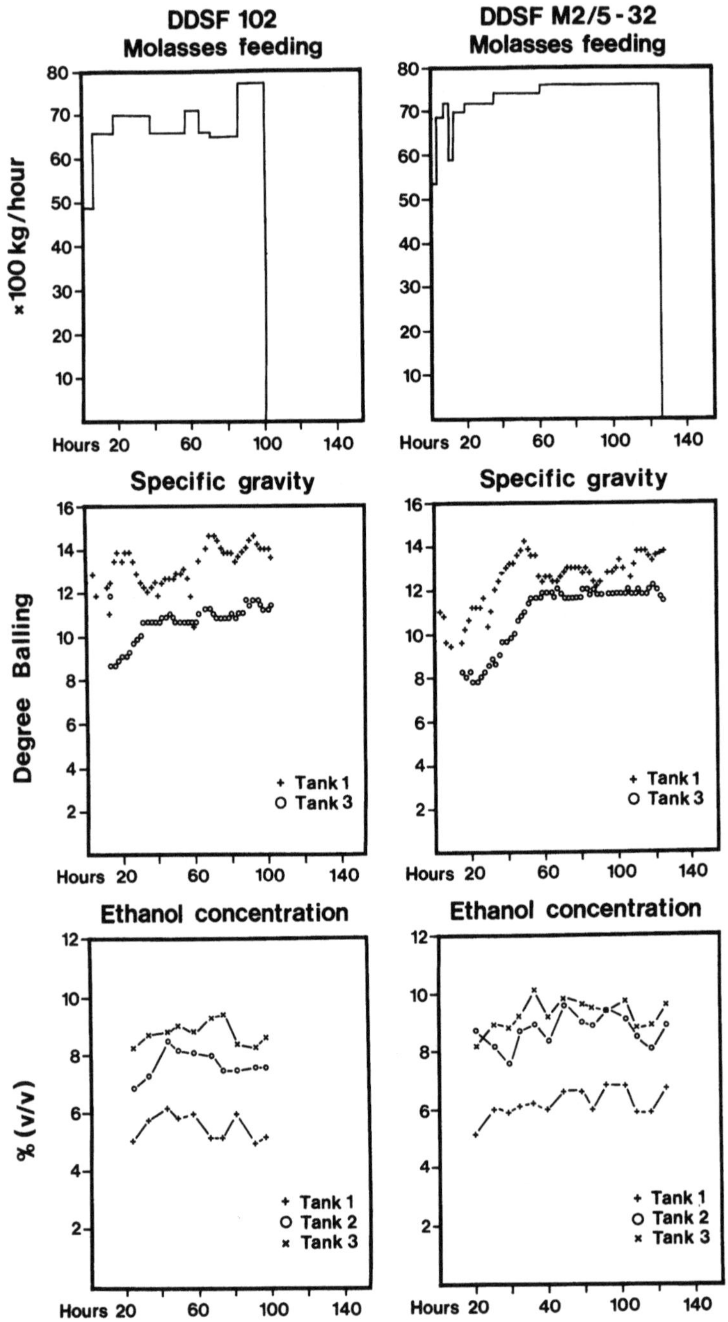

Fig. 3. Fluctuations in molasses feeding, specific gravity and ethanol concentration in the three fermentation vessels during fermentations with the yeast strains DDSF 102 and DDSF M2/5-32

2.3.4 Regulation of Fermentation

The regulation of the fermentation is achieved through altering the molasses dosage and by modifying the residence time by changing the level of filling in the fermenters. The flow through the system is almost constant during the fermentation, whereas the concentration of the molasses wort is adjusted to the current requirements.

Molasses dosage is constant the first 24 h until the recycled yeast appears in tank 1. Molasses feeding is then regulated according to the analyses taken every 2 h. The aim of the regulation is to avoid residual sugar in the last fermentation tank as well as obtaining a high ethanol production. If the sugar concentration in tank 3 is zero and the specific gravity has not been rising the previous 6–8 h, molasses dosage is increased. The specific gravity reflects the amount of the molasses added to the medium as the contribution from the yeast is constant through the fermentation. Furthermore the metabolic activity of the yeast shows in the fluctuations of the specific gravity: Rising metabolic activity results in a fall in specific gravity. Contrarily, a rising specific gravity with no prior increase in molasses dosage demands an increase in the fermentation time. This is done by raising the level of filling in the tanks, thus prolonging the residence time with up to 3 h. If this is not sufficient, addition of molasses is decreased.

Changes in substrate feeding influence the subsequent ethanol concentrations in the three fermentation tanks. In Fig. 3 are compared the fluctuations in both molasses feeding, specific gravity and the resulting ethanol concentrations during fermentations with both a production strain, DDSF 102 and a new hybrid DDSF M2/5–32.

The hybrid tolerates a more rapid initial rise in the molasses feeding and the total molasses concentration, reaches a higher value. DDSF M2/5–32 ferments faster, causing a slower rise in the specific gravity and higher ethanol concentrations than the reference strain. After 50 h, the specific gravity falls in the fermentation with DDSF M2/5–32 although the molasses afflux is constant. This implies a high metabolic activity. The rise in specific gravity for DDSF 102 in the last part of the fermentation is the result of a lowered metabolic activity.

The ethanol concentrations in the three fermentation tanks reflect the molasses added (Fig. 3). The final concentration reaches 10% for the hybrid and 9% for the reference. Furthermore, ethanol concentration in tank 2 is for DDSF M2/5–32, almost as high as the final concentration for DDSF 102. In the last 25 h, residual sugar was present in the fermentation with DDSF 102.

3 Evaluation of Results

3.1 Small-Scale Fermentations

A typical experimental trial of new hybrids includes 16 different yeast strains and 2 sets of a production strain used as an internal reference. All the tested new strains are compared to this strain.

The weight losses are computed for 2, 4, 8, 20, 27, and 44 h. Hybrids conferring a weight loss to the cultures significantly exceeding that of the reference are considered promising and retested in new fermentation experiments. Table 1 gives typical weight losses obtained with DDSF 102 and DDSF M2/5–32 in small-scale batch fermentations.

Evaluation by continuous small-scale fermentation of potential new production strains mainly takes into account the ability to produce high concentrations of ethanol under specific conditions. The higher the ethanol concentrations are in all three fermenter vessels the better.

Repeated tests of parental production strains and new yeast strains have shown that the described small-scale screening system is well suited to pick out distillers' yeasts with improved fermentation performance.

Table 1. Weight losses obtained with the yeast strains DDSF 102 and DDSF M2/5–32 in 100-ml batch fermentations. Fermentation temperature is 30 °C. Substrate is 40% (w/v) molasses diluted in water. pH is 5.0. The 20-h weight loss reflect differences in the fermentation rates. The final weight loss indicate a slightly better ethanol tolerance in DDSF M2/5–32

Strain	DDSF 102 Weight loss (g)		DDSF M2/5–32 Weight loss (g)	
	20 h	44 h	20 h	44 h
	7.57	8.65	8.03	8.87

3.2 Production-Scale Fermentations

Data collected from full-scale experiments are judged in a production context, the primary considerations being the total amount of ethanol and excess yeast formed. Table 2 exemplifies relevant results for the evaluation of a new production strain. These data compared with results regarding fermentation rate (Fig. 3), viability and ease of handling in the production environment determines the success of the new strain.

Table 2. Production parameters recorded from trials with a production strain DDSF 102 and a new distillers' yeast DDSF M2/5–32 created from DDSF 102. (Christensen 1987)

Strain	DDSF 102	DDSF M2/5–32
Production time, h	102	120
Ethanol produced, l	195,000	254,000
Ethanol produced per hour	1,915	2,116
Molasses used, t	705	920
Percent sugar, fermentable	46.3	44.9
Ethanol yield, l/t fermt. sugar	598	615

References

Alikhanyan SI, Nalbandyan GM (1971) Selection of wine yeast using mutagens. Sov Genet 2:1200–1205

Amerine MA, Ough CS (1980) Methods for analysis of musts and wines. Wiley, New York

Bose K, Ghose TK (1973) Studies on continuous fermentation of Indian cane sugar molasses by yeast. Process Biochem 8:23–25, 33

Brown SW, Oliver SG (1982) Isolation of ethanol-tolerant mutants of yeast by continuous selection. Eur J Microbiol Biotechnol 16:119–122

Burrows S (1970) Baker's yeast. In: Rose AH, Harrison JS (eds) The yeasts, vol 3. Academic, London, pp 349–420

Burrows S (1979) Baker's yeast. In: Rose AH (ed) Economic microbiology, vol 4. Academic, London, pp 32–64

Casey GP, Ingledew WM (1986) Ethanol tolerance in yeasts. In: Stewart GG, Russell I (eds) Critical reviews in microbiology, vol 13, issue 3. CRC, Boca Raton, pp 219–280

Christensen BE (1987) Breeding for ethanol-tolerant distillers' yeasts. In: Stewart GG, Russell I, Klein RD, Heibsch RR (eds) Biological research on industrial yeasts. Uniscience series. CRC, Boca Raton (in press)

Eschenbruch R, Cresswell KJ, Fischer BM, Thornton RJ (1982) Selective hybridization of pure culture wine yeasts. 1. Elimination of undesirable wine-making properties. Eur J Appl Microbiol Biotechnol 14:155–158

Fowell RR (1954) Biology staining schedules for first year students, 8th edn. H. K. Lewis, London, pp 27, 31. Referred to in: Harrison JS, Graham JCJ: Yeasts in the distillery practice. In: Rose AH, Harrison JS (eds) The yeasts, vol 3. Academic, London, pp 283–348

Guidoboni GE (1984) Continuous fermentation systems for alcohol production. Enzyme Microb Technol 6:194–200

Hacking AJ, Taylor IWF, Hanas CM (1984) Selection of yeast able to produce ethanol from glucose at 40 °C. Appl Microbiol Biotechnol 19:361–365

Inoue T (1982) Further studies on an improved micromethod for determination of total diacetyl in beer. Rep Res Lab Kirin Brew Co LTD 25:13–20

Irvali N, Bärwald G, Windisch S (1986) Einfluß der Stickstoffquelle auf die Leistung und Regenerierung der Hefe unter anaeroben Bedingungen. Die Branntweinwirtschaft, S 70–73

Ismail AA, Ali AAM (1971 a) Selection of high ethanol-yielding *Saccharomyces*. 1. Ethanol tolerance and the effect of training in *Saccharomyces cerevisiae*. Folia Microbiol 16:346–349

Ismail AA, Ali AAM (1971 b) Selection of high ethanol yielding *Saccharomyces*. II. Genetics of ethanol tolerance. Folia Microbiol 16:350–354

Korhola M (1983) Improvement of of yeast strains for added ethanol tolerance. In: Korhola M, Vaisanen E (eds) Gene expression in yeast. Proceedings of the Alco yeast symposium, Helsinki, pp 231–242

Nagodawithana TW, Steinkraus KH (1976) Influence of the rate of ethanol production and accumulation on the viability of *Saccharomyces cerevisiae* in rapid fermentation. Appl Environ Microbiol 31:158–162

Nagodawithana TW, Castellano C, Steinkraus KH (1974) Effect of dissolved oxygen, temperature, initial cell count and sugar concentration on the viability of *Saccharomyces cerevisiae* in rapid fermentations. Appl Microbiol 28:383–391

Offer G, Haldenwanger M (1984) Die Ausarbeitung eines biologischen Schnelltestes zur Beurteilung von Brennereirohstoffen – ein Beitrag für den wirtschaftlichen Brennereibetrieb. Die Branntweinwirtschaft 2–4, S 7

Panchall GJ, Stewart GG (1980) The effect of osmotic pressure on the production and excretion of ethanol and glycerol by a brewing yeast strain. J Inst Brew 86:207–210

Pierce JS (1970) Institute of brewing: analysis committee measurement of yeast viability. J Inst Brew 76:442–443

Reed G (1982) Production of fermentation alcohol as a fuel source. In: Reed G (ed) Prescott and Dunn's industrial microbiology. AVI, Westport, pp 835–859
Rose AH (1975) Growth and handling of yeasts. In: Prescott DM (ed) Methods in cell biology, vol XII. Yeast cells. Academic, London, pp 1–16
Rose D (1976) Yeasts for molasses alcohol. Process Biochem 12(3):10–12, 36
Rosén K (1978) Continuous production of ethanol. Process Biochem 13 (5):25–26
Seki T, Myoga S, Limtong S, Vedono S, Kumnuanta T, Taguchi J (1983) Genetic construction of yeast strains for high ethanol production. Biotechnol Lett 5:351–356
Somogyi M (1945) Determination of blood sugar. J Biol Chem 160:69–73
Thomas DS, Rose AH (1979) Inhibitory effect on growth and solute accumulation by *Saccharomyces cerevisiae* as effected by plasma-membrane lipid composition. Arch Microbiol 22:49–55
Thomas DS, Hossack JA, Rose AH (1978) Plasma membrane lipid composition and ethanol tolerance in *Saccharomyces cerevisiae*. Arch Microbiol 117:239–245
Thorne RSW (1961) Fermentation velocity of brewery yeasts. Brew Dig 36(7):38
Thornton RJ (1982) Selective hybridization of pure culture wine yeasts. II. Improvement of fermentation efficiency and inheritance of SO_2 tolerance. Eur J Appl Microbiol Biotechnol 14:159–164
Tressl R, Jakob R, Kossa T, Bronn WK (1976) Gaschromatographisch-massenspektrometrische Untersuchung flüchtiger Inhaltstoffe von Melasse. Die Branntweinwirtschaft, S 117–119
Zauner E, Bronn WK, Dellweg H, Tressl R (1979) Hemmwirkung von Nebenkomponenten der Rübenmelasse und von Pflanzenschutzmitteln auf die Atmung und Gärung von Hefe (*Saccharomyces cerevisiae*). Die Branntweinwirtschaft, S 154–163

Genetic Analysis and Manipulation of *Mucor* Species by DNA-Mediated Transformation

R. VAN HEESWIJCK, M. I. G. RONCERO, and L. P. JEPSEN

1 Introduction

1.1 Taxonomy of *Mucor*

The fungi belonging to the genus *Mucor* are wide-spread terrestrial moulds commonly isolated from soil or dung. They constitute the largest genus of the order Mucorales, class Zygomycetes, with a total of 60 species (Ingold 1978). *Mucor miehei* and *Mucor pusillus* may be considered as thermophilic members of the genus (Cooney and Emerson 1964) or as members of a separate genus, *Rhizomucor*, containing these two species and *Mucor tauricus* alone (Schipper 1978).

1.2 Industrial Use of *Mucor*

Mucor species are used as starter cultures in oriental fermentations based on rice, cassava and soya beans (Gray 1970; Hesseltine et al. 1985). They are of primary industrial interest because they excrete amylases (Adams and Deploey 1976), cellulases (Somkuti 1974), lipases (Somkuti and Babel 1968) esterases (Moskowitz et al. 1977) and proteases. Both *M. pusillus* (Arima et al. 1967) and *M. miehei* (Aunstrup 1966) produce acid proteases related to the proteolytic enzyme chymosin (EC 3.4.4.3) which are used as substitutes for the latter in a crucial step in the manufacturing of cheese: the clotting of milk as a result of κ-casein proteolysis (Foltmann 1967). Harboe (1984) has estimated that approximately 40% of the milk-clotting enzyme used in cheese manufacture is *Mucor miehei* protease. Other members of the genus *Mucor* which produce milk-clotting enzymes include *M. varians* (Bernardinelli et al. 1983), *M. bacilliformis* (Bottaro Castilla 1982), *M. circinelloides* (Jepsen 1986), and *M. racemosus* (Higashio and Yoshioka 1981).

1.3 Genetic Analysis and Manipulation

During the sexual cycle of *Mucor* in nature, the fusion of hyphae of opposite mating type results in the formation of zygospores, which are resting spores with dormancy periods of up to several months. Problems in obtaining formation and germination of zygospores under laboratory conditions have hitherto hindered the use of classical genetic methods in the analysis of mutants and in the generation of recombinants for the breeding of new strains. The traditional methods of mutagenesis, screening and selection, have been used with some success for strain improvement of *Mucor* (e.g. Higashio and Yoshioka 1981) but this approach is tedious and limited in extent by the finite size of the genome concerned. The ability to perform DNA-mediated transformation enables application of the modern techniques of molecular biology to the genetic analysis and manipulation of *Mucor* and introduces new possibilities for improvement of strains for various fermentation processes.

2 General Methods

2.1 Media

Complete medium, YPG (Bartnicki-Garcia and Nickerson 1962): 3.0 g yeast extract (Difco), 10.0 g peptone (Difco), 20.0 g glucose per litre distilled water. Minimal medium, YNB (Lasker and Borgia 1980): 0.5 g yeast nitrogen base without amino acids and ammonium sulphate (Difco), 1.5 g ammonium sulphate, 1.5 g glutamic acid per litre distilled water with 1–3% (w/v) glucose added post-sterilization.

For growth of auxotrophs the minimal medium is supplemented with amino acids at a final concentration of 100 μg/ml or with casamino acids at 2 mg/ml.

The media is adjusted to pH 4.5 with 1 *M* HCl unless restricted colony growth is required in which case the media is adjusted to pH 3.2. Acid hydrolysis of the agar at pH 3.2 is avoided by adding the 1 *M* HCl post-sterilization, or by preparing double strength solutions af agar and the other media components (pH 3.2) and autoclaving them separately.

2.2 Growth Conditions

Liquid cultures of mesophilic and thermopholic species are incubated at 28 °C and 40 °C respectively for optimal growth and germination of sporangiospores. Agar plate cultures are incubated at room temperature and 37 °C respectively as a matter of convenience.

Liquid cultures should constitute a maximum of 15% of the flask volume, and be shaken vigorously to maintain aerobic filamentous growth. If anaerobic conditions are required the cultures can be placed in a polycarbonate jar (Becton, Dickinson and Company) with a disposable hydrogen and carbon dioxide generator and incubated on a rotating table-top. *M. circinelloides* and other mesophilic species grow well under anaerobic conditions, producing large budding yeast-like cells. These cells form well-defined colonies on agar plates even at pH 4.5, but since they are multinucleate (Bartnicki-Garcia and Nickerson 1962) and tend to adhere to each other they are not ideal for use as single cell isolates. In our experience, sporangiospores of *M. miehei* and *M. pusillus* do not germinate in an anaerobic atmosphere and any growth of pregerminated cultures under these conditions is weak and filamentous.

2.3 Harvesting, Storage and Germination of Sporangiospores

1. Properties of Sporangiospores. Sporangiospores are asexual spores, the nuclei of which have been formed by a mitotic process. They constitute a stage of the life cycle suitable for use in laboratory manipulations. The sporangiospores range in size from 5–20 µm and can be uni- or multinucleate. The number of nuclei can be determined by Feulgen staining and is found to vary both with composition of the growth media and the strain used (van Heeswijck R unpublished 1986).

2. Harvesting of Sporangiospores. Sporangiospores are not produced in shaking liquid culture, but are profuse on the surface of solid media at pH 4.5. They are harvested in distilled water by scraping the surface of the culture gently with a glass rod, filtered through gauze or Monodur nylon mesh (pore size 22 µm) then washed in distilled water prior to further use. The concentration of sporangiospore suspensions is determined by haemocytometer counts and their viability by plating diluted aliquots on YPG medium pH 3.2. Cultures for sporangiospore production on YPG medium pH 4.5 are incubated at room temperature for 7 days (mesophiles) or 37 °C for 5 days (thermophiles), around which time the sporangiospores have greatest viability and germinate as synchronously as possible (see below).

3. Storage of Sporangiospores. Concentrated suspensions of sporangiospores can be stored in water at −20 °C, although some reduction in viability may occur. For long-term storage we use silica gel stocks (Davis and de Serres 1970) from which cultures are revived by shaking a small amount of the silica gel onto agar media. Sporangiospores can be harvested and stored overnight at 4 °C in water or growth medium with no detrimental effect on viability or synchrony of germination.

4. Germination of Sporangiospores. Germination of sporangiospores occurs upon transfer to fresh nutrient media at an appropriate growth temperature but can be very asynchronous, especially for mesophilic species. Synchronous cultures are desired for production of protoplasts (Sect. 4.1) and for selective enrichment of auxotrophic mutants (Sect. 5.3). It has not been possible to develop a method for specific induction of germination, however, the synchrony of germination can be optimized by using freshly harvested sporangiospores and a complete medium and it increases with increasing incubation temperature to a maximum at the optimal growth temperature (van Heeswijck 1984).

2.4 Replica Plating of Colonies

Toothpicks are used to transfer inocula from the compact sporulating mycelial colonies produced on pH 3.2 media in the dark. Once a grid pattern of colonies is established on a master plate it is most convenient to use a multi-pin replicator (9-cm diameter perspex base containing 240 stainless steel sewing pins) which is effectively sterilized by flaming in alcohol.

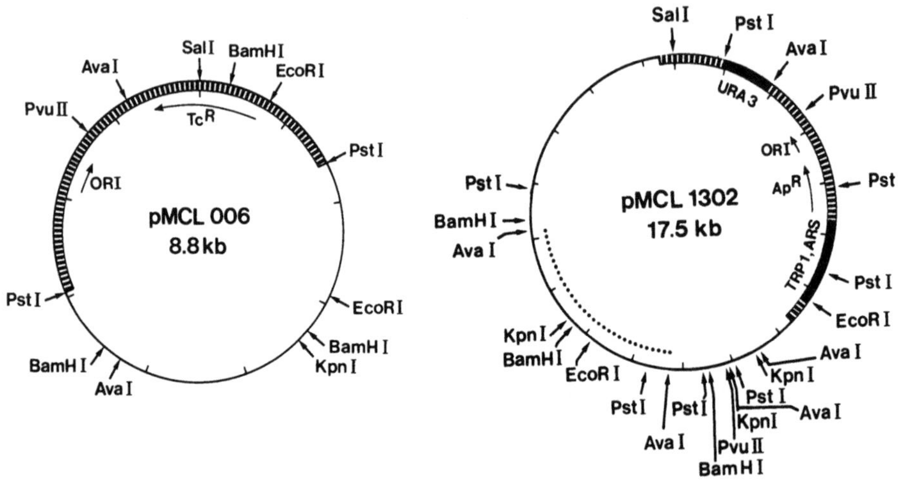

Fig. 1. Restriction maps of plasmids pMCL006 and pMCL1302. Plasmids are not drawn to the same scale. Location of restriction sites for AvaI, BamHI, EcoRI, KpnI, PstI, PvuII and SalI are shown. *Thick lines* are the vectors pBR322 and YRp17, respectively: *thick black lines* represent sequences derived from *S.cerevisiae; thick barred lines* represent sequences derived from *E. coli. Thin line* is the *Mucor* DNA insert

3 DNA-Mediated Transformation of *Mucor*

3.1 Plasmids pMCL006 and pMCL1302

Plasmid pMCL006 (Fig. 1) consists of pBR322 and a 4.4 kb Pst I fragment of Mucor DNA containing both a Leu^+ gene and a *Mucor* ars (autonomous replication sequence) (van Heeswijck 1986). It is a sub-clone of plasmid pMCL1302 (Fig. 1) which was selected from a *M. circinelloides* gene library by direct complementation of a Leu^- mutant of *M. circinelloides*. Both pMCL006 and pMCL1302 are capable of transforming the Leu^- mutant of *M. circinelloides* to prototrophy, and exist as autonomously replicating plasmid molecules within the transformants. They represent efficient shuttle vectors for the transfer of genetic information between *Eschericia coli, Saccharomyces cerevisiae,* and *M. circinelloides.*

3.2 Method for Transformation of *Mucor* with Plasmid DNA

The method developed for the transformation of a Leu^- mutant of *M. circinelloides* CBS 277.49 (strain R7B) to prototrophy with plasmids pMCL006 or pMCL1302 (van Heeswijck and Roncero 1984) is a modification of that used for transformation of *Neurospora crassa* (Case 1982):

1. Protoplasts are prepared as described in Sect. 4.1, washed gently three times in 0.5 *M* sorbitol, 10 m*M* morpholinopropanesulfonic acid (MOPS) pH 6.3, 50 m*M* $CaCl_2$, then resuspended in the same buffer at a concentration of ap-

proximately 10^7/ml. The soft pellet obtained during centrifugation (400 *g*, 5 min) is best retained if a swing-out rotor is used and the supernatant removed by aspiration. The protoplasts may be stored frozen at −70 °C until further use.

2. 200 µl of protoplasts is added to 25 µl of a solution containing plasmid DNA (0.1–1.0 µg) which has been preincubated with 1 mg heparin (Sigma) for 20 min on ice. 20 µl of 40% (w/v) polyethylene glycol (PEG) 4000 (Merck) in 10 m*M* MOPS pH 6.3, 50 m*M* $CaCl_2$ is added, and the mixture incubated for 30 min on ice.
3. 2.5 ml 40% (w/v) PEG 4000 in 10 m*M* MOPS pH 6.3, 50 m*M* $CaCl_2$ is added slowly and the mixture incubated for a further 25 min a room temperature, after which it is diluted by addition of 20 ml 0.5 *M* sorbitol, 10 m*M* MOPS pH 6.3, 50 m*M* $CaCl_2$ and centrifuged (400 *g*, 3 min).
4. The small, soft pellet is resuspended in 5 ml YPG medium pH 4.5, 0.5 *M* sorbitol and incubated for at least 30 min at room temperature to allow "recovery" of transformants before they are put under selective conditions.
5. After recentrifugation, the pellet is resuspended in 5 ml YNB medium pH 4.5, 0.5 *M* sorbitol. 1.0 ml aliquots are added to 9.0 ml of YNB medium pH 3.2 containing 3% glucose, 0.5 *M* sorbitol and 1% (w/v) agar (held molten at 50 °C), then overlayed onto agar plates containing YNB medium pH 3.2, 0.5 *M* sorbitol. Assessment of the number of potential colony forming units is made by plating diluted aliquots in agar overlays containing YPG medium pH 3.2, 0.5 *M* sorbitol.

After 2–3 days incubation at room temperture, prototrophic transformants become visible as small well-defined colonies which can be counted and transferred to fresh medium with toothpicks before they begin to sporulate. Transformation frequencies as high as 7800 Leu^+ colonies per µg DNA, or 3×10^{-3} per viable cell can be obtained with plasmid pMCL1302. The frequency obtained with pMCL006 is lower (2800 Leu^+ colonies per µg DNA) although the reason for this is not yet clear. Many factors can have an effect on the transformation frequency. The purity of the DNA is important and we prefer to use two to three cycles of CsCl gradient centrifugation during plasmid preparation, although "mini-prep" DNA can be used when only a small number of transformants is required. Omission of the "recovery" period or of the preincubation of DNA with heparin or the use of high plating densities all significantly reduce the transformation frequency. Attemps to transform Li^+-treated germlings according to the protocols of Ito et al. (1983) and Dhawale et al. (1984) have not been successful.

3.3 Properties of Transformants

The presence of PEG during transformation induces protoplast fusion (Sect. 4.4) increasing the possibility of heterokaryotic transformants. Repetitive cycles of single sporangiospore transfer produces homokaryotic isolates suitable for further analysis.

Transformants obtained with plasmids pMCL006 and pMCL1302 are mitotically unstable for the Leu^+ phenotype when grown under non-selective conditions and should thus be maintained on minimal YNB medium. If transformants of pMCL1302 are transferred to complete medium, only 5% of the viable sporangiospores produced retain the Leu^+ phenotype compared with 45% when growth is continued under selective conditions (van Heeswijck 1986). A total of two or three successive transfers to complete medium reduces the % Leu^+ to 0.8% and 0.056% respectively. The mitotic instability of the transformants stems from the existence of the plasmids as extrachromosomal autonomously replicating DNA molecules within the *Mucor* cells. Integration of the plasmids into the *Mucor* genomic DNA should produce mitotically stable transformants, but no such integrative transformants have yet been isolated.

4 Isolation, Regeneration and Fusion of Protoplasts from *Mucor*

Protoplasts are valuable tools in the manipulation of *Mucor* species and serve as the basis for DNA-mediated transformation and cell fusion experiments. They are relatively easy to prepare and readily undergo reversion to the normal cellular form.

4.1 Formation of Protoplasts from *Mucor*

The formation of protoplasts requires removal of the cell wall by digestion with a lytic enzyme(s). The sporangiospores of *Mucor* are not a suitable source of protoplasts due to the extreme resistance of the cell wall to lytic digestion (Jones et al. 1968). The high content of melanin [10% of dry weight according to Bartnicki-Garcia (1968)] is implicated in this resistance, as it is a known inhibitor of cell wall lytic enzymes. During germination, a vegetative wall is formed de novo underneath the sporangiospore wall, later emerging to become the germ tube wall (Bartnicki-Garcia et al. 1968). The lack of continuity between the spore and vegetative walls is correlated with marked differencies in their composition, including the absense of melanin in the latter. Chitin and chitosan become the main components; in *M. rouxii,* for example, they constitute 9.4% and 32.7% of the hyphal wall, respectively (Bartnicki-Garcia 1968). Chitinases are commercially available but are unable to hydrolyze chitosan. Chitosanases are not yet commercially available but are produced as extracellular enzymes by a variety of microorganisms including the actinomycete *Streptomyces* sp. no. 6 (Jones et al. 1968). The preparation of a lytic enzyme mixture from this microorganism is described in Sect. 4.2 and the method by which it can be used to produce protoplasts from germlings of Mucor is as follows (van Heeswijck 1984):

1. Sporangiospores (10^7/ml) are incubated in YPG medium pH 4.5 at 28 °C (mesophiles) or at 25 °C with a 2-h preincubation at 40 °C (thermophiles) until the majority of germ tubes are 35–50 µm in length. Conditions are used which give optimal synchronous germination (Sect. 2.3.4).

2. The suspension of germlings is collected on Monodur nylon cloth (mesh size 16 μm), washed extensively in 0.5 *M* sorbitol, 0.01 *M* sodium phosphate buffer pH 6.5 (to remove the majority of ungerminated spores) then resuspended in the same solution at a concentration of 10^7 cells/ml.
3. Streptozyme is added to give a final concentration of 0.5–1.0 mg/ml (approximately 1 unit chitosanase/ml) and the mixture incubated for 2–3 h at 23 °C.

Digestion of the cell wall and release of protoplasts (average size 7 μm) preferentially occurs at the hyphal apex. The total number of protoplasts can be up to three times the number of germlings treated, indicating that the cell membrane of the coenocytic hyphae is effectively resealed after each protoplast is released. The yield of protoplasts is highest when germlings have a germ tube length of 35–50 μm. Longer or shorter germ tubes result in a decrease in the yield of protoplasts and a decrease in their homogeneity (as evidenced by a greater variation in size, refractility and apparent membrane integrity).

The optimal concentration of sorbitol is 0.35–0.5 *M*. At concentrations of 0.65 *M* or above no cell wall degradation is observed, and at concentrations of 0.3 *M* or lower, the action of streptozyme is accompanied by extensive cell lysis. Other sugar alcohols such as mannitol can effectively replace sorbitol as osmotic stabilizer, but solutions of salts [KCl, $MgCl_2$, $(NH_4)_2SO_4$ or $MgSO_4$] are ineffective and apparently inhibit the activity of the streptozyme preparation. No commercially available polysaccharases have any lytic effect on Mucor hyphae when used alone, however preparations of snail gut digestive enzymes [e.g. glusulase (Endo Lab. Inc.)] and Novozym 234 (Novo Industries) increase the yield of protoplasts obtained when used in combination with a low concentration of streptozyme. Since the snail gut digestive enzymes contain significant nuclease activity, use of Novozym 234 is preferred, and it can be routinely added at a final concentration of 1.0 mg/ml to the protoplast mixture. The addition of chitinase has no effect on protoplast yields, presumably due to the presence of significant chitinase activity in the streptozyme preparation itself (Sect. 4.2).

Treatment of the mesophilic species *M. circinelloides, M. racemosus,* and *M. rouxii* with streptozyme as described above results in the formation of high yields of stable protoplasts suitable for use in DNA-mediated transformation and protoplast fusion experiments. When using thermophilic species, growth at low temperature (25 °C) is critical for subsequent formation of stable protoplasts of homogenous morphology. Growth at 40 °C results in at least a three fold reduction in protoplast yields, the morphological homogeneity and stability of the protoplasts is decreased and extensive cell lysis is observed. In order to obtain synchronous growth and germination at the sub-optimal temperature of 25 °C the sporangiospores from thermophilic species are preincubated at 40 °C for 2 h. During such a short time at 40 °C the spores begin to swell but the emergence of germ tubes does not occur until some hours later, i.e. during incubation at the lower temperature.

The formation of protoplasts from *M. racemosus, M. miehei* and *M. pusillus* has also been described by Genthner and Borgia (1978) and Ohnuki et al. (1982).

4.2 Preparation of a Lytic Enzyme from *Streptomyces* sp. no. 6 (Streptozyme)

1. Streptomyces sp. no. 6 (obtained from D. Jones, the Macaulay Institute for Soil research, Aberdeen, Scotland) is grown in the medium of Skujins et al. (1965) containing chitin and chitosan as sole C-sources: 1.05 g $K_2HPO_4 \cdot 3H_2O$, 0.2 g KH_2PO_4, 0.5 g $(NH_4)_2SO_4$, 5.0 g chitin (Sigma), 1.0 g chitosan (Sigma) per litre distilled water. The chitosan is first dissolved in 30 ml 1% (w/v) acetic acid, neutralized and precipitated with NaOH, then washed extensively with water. The medium is adjusted to pH 7.0, autoclaved then completed with addition of 1 ml of each of 1 *M* MgSO4, 0.1 *M* $CaCl_2$, and 1 mg/ml $ZnSO_4$.
2. After 7–8 days of incubation of room temperature with shaking the cultures are centrifuged (5000 *g*, 30 min), the supernatant is brought to 95% saturation with ammonium sulphate (650 g/l) at 40 °C and left stirring gently overnight.
3. The precipitate is collected by centrifugation (5000 *g*, 60 min), resuspended in a minimal volume cold 20 m*M* sodium phosphate buffer pH 6.5 (approximately 20 ml per litre of production culture), and dialyzed extensively the same buffer at 4 °C. Dialysis against water should be avoided as it leads to extensive precipitation of protein and loss of activity.
4. After centrifugation to remove any insoluble material the preparation is lyophilised and stored at −20 °C. Up to 200 mg protein per litre of culture can be obtained with specific activities of up to 2.6 units chitosanase/mg protein. Chitosanase activity is determined according to Price and Storck (1975) using the indole method of Dische (1955) to assay for the release of hexosamine.

   The lytic enzyme(s) can also be induced by growth of the *Streptomyces* in media-containing *Mucor* cell walls, or by incubation in a medium containing 0.05 *M* D-glucosamine (Price and Storck 1975) but these procedures are more complicated and give streptozyme preparations lower in yield and specific activities.

4.3 Regeneration of Protoplasts

The protoplasts of mesophilic and thermophilic species regenerate at frequencies of up to 40% and 6% respectively when plated onto solid media in 10-ml overlays containing 1% agar, 0.5 *M* sorbitol and YNB or YPG medium, pH 3.2. Direct spreading of protoplast suspensions onto sorbitol containing agar with considerable surface moisture allows equivalently good regeneration but is not preferred, since cells clump and float on the plate surface, reducing the accuracy of colony counts and hindering the isolation of single colonies. Another disadvantage is that any drying of the plate surface significantly reduces the regeneration frequencies obtained.

4.4 Fusion of Protoplasts

Protoplast fusion provides a simple and rapid procedure for the formation of heterokaryons, which are useful in complementation studies. Nuclear fusion has not yet been observed in heterokaryons of *Mucor*, nevertheless protoplast fusion remains a potential means of empirical breeding and strain improvement.

1. Protoplasts are prepared as described in Sect. 4.1, washed once in 0.01 *M* sodium phosphate buffer pH 6.5, 0.5 *M* sorbitol, then resuspended in the same at a concentration of 5×10^6–5×10^7/ml.
2. 1 ml of protoplasts from two different auxotrophs are mixed then centrifuged at 400 g/5min. The pelleted protoplasts are resuspended in 5 ml of 45% (w/v) PEG 4000 containing 60 m*M* $CaCl_2$, and incubated at room temperature for 30 min.
3. 30 ml of 0.01 *M* sodium phosphate buffer pH 6.5, 0.5 *M* sorbitol is added, the protoplasts centrifuged, then respended in 10 ml of the same. 1 ml aliquots are plated as described above (Sect. 4.3) onto minimal (YNB) medium and complete (YPG) medium pH 3.2.

   The frequency of fusion is assessed by determination of the ratio of the number of colonies that arise as a result of nutritional complementation on minimal medium when compared to the number growing on complete medium. Using *M. circinelloides* we have obtained fusion frequencies of up to 1.7% or 4.6×10^{-3}/protoplast pair. Similar results have been reported by Lasker and Borgia (1980), while Ohnuki et al. (1982) have obtained intra-specific and inter-specific heterokaryon formation with *M. pusillus* and *M. miehei.*

5 Isolation of Marker Genes for Use in DNA-Mediated Transformation

Any efficient transformation system requires a method by which transformants can be selected from the total population of regenerating cells. This is accomplished by inclusion of a selectable marker gene on the transformation vector.

5.1 Selectable Marker Genes Coding for Resistance to Antibiotics

Antibiotic resistance genes have been isolated from a number of organisms and used successfully as marker genes in a variety of transformation systems (Table 1). Their utility in the transformation of *Mucor* species may be limited in some cases by the high level of natural resistance to the antibiotic concerned (MIG Roncero and LP Jepsen, unpublished results).

5.2 Selectable Marker Genes Coding for Metabolic Enzymes

Genes coding for metabolic enzymes have proved successful as selectable marker genes in the transformation of a number of species of filamentous fungi, including Mucor (van Heeswijck and Roncero 1984). Their use requires that the receptor cell is deficient in the activity to be selected for. This is normally achieved by using

Table 1. Sensitivity of *Mucor* to selected antibiotics

Antibiotic	Concentration	Growth[a] of		Examples of use in selection systems
		M. circinelloides	*M. Miehei*	
G 418	2 mg/ml	++	–	Selection of transformants of *N. crassa* at 2 mg/ml (Bull and Wooton 1984)
Neomycin	500 μg/ml 2 mg/ml	++ +/–	++ –	Selection of transformants of *Ustilago maydis* at 4 mg/ml (Banks 1983)
Hygromycin B	400 μg/ml 800 μg/ml	+ +/–	– –	Selection of transformants of *Saccharomyces cerevisiae* at 200 μg/ml (Gritz and Davies 1983)
Oligomycin	3 μg/ml 20 μg/ml	++ +/–	– –	Selection of transformants of *A. nidulans* at 3 μg/ml (Ward et al. 1986)
Benomyl	10 μg/ml	++	+	Selection of transformants of *N. crassa* at 0.5 μg/ml (Orbach et al. 1986)
Methotrexate	1 mg/ml	++	+	Selection of transformants of mouse fibroblasts at 0.4 μ*M* (O'Hare et al. 1981)

[a] ++: normal growth; +: slightly inhibited growth; +/–: strongly inhibited growth, but with some background and/or frequent appearance of resistant colonies; –: no growth.

auxotrophic mutants and selecting for transformation mediated reversion to prototrophy. In some cases, however, even wild-type strains may be deficient in a selectable activity, as with the amdS gene of *Aspergillus nidulans* (coding for an acetamidase) that can be used in the transformation of wild-type strains of *Aspergillus niger* and *Cochliobolus heterostrophus* (Kelly and Hynes 1985; Turgeon et al. 1985).

The isolation of homologous genes suitable for use as selectable markers can proceed by in vitro cloning techniques of cDNA synthesis, hybridization to oligonucleotide probes and hybridization to heterologous gene probes, or the screening of genomic libraries of specific mutations. Using DNA from a genomic library of *M. circinelloides* constructed in the yeast-*E. coli* shuttle vector YRp17, we have attempted to complement mutations in *E. coli* (argE3, his-4, ilvE12, ilv-318, leuB6, proA2, pro-82, thr-1, thr-300) and *S. cerevisiae* (arg4, his3A, his4, hom3, ilv2, ilv3, ilv5, leu2, lys1, ura4) with no success. This suggests the existence of barriers to the expression of at least some Mucor genes in these heterologous systems. A Leu^+ gene was consequently selected from this library by direct complementation of a *M. circinelloides* Leu^- mutant (van Heeswijck and Roncero 1984). The high frequency of transformation which can be obtained in *M. circi-*

nelloides (Sect. 3), and the ease with which unmodified plasmids can be recovered from the transformant DNA (Sect. 6.3) makes this approach the most effective for cloning this type of selectable marker in *Mucor* species.

5.3 Isolation and Enrichment of Auxotrophic Mutants of *Mucor*

The cloning of genes by direct complementation requires the isolation of mutants with suitable phenotypes. The following method has been developed for the isolation of auxotrophs of *Mucor*, specifically *M. circinelloides* CBS 277.49 and *M. miehei* CBS 370.65 (Roncero 1984).

1. Approximately 10^8 wild-type sporangiospores are spread on minimal (YNB) medium pH 4.5 supplemented with 2 mg/ml casamino acids, then immediately irradiated with a dose of ultraviolet light sufficient to give a survival rate of 0.5%–2%.
2. The plates are incubated at 28 °C (mesophiles) or 40 °C (thermophiles) for 7 days resulting in confluent mycelial growth and sporulation. The spores are harvested and stored at −20 °C until further use.
3. 25 ml of YNB medium pH 4.5 is inoculated with 10^8 sporangiospores and incubated with shaking at the appropriate temperature for 4–5 h until approximately 90% of the spore population has started to germinate.
4. The germlings are collected by centrifugation (6000 *g*, 10 min) and resuspended in an equal volume of YNB lacking ammonium sulphate (to avoid precipitation of N-glycosylpolifungin, NGP), and then NGP (dissolved in water at 10,000 units/ml) is added to give a final concentration of 1000 units/ml.
5. After 3 h of incubation the germlings are washed three times with water, diluted and plated on solidified YNB medium pH 3.2 supplemented with 2 mg/ml casamino acids to give 50–100 colonies per plate.
6. Plates are incubated for 7 days after which colonies are counted and replica plated onto YNB medium with and without casamino acids. The phenotype of any putative auxotrophs is determined by transferring them to a variety of supplemented minimal media, each one containing 4–5 different amino acids as described by Holliday (1956). The nutritional requirement of a particular mutant is identified by growth on one or more of these combinations.

The method described involves use of the polyene antibiotic N-glycosyl-polifungin (NGP) as a counter selective agent. Since germinating spores are more sensitive to NGP than ungerminated spores, a preferential killing of prototrophs occurs when mutagenized cultures are exposed to this antibiotic whilst growing in minimal medium.

NGP is synthesized by Prof. Borowski and collaborators at the Department of Drug Technology and Biochemistry of the Gdansk Institute of Technology, Poland (Falkowski et al. 1973). It has the advantage of being water-soluble unlike most other polyene antibiotics (e.g. nystatin), and has proved successful for selection of auxotrophic and temperature-sensitive mutants in *Aspergillus nidulans* (Bal et al. 1974) and *S. cerevisiae* (Polaina and Conde 1981) as well as *Mucor*. The method described above involves only one round of treatment with NGP and gives mutant yields of 4.4% for *M. circinelloides* and 0.36% for *M. miehei*. Additional (successive) cycles of counter selection might be advisable when specific and rare phenotypes of mutants are needed. Interestingly, while at least six different phenotypic classes of auxotrophs have been obtained from *M. circinelloides*, only lysine-requiring auxotrophs have been obtained from *M. miehei*, indicating perhaps aneuploidy or partial heterozygosity of the parental strain used. This cannot be explained by a relative higher resistance of the Lys^- mutants to NGP, since two such auxotrophs have been isolated without any exposure to this agent. Auxotrophic mutants of *M. racemosus, M. miehei,* and *M. pusillus* have also been isolated by Peters and Sypherd (1978) and Ohnuki et al. (1982).

6 Analysis of Transformants

Standard techniques of Southern and Northern hybridisation and Western immuno-blots can be used for the analysis of *Mucor* transformants and the proteins they manufacture. Methods for the isolation of RNA and DNA are described below. The recovery of plasmid DNA is of particular relevance to the cloning of *Mucor* genes by direct complementation of mutants.

6.1 Isolation of RNA

1. 100 ml liquid medium is inoculated with 10^6–10^7 spores at the appropriate growth temperature (Sect. 2).
2. The mycelium is harvested during the exponential growth phase by filtration on Monodur nylon mesh (pore size 100 μm), washed well with cold water and then as much liquid as possible is removed by suction.
3. The mycelium is ground with crushed glass in liquid nitrogen using a mortar and pestle. When the homogenate has a "soft-ice"-like appearance, 4 ml of extraction buffer (10 m*M* TRIS/HCl pH 8.0, 1 m*M* EDTA) per gram of mycelium is added. After thawing, SDS is added to a final concentration of 1% (w/v) and the homogenate is spun at 4000 *g* for 5 min (this step should be performed as fast as possible).
4. The supernatant is extracted twice with an equal volume of phenol-chloroform-isoamylalcohol (25:24:1), once with chloroform-isoamylalcohol (24:1) and the nucleic acid precipitated by addition of 0.1 vol 3 *M* sodium acetate pH 4.8 and 2.5 vol 96% ethanol.
5. After centrifugation at 10,000 *g* for 10 min, the nucleic acid pellet is dissolved in water (recentrifugation may be necessary in order to remove water-insoluble substances) and made 1 *M* in sodium chloride. The solution is frozen in liquid nitrogen and allowed to thaw in the refrigerator.
6. After centrifugation at 10,000 *g* for 10 min, the resulting RNA pellet is dissolved in water, ethanol precipitated, then redissolved in water and the OD_{260} and OD_{280} measured. Yields of about 0.1% of cell dry weight can be expected. RNA is kept at −20 °C in water or 67% ethanol, 0.08 *M* sodium acetate. If

RNA is to be used as template for in vitro translations, or a rare messenger is to be detected by Northern hybridization, affinity chromatography on poly(U) sepharose is necessary.

6.2 Isolation of DNA

The method used is a modification of that of Yelton et al. (1984) developed for *A. nidulans:*

1. 100 ml YNB medium pH 4.5 is inoculated with 10^7 sporangiospores and incubated for 16–20 h at the appropriate growth temperature (Sect. 2).
2. Cultures are collected using a Millipore filtration funnel, washed twice with cold water and then as much liquid as possible is removed by suction.
3. The mycelial mats are frozen in liquid nitrogen and ground to a fine powder in a mortar with a pestle. 10 ml of 50 m*M* EDTA pH 8.5, 0.2% SDS is added, mixed well for approximately 1 min at room temperature, then incubated at 68 °C for 15 min.
4. After cooling to room temperature the cell lysate is centrifuged (12,000 *g*, 15 min, room temp.), the supernatant removed with a pipette and incubated for 1 h on ice with addition of 600 µl 8 *M* potassium acetate pH 4.2.
5. After centrifugation (25,000 *g*, 15 min, 4 °C) the supernatant obtained is added to an equal volume of isopropanol at room temperature then recentrifuged (12,000 *g*, 30 min, room temp.).
6. The nucleic acid-containing pellet is resuspended in 4.0 ml 10 m*M* TRIS pH 7.6, 1 m*M* EDTA, then incubated for 1 h at 37 °C with 40 µl of RNase A (10 mg/ml, preheated at 90 °C for 10 min). If the DNA is to be digested with restriction enzymes prior to analysis it is purified by CsCl-density gradient centrifugation (p = 1.66 g/ml) otherwise it is simply extracted with an equal volume of phenol prior to ethanol precipitation. The average yield of DNA obtained is 50 µg/100 ml culture.

6.3 Recovery of Plasmid DNA

In order to recover plasmids from *Mucor* transformants, total DNA (isolated as described in Sect. 6.2) can be used in the transformation of *E. coli* with subsequent selection for a relevant marker gene (e.g. Ap^R or Tc^R–ampicillin or tetracycline resistance). The plasmids can then be isolated in an unmodified form from the cultures of the *E. coli* transformants by conventional methods. The *Mucor* DNA preparations in themselves inhibit the plasmid-mediated transformation of *E. coli* by 70–99%, depending on the method of DNA purification (van Heeswijck and Roncero 1984), nevertheless when using 1 µg of DNA per transformation, up to four *E. coli* transformants per µg can be obtained.

An enriched fraction of plasmid DNA can be isolated from total *Mucor* DNA by CsCl-density gradient centrifugation ($p = 1.55$ g/ml) in the presence of ethidium bromide (EtBr) (600 µg/ml). This fraction presumably contains covalently closed plasmid circles and is located in the region of the gradient below the flourescent band containing the chromosomal DNA. Its position can be determined by centrifuging parallel gradients containing the transforming plasmid isolated form *E. coli*. Using this enriched plasmid fraction, up to 700 *E. coli* transformants can be obtained from the *Mucor* DNA isolated from a 100-ml culture.

7 Genetic Manipulation of *Mucor* by DNA-Mediated Transformation Concluding Remarks

The availability of a DNA-mediated transformation system facilitates the cloning of *Mucor* genes, particularly where mutants are available whose phenotypes allow direct selection of specific genes from genomic libraries. Structural and/or regulatory regions of the cloned genes can then be analyzed, altered in vitro by use of techniques such as site-directed mutagenesis and then reintroduced into *Mucor*. Maintenance of the transforming DNA as an autonomously replicating plasmid facilitates determination of the effects of the altered DNA sequences since isogenic transformants can be readily isolated. Introduction and expression of the altered homologous genes, or alternatively of heterologous genes isolated from other organisms, should enable manipulation of the type and/or amount of specific substances produced by *Mucor* under any given set of fermentation conditions.

References

Adams PR, Deploey JJ (1976) Amylase production by *Mucor miehei* and *Mucor pusillus*. Mycologia 68:934–938

Arima K, Iwasaki S, Tamura G (1967) Milk clotting enzyme from microorganisms. Part I. Screening test and the identification of the potent fungus. Agric Biol Chem 31:540–545

Aunstrup K (1966) Fremgangsmåde til fremstilling af et mælkekoagulerende enzym. Patent 110031 Danmark Direktoratet for Patent -og Varemærkevæsenet København

Bal J, Balbin E, Pieniazek NJ (1974) Method for isolating auxotrophic mutants in *Aspergillus nidulans* using N-glycosyl-polifungin. J Gen Microbiol 84:111–116

Banks GR (1983) Transformation of *Ustilago maydis* by a plasmid containing yeast 2-micron DNA. Curr Genet 7:73–77

Bartnicki-Garcia S (1968) Cell wall chemistry, morphogenesis and taxonomy of fungi. Annu Rev Microbiol 22:88–108

Bartnicki-Garcia S, Nickerson WJ (1962) Induction of yeastlike development in *Mucor* by carbon dioxide. J Bacteriol 84:829–840

Bartnicki-Garcia S, Nelson N, Cota-Robles E (1968) Electron microscopy of spore germination and cell wall formation in *Mucor rouxii*. Arch Microbiol 63:242–255

Bernardinelli SE, Bottaro Castilla HR, Waehner RS, Muse J, Fraile ER (1983) Produccion y propiedades de la enzima coagulante de leche. Rev Argent Microbiol 15:95–104

Bottaro Castilla HR, Waehner RS, Meinardi CA, Zalazar CA, Fraile ER (1982) Produccion de un concentrados de protease acida de *Mucor bacilliformis*. Rev Argent Microbiol 14:115–118

Bull JH, Wooton JC (1984) Heavily methylated amplified DNA in transformants of *Neurospora crassa*. Nature 310:701–704

Case ME (1982) Transformation of *Neurospora crassa* utilising recombinant plasmid DNA. In: Hollaender A, De Moss RD, Kaplan S, Kanisky J, Savage D, Wolfe RS (eds) Genetic engineering of microorganisms for chemicals. Plenum, New York, p 87

Cooney DG, Emerson R (1964) Thermophilic fungi. Freeman, San Francisco

Davis RH, de Serres FJ (1970) Genetic and microbiological research techniques for *Neurospora crassa*. In: Colowick SP, Kaplan NO (eds) Methods Enzymol, vol XVIIA. Academic, New York, pp 79–143

Dhawale SS, Paietta JV, Marzluf GA (1984) A new rapid and efficient transformation procedure for *Neurospora*. Curr Genet 8:77–79

Dische Z (1955) New color reactions for determination of sugars in polysaccharides. In: Glick D (ed) Methods of biochemical analysis, vol 2. Interscience, New York, pp 313–358

Falkowski L, Bobrowski M, Buluk H, Bylec E, Cybulska B, Golik J, Kolodziejczyk P, Pawlak J, Rudowski A (1973) Antibiotics containing polyene macrolide groups. German patent 2239891

Foltmann B (1967) A review on prorennin and rennin. C R Trav Lab Carlsberg 35:143–231

Genthner FJ, Borgia PT (1978) Spheroplast fusion and heterokaryon formation in *Mucor racemosus*. J Bacteriol 134:349–352

Gray WD (1970) The use of fungi as food and in food processing. CRC, Chemical Rubber Company Butterworths, London

Gritz L, Davies J (1983) Plasmid encoded hygromycin B resistance. The sequence of hygromycin B phosphotransferase gene and its expression in *Eschericia coli* and *Saccharomyces cerevisiae*. Gene 25:179–188

Harboe MK (1984) Commercial aspects of aspartic proteases. In: Kostka V (ed) Aspartic proteinases and their inhibitors. De Gruyter, Berlin, pp 537–550

Hesseltine CW, Featherston CL, Lombard GL, Dowell VR Jr (1985) Anaerobic growth of molds isolated from fermentation starters used for food in Asian Countries. Mycologia 77:390–400

Higashio K, Yoshioka Y (1981) Preparation and some properties of crude enzymes from Mucor racemosus no. 50 and its mutants (studies on milk clotting enzyme from microorganisms, part II). Nippon Nogeikagaku Kaishi 55:573–581

Holliday R (1956) A new method for the identification of biochemical mutants of microorganisms. Nature 178:987

Ingold CT (1978) The biology of *Mucor* and its allies. Studies in biology, no 88. Edward Arnold

Ito H, Fukuda Y, Murata K, Kimura A (1983) Transformation of intact yeast cells treated with alkali cations. J Bacteriol 153:163–168

Jepsen LP (1986) Studies on the milk clotting enzymes produced by members of the genus *Mucor*. Cand scienthesis, Copenhagen University

Jones D, Bacon JSD, Farmer VC, Webley DM (1968) Lysis of cell walls of *Mucor ramannianus* Moller by a *Streptomyces* sp. J Microbiol Serol 83:173–182

Kelly MK, Hynes MJ (1985) Transformation of *Aspergillus niger* by the amdS gene of *Aspergillus nidulans*. EMBO J 4:475–479

Lasker BA, Borgia PT (1980) High frequency heterokaryon formation by *Mucor racemosus*. J Bacteriol 141:565–569

Moskowitz GR, Cassaigne R, West IR, Shen T, Feldman LI (1977) Hydrolysis of animal fat and vegetable oil with *Mucor miehei* esterase. Properties of the enzyme. J Agric Food Chem 25:1146–1150

O'Hara K, Benast C, Breathnach R (1981) Transformation of mouse fibroblasts to methotrexate resistance by a recombinant plasmid expressing a procaryotic dihydrofolate reductase. Proc Natl Acad Sci USA 78:1527–1531

Ohnuki T, Etoh Y, Beppu T (1982) Intraspecific and interspecific hybridisation of *Mucor pusillus* and *Mucor miehei* by protoplast fusion. Agric Biol Chem 46:451–458

Orbach MJ, Porro EB, Yanofsky C (1986) Cloning and characterisation of the gene for β-tubulin from a benomyl-resistant mutant of *Neurospora crassa* and its use as a dominant selectable marker. Mol Cell Biol 6:2452–2461

Peters J, Sypherd PS (1978) Enrichment of mutants of *Mucor racemosus* by differential freeze-killing. J Gen Microbiol 105:77–81

Polaina J, Conde J (1981) Use of the polyene antibiotic N-glycosyl polifungin in counter-selecting yeast mutants. Mutat Res 91:111–114

Price JS, Storck R (1975) Production, purification and characterisation of an extracellular chitosanase from *Streptomyces*. J Bacteriol 124:1574–1585

Roncero MIG (1984) Enrichment method for the isolation of auxotrophic mutants of *Mucor* using the polyene antibiotic N-glycosyl-polifungin. Carlsberg Res Commun 49:685–690

Schipper MAA (1978) 1. On certain species of *Mucor* with a key to all accepted species. 2. On the genera *Rhizomucor* and *Parasitella*. Stud Mycol 17

Skujin JJ, Potgieter, Alexander M (1965) Dissolution of fungal cells by a *Streptomycete chitinase* and β(1–3)glucanase. Arch Biochem Biophys 111:358–364

Somkuti GA (1974) Synthesis of cellulase by *Mucor pusillus* and *Mucor miehei*. J Gen Microbiol 81:1–6

Somkuti GA, Babel FJ (1968) Lipase activity of *Mucor pusillus*. Appl Microbiol 16:617–619

Turgeon BG, Garber RC, Yoder OC (1985) Transformation of the fungal maize pathogen *Cochliobolus heterostrophus* using the *Aspergillus nidulans* amdS gene. Mol Gen Genet 201:450–453

Van Heeswijck R (1984) The formation of protoplasts from *Mucor* species. Carlsberg Res Commun 49:597–609

Van Heeswijck R (1986) Autonomous replication of plasmids in *Mucor* transformants. Carlsberg Res Commun 51:433–443

Van Heeswijck R, Roncero MIG (1984) High frequency transformation of *Mucor* with recombinant plasmid DNA. Carlsberg Res Commun 49:691–702

Ward M, Wilkinson B, Turner G (1986) Transformation of *Aspergillus nidulans* with a cloned oligomycin-resistant ATP synthase subunit 9 gene. Mol Gen Genet 202:265–270

Yelton MM, Hamer JE, Timberlake WE (1984) Transformation of *Aspergillus nidulans* by using a trpC plasmid. Proc Natl Acad Sci USA 81:1470–1474

Final Product

Malt and Hop Flavanoids in Pilsner Beer

J. A. Delcour

1 Importance

In brewing research there has been a considerable degree of interest in beer flavanoids. These substances (see Fig. 1) do, in fact, contribute to beer haze formation (Gardner and McGuinness 1977; Dadic 1984) by reactions and interactions with protein material.

Earlier feelings were that these compounds would contribute to the flavour of the beverage but more recent findings point out that they are of no impact on the flavour (Erdal et al. 1980, 1983; Erdal 1986; Delcour et al. 1984, 1985a, b; von Wettstein et al. 1985) and flavour stability (Erdal et al. 1983; Erdal 1986).

These substances originate from the raw materials (namely barley and hops) and several of them have been identified or synthesized in the free phenolic form (Fletcher et al. 1977; Eastmond 1974; Foo and Porter 1983; Hemingway and Foo 1983; Delcour et al. 1983, 1985c; Delcour and Vercruysse 1986). Beer also contains flavonols and flavonol glycosides, but their concentration is much lower and their impact upon beer haze is much less pronounced than that of the flavanoids (Vancraenenbroeck et al. 1983).

In view of the above, progress has been made towards the identification of these substances in beer and its raw materials as well as towards their quantitative estimation in beer.

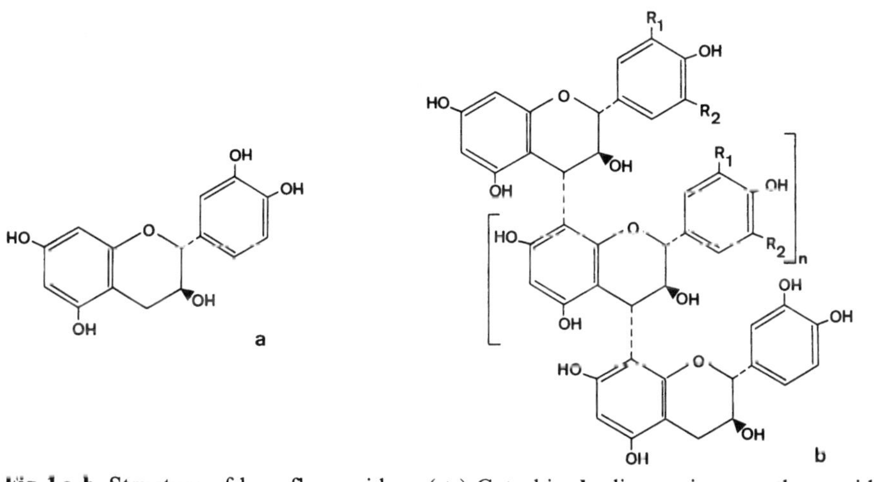

Fig. 1 a, b. Structure of beer flavanoids. **a** (+)-Catechin; **b** oligomeric proanthocyanidins, $R_1 = R_2 = H$ propelargonidin; $R_1 = OH$, $R_2 = H$ procyanidin; $R_1 = R_2 = OH$ prodelphinidin; $n = 0$ dimer; $n = 1$ trimer

2 Identification of Flavanoids in Barley, Malt, Hops and Beer

2.1 Barley and Malt Flavanoids

Table 1 lists the flavanoids that have been identified in both barley and malt by comparison of their chromatographic behaviour with that of the reference substances and by ^{1}H nmr of their peracetates. The suspected presence of procyanidin B6 in malt (Jerumanis 1979) has so far not been confirmed. At the dimer and trimer level in barley and malt only these flavanoids have been described that have a 2,3-trans stereochemistry in both their "upper" and "lower" units, and prodelphinidins as well as procyanidins are present in these raw materials.

Table 1. Identification of barley (B) and malt (M) flavanoids

Flavanoid	Identification	References	B/M
(+)-Catechin	Nylon 66	1	B
	Sephadex G-25 S	2, 3	B
	RP-HPLC[a]	4,5	B
(−)-Epicatechin	Nylon-66	1	B
(+)-Gallocatechin	Nylon-66	1	B
(−)-Epigallocatechin	Nylon-66	1	B
Procyanidin B3	Sephadex G-25 S	3	B
	RP-HPLC	4, 5	B
	^{1}H nmr[b]	6–9	B
Prodelphinidin B3	RP-HPLC	4, 5	B
	^{1}H nmr	6–9	B
Procyanidin C2	^{1}H nmr	6–9	B
(4,8:4,8)-all-trans-(+)-Gallocatechin-bi-[(+)-catechin]	^{1}H nmr	6–9	B
(4,8:4,8)-all-trans-bi-[(+)-Gallocatechin]-(+)-catechin	^{1}H nmr	6–9	B
(4,8:4,8)-all-trans-(+)-Catechin-(+)-gallocatechin-(+)-catechin	^{1}H nmr	8, 9	B
(+)-Catechin	Sephadex G-25 S	3	M
Procyanidin B3	Sephadex G-25 S	3	M
	RP-HPLC	10	M
Procyanidin B6	RP-HPLC	10	M
Procyanidin C2	RP-HPLC	10	M

[a] RP-HPLC = reversed-phase high-performance liquid chromatography.
[b] ^{1}H nmr = proton nuclear magnetic resonance.
[1] Vancraenenbroeck et al. (1970); [2] Vancraenenbroeck et al. (1974); [3] McMurrough (1979); [4] McMurrough (1981); [5] McMurrough et al. (1983); [6] Mulkay et al. (1981a); [7] Mulkay et al. (1981b); [8] Outtrup (1981); [9] Outtrup and Schaumburg (1981); [10] Jerumanis (1979).

2.2 Hop Flavanoids

The reported identifications of hop flavanoids are summarised in Table 2. Jerumanis (1979) concludes the presence of prodelphinidins in hops. McMurrough, who initially believed that hops contain no prodelphinidins (1979), later (1981) found that the ratio of cyanidin to delphinidin in the hydrolysate of whole hops varies from 1.2 to 6.2 and averaged 2.5, the delphinidin most probably originating from the polymeric flavanol fraction. The dimeric and trimeric procyanidins in hops consist of (+)-catechin as well as of (−)-epicatechin units.

Table 2. Identification of hop flavanoids

Flavanoid	Identification	References
(+)-Catechin/(−)-Epicatechin	Sephadex G-25 S	1–4
	RP-HPLC	5, 6
(+)-Gallocatechin	Sephadex G-25 S	
	Nylon 66	1–3
Procyanidin B1	^{1}H nmr	1
	Sephadex G-25 S	4
	RP-HPLC	6
Procyanidin B2	Sephadex G-25 S	4
	RP-HPLC	5, 6
Procyanidin B3/procyanidin B4	^{1}H nmr	1
	Sephadex G-25 S	4
	RP-HPLC	5, 6
Procyanidin B6	RP-HPLC	5
Procyanidin C1/procyanidin C2	RP-HPLC	5

[1] Gorissen (1969); [2] Vancraenenbroeck et al. (1969); [3] Vancraenenbroeck et al. (1974); [4] McMurrough (1979); [5] McMurrough (1981); [6] Jerumanis (1979).

2.3 Beer Flavanoids

Literature data comprising the evidence for the presence of individual flavanoids in beer is presented in Table 3. Most frequently encountered is the dimeric procyanidin B3. The dimeric prodelphinidin B9 is believed to arise from isomerization of prodelphinidin B3 (Delcour and Tuytens 1984).

Since no flavanoids with (−)-epicatechin units have been detected in barley or malt, the procyanidins B1, B2, B4, B5, A2, and C1 must originate from hops. In view of the data listed above (Tables 1 and 2) prodelphinidin B3 most probably stems from barley.

Table 3. Identification of beer flavanoids

Flavanoid	Identification	References
(+)-Catechin	PC[a]	1
	GC[a]	2–5
	TLC[a]	6
	RP-HPLC	7, 8
(−)-Epicatechin	PC	1
	GC	4, 5
	TLC	6
	RP-HPLC	7–9
(+)-Gallocatechin	GC	3–5
	Nylon 66	6
(−)-Epigallocatechin	Nylon 66	6
Procyanidin B1	Sephadex G-25 S	6
Procyanidin B2	GC	5
	Sephadex G-25 S	6
	RP-HPLC	7
Procyanidin B3	GC	2, 5, 10
	Sephadex G-25 S	6
	RP-HPLC	7–9, 11
	^{1}H nmr	12
Procyanidin B4	Sephadex G-25 S	6
	RP-HPLC	7
Procyanidin B5/procyanidin B6/ procyanidin A2	GC	5
Prodelphinidin B3	RP-HPLC	9
	^{1}H nmr	12
Prodelphinidin B9	^{1}H nmr	12
Procyanidin C1/procyanidin C2	RP-HPLC	7

[a] PC = paper chromatography, GC = gas chromatography, TLC = thin layer chromatography.
[1] Gramshaw (1967); [2] McGuinness et al. (1975); [3] Gracey and Barker (1976a,b); [4] Drawert et al. (1977); [5] Vancraenenbroeck et al. (1979); [6] Vancraenenbroeck et al. (1977); [7] Jerumanis (1979); [8] McMurrough et al. (1983); [9] Kirby and Wheeler (1980); [10] Eastmond (1974); [11] Delcour et al. (1981); [12] Delcour and Tuytens (1984).

3 Analysis of Beer Flavanoids

Several assays have been developed for the quantitative estimation of polyphenolics in beer. Usually, four different approaches have been chosen.

In a first series of methods the attempt has been made to estimate "anthocyanogens" or the so-called proanthocyanidins by reaction of a phenolic beer extract with an alcoholic hydrochloric acid solution and measurement of the E_{545}-value of the resulting solutions under standardized conditions. Secondly, the reaction of ferric citrate with o-di- and o-trihydroxy phenolic moieties under alkaline conditions has been accepted by the European Brewery Convention (Analytica 1975). Thirdly, Singleton and Rossi (1965) have developed a method based upon colourimetry of total phenolics with phosphomolybdic-phosphotungstic acid and they have applied this method to wine. This procedure and a procedure to estimate the nonflavonoid phenols in the same beverage (Kramling and Singleton 1969; Singleton 1974) are discussed in a chapter in this series by Dr. Singleton but at our laboratory have proved to be less useful in the case of beer (Delcour and Janssens de Varebeke 1985). Finally and more recently, a colourimetric assay (based upon the reaction of their A-rings with the chromogen p-dimethylaminocinnamaldehyde) has been developed for flavanoids in beer (Delcour and Janssens de Varebeke 1985).

We here report on all of these analysis methods except those developed by Dr. Singleton.

3.1 Analysis of Beer Anthocyanogens

3.1.1 Introduction

The term anthocyanogens denotes all substances in wort and beer that yield red pigments (anthocyanidins) on heating with an alcoholic hydrochloric acid solution in the presence of molecular oxygen (see Fig. 2). The term covers all dimeric and higher oligomeric flavanoids. In current polyphenol terminology these substances are called proanthocyanidins. Since in the anthocyanogen assay the monomeric flavanoids such as (+)-catechin and (−)-epicatechin do not react to cyanidin, they are not included in the analysis. It should moreover be taken into account that the yield of cyanidin released is clearly dependent upon the degree

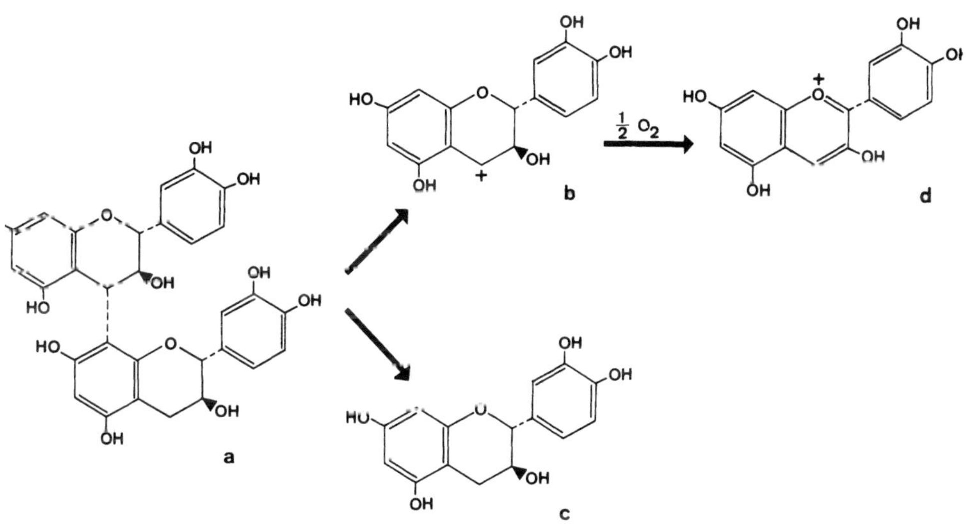

Fig. 2a–d. The anthocyanogen reaction. **a** Procyanidin B3; **b** flavanyl carbenium ion; **c** (+)-catechin; **d** cyanidin

of polymerization of the proanthocyanidins. It is also obvious that the actual analytical result will be dependent upon the proanthocyanidin used for the preparation of the calibration plot (Drawert et al. 1979).

In order to also estimate the catechin fraction in beer McFarlane et al. (1963) have proposed two colourimetric assays. Their method has been modified by Narziss and Kessler (1971). Dadic (1971) has estimated anthocyanogens and catechins separately in beer by binary analysis of the absorbancies at two wavelengths (455 and 545 nm) of the coloured pigments derived both from catechins and anthocyanogens on heating with alcoholic hydrochloric acid. These methods have all failed to gain widespread acceptance, mainly due to the amount of labour involved in their application.

The methods of Harris and Ricketts (1959) and Jerumanis (1973) are easier to be used for routine analysis. At this laboratory we currently apply the procedure by Jerumanis (1973).

3.1.2 Method (Jerumanis 1973)

Reagents. All reagents are of analytical grade unless otherwise specified.
- n-butanol;
- polyclar AT (PVPP), Gaf Chemicals;
- N-methylpyrrolidone: to 1.0 l of the commercial product are added 5 drops of a concentrated hydrochloric acid solution. The reagent is subsequently purified by vacuum distillation at about 150 °C;
- chromogen solution: 15.4 mg of $FeSO_4 \cdot 7\,H_2O$ is dissolved in 90 ml of a concentrated hydrochloric acid solution and the volume is adjusted to 100.0 ml with distilled water. The resulting solution is carefully mixed into 200 ml of purified N-methylpyrrolidone while cooling under tap water to avoid colouration of the reagent. The reagent is then stored at ambient temperature during 8 h and afterwards at 4 °C.

Procedure. Polyclar AT (100 mg) is weighed in a graduated glass centrifuge tube (15 ml). Degassed beer[1] (2.0 ml) is added and the volume adjusted to 8.0 ml with distilled water.

The contents of the tube are now thoroughly mixed by shaking (1 min). The mixture is further diluted to 15.0 ml with distilled water and centrifuged so that all PVPP precipitates. The supernatant is then discarded and the adsorbent washed with 10.0 ml of distilled water by renewed shaking (30 s). The volume is brought to 15.0 ml and the PVPP is again precipitated by centrifugation. After decantation, the anthocyanidin pigments are formed as follows: add to the PVPP 7.0 ml of the chromogen reagent and suspend the polymer with a small glass rod. The tube is then heated in a boiling water bath for 30 min. The anthocyanidin solution is then cooled and diluted to 10.0 ml with n-butanol. After mixing, the contents are filtered on a paper filter and their E_{550s}-values measured in a 1-cm cell.

At the same time a blank is prepared by performing all the above manipulations with water instead of beer, and its E_{550b}-value is recorded as well.

[1] Do not degas the beer by filtration since paper adsorbs polyphenolic substances.

We find it useful to obtain the E_{550s}-value as the average of two separate estimations.

The concentration of anthocyanogens in the beer can then be calculated as

$$\text{Anthocyanogens (mg/l)} = (E_{550s} - E_{550b}) \times 162.5.$$

Analytical data for 19 Belgian Pilsner beers by this method varied between 2.0 and 60.0 mg/l with an average at 33.8 mg/l (Delcour and Janssens de Varebeke 1985).

3.2 Analysis of Total Polyphenolics

3.2.1 Introduction

Jerumanis (1973) has described two methods for the estimation of phenolics in beer. In the assays the colour produced in the reaction of the polyphenols with ammonium ferric citrate under alkaline conditions is measured.

Structural requirements for the reaction to occur are that the phenolic compounds have at least two o-hydroxyl groups. The molar colour yield of a prodelphinidin (having three o-hydroxyl groups on the B-ring) will be higher than that of the corresponding procyanidin. All results are expressed as (+)-catechin equivalents.

Since in most oligomeric beer proanthocyanidins both the "upper" and the "lower" flavan units have B-rings that will react with the chromogen reagent, the molar yield will not only depend on the hydroxylation pattern of the particular B-rings but also on the degree of polymerization of the flavanoids.

In a first assay the author (Jerumanis 1973) measures the beer polyphenolic content as the difference in colour yield obtained with beer before and after adsorption of the polyphenolics on Polyclar AT. In this way interferences in the colour development that are due to melanoidins, reductones, cysteine and ascorbic acid are avoided.

The procedure is, however, tedious and the European Brewery Convention (Analytica 1975) has therefore accepted a simplified version of the method, although interference by the named compounds can in fact occur in dark beers (Jerumanis 1973).

3.2.2 Method

Reagents. All reagents are of analytical grade unless otherwise specified.

- Carboxymethylcellulose (CMC/EDTA): a 1% solution of the sodium salt of carboxymethylcellulose (low viscosity, BDH 27649) with the addition of 0.2% EDTA. Each month a fresh solution should be prepared by slowly adding 10 g of pure carboxymethylcellulose (sodium salt) and 2 g of pure disodium ethylene diamine tetra acetate to about 500 ml of distilled water with stirring. When all the solid material has dispersed and dissolved (1–3 h standing, stirring or maceration will speed up the process) transfer quantitatively to a 1,000-ml measuring flask and bring to volume with distilled water. Where necessary, the solution should be clarified by centrifuging;

- ferric reagent: the solution must be completely clear, so that a fresh solution of 3.5% green ammonium ferric citrate (16% Fe) (BDH 27882) should be prepared each week (or more frequently);
- ammonia reagent: pure concentrated ammonia diluted with two volumes of distilled water.

Procedure. The beer is degassed by shaking.[1] Hazy beers are clarified by centrifuging, not by filtration. The solutions to be measured in the spectrophotometer should be visually checked for clarity before reading. It is essential in the test and in the blanks to mix thoroughly after the addition of each reagent.

Main test. 10 ml of beer and 8 ml of CMC/EDTA reagent are pipetted into a 25-ml graduated flask with a ground glass stopper and are well mixed; then 0.5 ml of ammonia reagent is added and the solution is thoroughly mixed again. After making up to 25 ml the solution is mixed once more. After 10 min the extinction in a 10 mm cell is measured at 600 nm against a blank prepared as below.

Blank. 10 ml of the beer and 8 ml of CMC/EDTA reagent are mixed in a 25-ml graduated flask and, after adding 0.5 ml of ammonia reagent, mixed again. After this the solution is brought to 25 ml with distilled water, well mixed again and stood for 10 min before measurement.

Dilution of Samples. When it is desired to dilute the sample for analysis, a 50-ml graduated flask is employed and, after the addition of the CMC/EDTA and ferric citrate reagents, 25 ml of distilled water is added, well mixed, and followed by the ammonia reagent.

The concentration of polyphenols in mg/l (+)-catechin equivalents is given by

$$\text{Total polyphenolics (mg/l)} = E_{600} \times 820.$$

In the case of 19 Belgian Pilsner beers, analytical data by this method varied between 204 and 47 mg/l with an average of 138 mg/l (Delcour and Janssens de Varebeke 1985).

3.3 Analysis of Flavanoids

3.3.1 Introduction

The p-dimethylaminocinnamaldehyde reagent has been used by McMurrough to monitor chromatographic separations of flavanoids from barley, hops (McMurrough 1979; McMurrough and McDowell 1978; McMurrough et al. 1983), worts and beers (McMurrough 1979; McMurrough et al. 1983).

It has been established by McMurrough and McDowell (1978) that phenolics react with the chromogen if they contain m-oriented di- or tri-hydroxy substituted benzene rings. The structural requirements for reaction by flavanoids are that they have a single bond at the 2,3-position. The method will thus not estimate the

[1] Do not degas the beer by filtration since paper adsorbs polyphenolic substances.

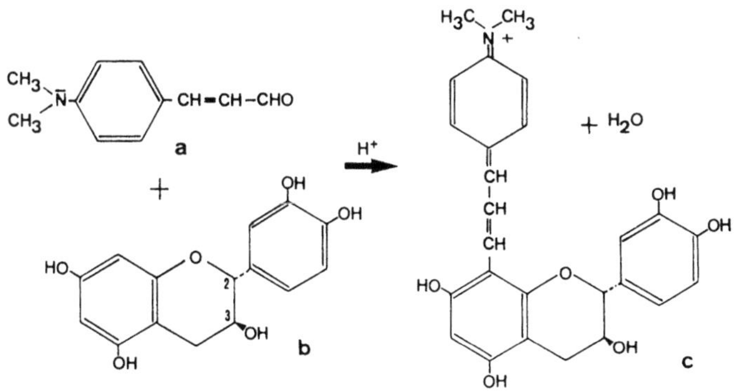

Fig. 3 a–c. Condensation reaction of p-dimethylaminocinnamaldehyde (**a**) with (+)-catechin (**b**) yielding the coloured pigment (**c**)

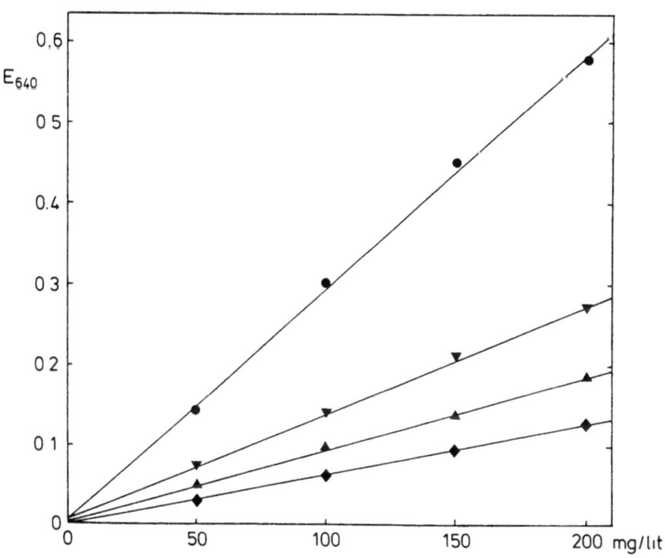

Fig. 4. p-Dimethylaminocinnamaldehyde colour yield of different solutions of phenolics in 5% ethanol under the assay conditions. (●) (+)-Catechin; (▼) procyanidin B4; (▲) procyanidin B3; (◆) a mixture of trimeric and tetrameric procyanidins mainly containing procyanidin C2

flavonols and flavonol glycosides present in beer since the nucleophilicity of their A-rings is reduced to a great extent.

By analogy with the reaction of flavanoids with vanillin (Ribéreau-Gayon 1968), the condensation with p-dimethylaminocinnamaldehyde can be visualized as shown in Fig. 3. Delcour and Janssens de Varebeke (1985) have developed a colourimetric assay for flavanoids in beer based upon the application of this chromogen.

As shown in Fig. 4, the colour yield by this method varies for different phenolics. Not only the degree of polymerization affects the response of flavanoids to the chromogen but also the 2,3-stereochemistry of the heterocyclic rings. Thus, a 2,3-cis configuration in one of the heterocyclic rings at the dimer level (procyanidin B4, "lower" unit) results in a greater reactivity than that of the all-trans dimer (procyanidin B3). This phenomenon is also observed at the monomer level, where (−)-epicatechin gives a greater colour yield than (+)-catechin.

The method described here has been published before (Delcour and Janssens de Varebeke 1985) and has recently been accepted by the European Brewery Convention Analysis Committee as one of the Analytica (1987) methods.

3.3.2 Method

Reagents. All reagents are of analytical grade unless otherwise specified.

- chromogen reagent: p-dimethylaminocinnamaldehyde (Janssen Chimica, 1,000 g) is dissolved in a previously cooled mixture of 250 ml concentrated hydrochloric acid and 750 ml methanol and the solution is diluted to a final volume of 1 l with methanol. Fresh reagent is prepared once a week and the reagent is stored in the dark (McMurrough et al. 1983).

Procedure. Degassed[1] beer (10.00 ml) is diluted to 100 ml with water and 1.00 ml of this solution is pipetted into a test tube. Chromogen reagent (5.00 ml) is added, and the absorbance is read at 640 nm after 10 min (E_{640s}). A blank is prepared by mixing 5.00 ml of the chromogen reagent with 1.00 ml water and measurement of the E_{640b} value.

The content of flavanoids in the beer can be read from the calibration plot (see Fig. 4) or from the following equation:

$$\text{Flavanoids mg/l} = (E_{640s} - E_{640b}) \times 335.$$

Ascorbic acid, cysteine and beer proteins do not interfere in this method. Since in this assay beer is analyzed that has been diluted by a factor 10, no corrections are necessary for beer colour. Although we initially considered such influences they proved to be negligible.

3.4 Comparison of the Methods

Figure 5 shows how the results of p-dimethylaminocinnamaldehyde assay compare with those of the Jerumanis (1973) methods for total polyphenolics and anthocyanogens.

In the case of 19 commercial Belgian Pilsner beers linear regression equations relating the concentration of phenolics measured by the Analytica (1975) or by the anthocyanogen assay (Jerumanis 1973) to the concentration of flavanoids estimated with the p-dimethylaminocinnamaldehyde reagent were as follows (Del-

[1] Do not degas the beer by filtration since paper adsorbs polyphenolic substances.

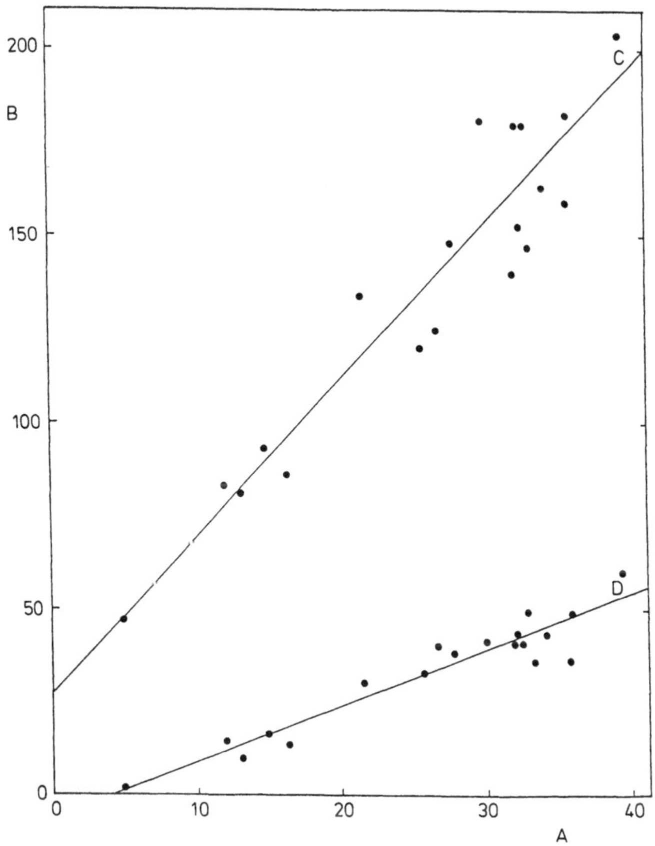

Fig. 5. Scatter plot of the results of polyphenolics estimations for 19 different Belgian Pilsner beers. *A* p-dimethylaminocinnamaldehyde reactive substances (catechin equivalents, mg/l and *B* corresponding results for total polyphenolics (EBC method, *C* catechin equivalents, mg/l and anthocyanogens (*D* leucocyanidin equivalents, mg/l)

cour and Janssens de Varebeke 1985)

$$\text{Total polyphenolics (mg/l)} = 28 + 4.194 \times \text{flavanoids (mg/l)} \quad r^2 = 0.903$$

$$\text{Anthocyanogens (mg/l)} = -5.9 + 1.512 \times \text{flavanoids (mg/l)} \quad r^2 = 0.916$$

Considering the totally different reactions involved we were quite surprised to find a reasonable linear correlation between the results by the different methods.

The p-dimethylaminocinnamaldehyde method yields analytical results that are more in line with recent high-performance liquid chromatography estimations for flavanoids in beer (Kowaka et al. 1983; McMurrough et al. 1983) than are the results by the classic Analytica (1975) method.

Contrary to this procedure, the flavanoid assay does not measure the flavonols, but this is rather an advantage than a disadvantage if one wishes to analyze for those substances responsible for beer haze formation.

Of further interest is that proanthocyanidin-free beers brewed from proanthocyanidin-free malt and tannin-free hop extract yield negligible concentrations of anthocyanogens and flavanoids, but still about 50 mg/l of total polyphenolics when analyzed according to the above methods.

4 Significance of the Analytical Results

Much effort has been put into finding a relationship between the level of total polyphenolics, anthocyanogens or flavanoids in beer and the liability of different beer types to form haze. However, little success has been achieved in this respect since for different beers the haze development depends not only on the concentration of the polyphenolics but also on the beer proteins, molecular oxygen and the level of reducing substances present.

Better results have been obtained when comparing the liability to haze formation of one particular beer after treatment with different levels of polyphenol adsorbents such as PVPP and the resulting analytical data for total polyphenolics and anthocyanogens. From this it follows that analysis for phenolics is of importance for the prediction of the colloidal stability of a particular beer.

As already stated, the molar colour yield of different polyphenolics by the above assays is dependent upon such factors as their hydroxylation pattern (B-rings) and their degree of polymerization. Since, moreover, the haze potential of proanthocyanidins varies with the same structural characteristics, it has been attempted to relate the levels of specific proanthocyanidins to the haze stability of the resulting beers. Here again the main conclusion is that for one particular beer this can be done, but that there is no unambiguous relationship between the haze stability of different beers and their content of specific proanthocyanidins (Vancraenenbroeck et al. 1983).

5 Analysis of beer Flavanoids by High-Performance Liquid Chromatography

The future of the analysis of flavanoids from hops and malt in beer depends upon the development and the general application of means to estimate specific compounds. Due to the fact that polyphenolics are only minor constituents in beer, it will always be necessary to first remove them from the beverage by extraction or adsorption. From there on can analyze for these substances by gas chromatography or by high-performance liquid chromatography.

5.1 Gas Chromatography

In the case of gas chromatography of flavanoids, the main problems are the need for a complete and reproducible derivatization of the substances, the high temperatures needed in order to obtain chromatographic separations and the unknown stability of the derivatives at high temperatures.

Several methods have been published in the literature (McGuinness et al. 1975; Gracey and Barker 1976a, b; Vancraenenbroeck et al. 1979), but the applications of high-performance liquid chromatography have made the use of gas chromatography outdated.

5.2 High Performance Liquid Chromatography

Jerumanis (1979) has obtained satisfactory separations of beer proanthocyanidins with a reversed-phase gradient elution high-performance liquid chromatography system. The isolation of the phenolic extract by the author is by adsorption of the compounds onto polyamide.

Since it has been shown by Van Roey (1979) that the desorption of dimeric and trimeric procyanidins from polyamide is incomplete, and since the higher the degree of polymerization the more the polyphenolics are irreversibly adsorbed some authors (Kirby and Wheeler 1980; Delcour et al. 1981) have proposed to apply Sephadex LH-20 as adsorbent for polyphenolics from beer. This technique had been applied earlier by Lea (1978) and Lea et al. (1979) for the removal of phenolics from cider and wine.

The use of Sephadex LH-20 has the advantage that the gel can adsorb relatively small quantities of polyphenolic material from relatively large volumes of aqueous solutions and yet permit quantitative recovery of such adsorbed material by subsequent elution with 60% acetone. Furthermore, no depolymerization of proanthocyanidins is observed on this gel, as seems to be the case on Nylon 66 (Jerumanis 1979).

We here propose to apply the Sephadex LH-20 adsorption and desorption technique of Delcour et al. (1981) and to perform the separation of the resulting flavanoids as outlined by Jerumanis (1979).

5.2.1 Isolation of Proanthocyanidins from Beer

Samples of beer (0.5 l) are degassed and percolated through a column of Sephadex LH-20 (195 × 50 mm, previously swollen in water) at a flow rate of 250 ml/h. The column is eluted with distilled water (2.25 l) until the E_{280}-value is less than 0.3 when the elution of most of the protein fractions is finished. An acetone-water mixture (60%, 3.0 l) is then used as the eluting solvent and recovered. The acetone is removed by vacuum evaporation at 30 °C. The resulting solutions are frozen with liquid air and subjected to lyophilization. The powder obtained is then submitted to high-performance liquid chromatography.

5.2.2 High Performance Liquid Chromatography

Jerumanis (1979) separates beer polyphenolics in a gradient of acetic acid in water on a reversed-phase Sil C18 HL 10 μ column (40 × 4 mm, RSL, Belgium). A pre-column (10 × 2.1 mm, CoPell ODS, Chrompack, The Netherlands) is used in the method. The elution solvents are twice-distilled water and an acetic acid solution (10%).

The solutions are degassed by filtration on a Gelman 0.45 μ filter. The gradient is linear from 2% acetic acid to 10% acetic acid in 100 min. After use the columns are washed in a 10% solution of acetic acid in methanol. The flow rate is 1.0 ml/min and detection is by measurement of the E_{280}-values. Under these chromatographic conditions the elution order of some known beer flavanoids is procyanidin B3, procyanidin C2, procyanidin B4, (+)-catechin, procyanidin B2, (−)-epicatechin and procyanidin Cl.

For quantitative work, all polyphenol absorption values can be related to (+)-catechin, since on a unit weight basis the ultraviolet absorbance at 280 nm of procyanidins is identical, giving a molar extinction for the whole series of E = 12.5 *M* where *M* is the molecular weight (Lea 1978). In the case of prodelphinidins this approach will, however, lead to slightly erroneous values, in which case it seems appropriate to apply the procedure by Jerumanis (1985).

Acknowledgements. Dr. J. Jerumanis and Dr. J. S. Pierce, Editor of the Journal of the Institute of Brewing, are gratefully thanked for their permission to use data published earlier.

References

Analytica EBC (1975) Schweiz Brau Rundsch

Analytica EBC (1987) Schweiz Brau Rundsch

Dadic M (1971) Analytical methods for polyphenols in brewing. Part II. A new method for determination of anthocyanogens and catechins (tanninogens) in wort and beer. Proc Am Soc Brew Chem, pp 159–170

Dadic M (1984) Beer stability–a key to success in brewing. The award of merit lecture. Mast Brew Assoc Am Tech Q 21:9–26

Delcour JA, Janssens de Varebeke D (1985) A new colourimetric assay for flavanoids in Pilsner beers. J Inst Brew 91:37–40

Delcour JA, Tuytens GM (1984) Structure elucidation of three dimeric proanthocyanidins isolated from a commercial Belgian Pilsner beer. J Inst Brew 90:153–161

Delcour JA, Vercruysse SAR (1986) Direct synthesis of the barley proanthocyanidins prodelphinidin B3, prodelphinidin C2 and two trimeric proanthocyanidins with a mixed prodelphindin-procyanidin stereochemistry. J Inst Brew 92:244–249

Delcour JA, Verhulst LL, Geys R, Deliever H, Dondeyne P (1981) Trimeric and oligomeric flavanoids in beer: a joint application of Sephadex LH-20 and reversed-phase gradient elution high-performance liquid chromatography. J Inst Brew 87:391–393

Delcour JA, Ferreira D, Roux DG (1983) Synthesis of condensed tannins. Part 9. The condensation sequence of leucocyanidin with (+)-catechin and with the resultant procyanidins. J Chem Soc Perkin I, pp 1711–1717

Delcour JA, Vandenberghe MM, Dondeyne P, Schrevens EL, Wijnhoven J, Moerman E (1984) Flavour and haze stability differences in unhopped and hopped all-malt Pilsner beers brewed with proanthocyanidin-free and with regular malt. J Inst Brew 90:67–72

Delcour JA, Schoeters MM, Meysman EW, Dondeyne P, Schrevens EL, Wijnhoven J, Moerman E (1985a) Flavour and haze stability differences due to hop tannins in all-malt Pilsner beers brewed with proanthocyanidin-free malt. J Inst Brew 91:88–92

Delcour JA, Schoeters MM, Dondeyne P, Schrevens EL, Wijnhoven J, Moerman E (1985b) Flavour and haze stability differences due to hop and malt tannins in all-malt Pilsner beers brewed with proanthocyanidin-free and with regular malt. J Inst Brew 91:88–92

Delcour JA, Serneels EJ, Ferreira D, Roux DG (1985c) Synthesis of condensed tannins. Part 13. The first 2,3-trans-3,4cis-procyandins: sequence of units in a "trimer" of mixed stereochemistry. J Chem Soc Perkin I, pp 669–676

Drawert F, Leupold G, Lessung V (1977) Gaschromatographische Analyse von phenolischen Verbindungen im Bier. Brauwissenschaft 30:13–18

Drawert F, Graf A, Bhiwapurker S, Leupold G, Beier J (1979) Zur Problematik der Analyse phenolischer Verbindungen. Proc Eur Brew Conv Congr, Berl, pp 335–340

Eastmond R (1974) The separation and identification of a dimer of catechin occurring in beer. J Inst Brew 80:188–192

Erdal K (1986) Proanthocyanidin-free barley–malting and brewing. J Inst Brew 92:220–224

Erdal K, Ahrenst-Larsen B, Jende-Strid B (1980) Use of proanthocyanidin-free barley in beer brewing. In: Inglett GE, Munck L (eds) Cereals for food and beverages. Academic, New York, pp 365–379

Erdal K, Outtrup H, Ahrenst-Larsen B (1983) The role of proanthocyanidins in beer flavour and flavour stability. Proc Eur Brew Conv Congr, London, pp 557–564

Fletcher AC, Porter LJ, Haslam E, Gupta RJ (1977) Plant proanthocyanidins. Part 3. Conformational and configurational studies of natural procyanidins. J Chem Soc Perkin I, pp 1628–1637

Foo LY, Porter LJ (1983) Synthesis and conformation of procyanidin diastereoisomers. J Chem Soc Perkin I, pp 1535–1543

Gardner RJ, McGuinness JD (1977) Complex phenols in brewing–a critical survey. Mast Brew Assoc Am Tech Q 14:250–261

Gorissen H (1969) Les composés phénoliques du houblon. Doctoral thesis, Université Catholique de Louvain, Leuven

Gracey DEF, Barker RL (1976a) Studies on beer haze formation. I. Gas chromatographic profiles of beer: polyamide resin adsorbates. J Inst Brew 82:72–77

Gracey DEF, Barker RL (1976b) Studies on beer haze formation. II. Dimeric flavanoids observed in profiles of beer: Nylon 66 adsorbates. J Inst Brew 82:78–83

Gramshaw JW (1967) Phenolic constituents of beer and brewing materials, I. Phenolic and nitrogenous components removed from beer by polyamide resins. J Inst Brew 73:258–270

Harris G, Ricketts RW (1959) Studies on non-biological hazes of beers. VIII. Rapid estimation of anthocyanogens in beer. J Inst Brew 65:331–333

Hemingway RW, Foo LY (1983) Condensed tannins: quinone methide intermediates in procyanidin synthesis. J Chem Soc Chem Commun, pp 1035–1036

Jerumanis J (1973) Mise au point des méthodes de dosage des polyphénols en brasserie. Bull Assoc R Anc Etud Brass Univ Louv 69:1–16

Jerumanis J (1979) Séparation et identification de flavanoides par chromatographic liquide à haute performance. Proc Eur Brew Conv Congr, Berlin, pp 309–319

Jerumanis J (1985) Quantitative analysis of flavanoids in barley, hops and beer by high-performance liquid chromatography. J Inst Brew 91:250–252

Kirby W, Wheeler RE (1980) The extraction of beer polyphenols and their assay by high-performance liquid chromatography. J Inst Brew 86:15–17

Kowaka K, Fukuoka Y, Kawasaki H, Asano K (1983) The true value of aroma hops in brewing. Proc Eur Brew Conv Congr, Lond, pp 71–78

Kramling TE, Singleton VL (1969) An estimation of the non-flavonoid phenols in wines. Am J Enol Vitic 20:86–92

Lea AGH (1978) The phenolics of ciders: oligomeric and polymeric procyanidins. J Sci Agric 29:471–477

Lea AGH, Bridle P, Timberlake CF, Singleton VL (1979) The procyanidins of white grapes and wines. Am J Enol Vitic 30:289–300
McFarlane WD, Sword PF, Blinoff G (1963) Catechin–tannins in brewing. Proc Eur Brew Conv Congr, Brussels, pp 174–180
McGuinness JD, Laws DRJ, Eastmond R, Gardner RJ (1975) The estimation and significance of catechin and dimeric catechin in beer. J Inst Brew 87:237–241
McMurrough I (1979) Chromatographic procedures for measuring polyphenol haze precursors. Proc Eur Brew Conv Congr Berl, pp 321–324
McMurrough I (1981) High-performance liquid chromatography of flavanoids in barley and hops. J Chromatogr 218:683–693
McMurrough I, McDowell J (1978) Chromatographic separation and automated analysis of flavanols. Anal Biochem 91:92–100
McMurrough I, Hennigan GP, Loughrey MJ (1983) Contents of simple, polymeric and complexed flavanols in worts and beers and their relationship to haze formation. J Inst Brew 89:15–23
Mulkay P, Touillaux R, Jerumanis J (1981 a) Proanthocyanidins of barley: separation and identification. J Chromatogr 208:419–423
Mulkay P, Touillaux R, Jerumanis J (1981 b) Les prodelphinidines de l'orge: séparation, identification et influence sur la stabilité colloidale de la bière. Cerevisia 6:29–33
Narziss L, Kessler H (1971) Ein Beitrag zur Veränderung der Malzgerbstoffe während des Mälzungs- und Brauprozesses. Modifizierte Methode zur Trennung der Catechine und Anthocyanogene bzw. Leucocyanidine nach McFarlane. Brauwissenschaft 24:14–17
Outtrup H (1981) Structure of prodelphinidins in barley. Proc Eur Brew Conv Congr, Copenhagen, pp 323–333
Outtrup H, Schaumburg K (1981) Structure elucidation of some proanthocyanidins in barley by 270MHz NMR spectroscopy. Carlsberg Res Commun 46:43–52
Ribéreau-Gayon P (1968) Les composés phénoliques des végétaux. Dunod, Paris
Singleton VL (1974) Analytical fractionation of the phenolic substances of grapes and wine and some practical uses of such analyses. In: Webb DA (ed) Chemistry of winemaking. Am Chem Soc, Wash, pp 184–211
Singleton VL, Rossi JA (1965) Colorimetry of total phenolics with phosphomolybdic-phosphotungstic acid reagents. Am J Enol Vitic 16:144–158
Vancraenenbroeck R, Callewaert W, Gorissen H, Lontie R (1969) Séparation chromatographique et identification de flavonols et de flavanes du houblon. Proc Eur Brew Conv Congr, Interlaken, pp 29–43
Vancraenenbroeck R, Gorrison H, Debeer L, Lontie R (1970) Séparation chromatographique et identification de flavanoides de l'orge. Bull Assoc R Anc Etud Brass Louvain 66:73–89
Vancraenenbroeck R, Jerumanis J, De Clerck J (1974) Les polyphénols en brasserie. Bull Assoc R Anc Etud Brass Univ Louv 70:57–70
Vancraenenbroeck R, Gorissen H, Lontie R (1977) Separation of beer flavanoids by column chromatography. Proc Eur Brew Conv Congr, Amsterdam, pp 429–439
Vancraenenbroeck R, Kara-Zaitri M, Devreux A (1979) Influence de la teneur en proanthocyanidines dimères et en catéchines de la bière sur la stabilité colloidale. Proc Eur Brew Conv Congr, Berl, pp 293–307
Vancraenenbroeck R, De Brackeleire C, Willems Y, Devreux A (1983) Détermination quantitative des flavonols dans la bière. Cerevisia 8:13–19
Van Roey G (1979) Invloed van enkele parameters der gaschromatografie op de kwantitatieve bepaling van dimerisch catechine in bier. Thesis, Katholieke Universiteit te Leuven, Leuven
Von Wettstein D, Nilan RA, Ahrenst-Larsen B, Erdal K, Ingversen J, Jende-Strid B, Kristiansen KN, Larsen J, Outtrup H, Ullrich SE (1985) Proanthocyanidin-free barley in brewing: progress in breeding for high yield and research tool in polyphenol chemistry. Mast Brew Assoc Am Tech Q 22:41–52

Analytical Measurement of Volatile Sulphur Compounds in Beer

T. PEPPARD

1 Introduction

Many sulphur compounds, particularly those of relatively low molecular weight, are powerful odorants even at very low concentration. Consequently the presence of even trace amounts of such compounds in foods and beverages can have a profound effect on flavour. In the case of beer, these effects are mostly deleterious, although small amounts of certain sulphur compounds are generally considered beneficial to the flavour of the product.

Most of the sulphur-containing material present in beer is composed of involatile constituents such as amino acids, proteins and inorganic sulphate. These substances are not directly responsible for sulphury flavours in beer, but are important as precursors of the low molecular weight sulphur compounds which are. Volatile sulphur compounds usually account for less than 1% of the total quantity of sulphur-containing substances in beer. The actual quantities of materials responsible for sulphury flavours in beer are therefore exceedingly small, frequently being less than 1 μg/l (ppb). Consequently, it is only within the last 15 years or so, with the advent of mild concentration techniques and highly sensitive and selective methods of detecting, identifying and quantifying volatile sulphur compounds, that brewing scientists have been able to address problems relating to sulphury flavours in beer. It is the aim of this chapter to discuss the most important methods currently available for the analytical measurement of volatile sulphur compounds in beer. The nature of the volatile sulphur compounds present in beer is discussed briefly in Sect. 2 below. However, aspects concerning the sensory characterization of sulphury flavours in beer (Arkima et al. 1982; Peppard 1985 b) are considered beyond the scope of the present work.

2 Volatile Sulphur Compounds in Beer

The first reported link between volatile sulphur compounds and undesirable odours in beer was made over 80 years ago by Frew (1898), who demonstrated the presence of hydrogen sulphide in the gases evolved during primary fermentation. Since then, over 40 volatile sulphur-containing constituents of beer have been identified, and these have been listed by Nykänen and Suomalainen (1983). In addition, a number of authors have discussed the origin and control of sulphury flavours in beer (Wainwright 1972; Peppard 1984; Narziss et al. 1985).

The types of volatile sulphur compounds detected in beer are surprisingly varied, and include hydrogen sulphide, sulphur dioxide, thiols, sulphides, polysulphides, thioesters, thiazoles and thiophenes. The origins of these substances include both of the principal raw materials of brewing, namely malt and hops, as well as a number of important steps in the brewing process, but particularly that of fermentation. Additionally, sulphur compounds can arise through microbial infection, the onset of staling reactions and the exposure of beer to light.

One of the most important sulphur compounds derived from malt is dimethyl sulphide (DMS), which has been shown to be a major contributor to the flavour of many lager-type beers brewed from lightly kilned malts. On the other hand, ales and other beers brewed from dark or highly kilned malts have been shown to contain significant levels of the products of Maillard browning reactions, including sulphur-containing compounds such as thiazoles and thiophenes. Sulphur compounds derived from hops originate in the main from the essential oil of hops. This has been shown to contain thioesters, sulphides and polysulphides, and also, in the case of hops contaminated with the fungicide elemental sulphur, a variety of sulphur-containing terpenoid constituents. During the course of a normal fermentation, yeast excretes a range of sulphur compounds into the surrounding medium. These include mainly hydrogen sulphide and sulphur dioxide, but also methionol and a number of related compounds. Hydrogen sulphide, DMS and other low molecular weight thiols and sulphides can also be formed through the action of numerous spoilage organisms. However, the final aseptic packaging of a beer of sound flavour does not necessarily guarantee that the subsequent development of sulphury flavour will not still occur. This can happen by two principal means, namely staling and (in the case of bottled beer) the effect of light. The formation of simple alkanethiols, methional, and dimethyl tri- and tetrasulphides has been reported to occur during the staling of beer, whilst exposure of beer (particularly when bottled in clear or green glass) to light leads to the production of methanethiol and 3-methyl-2-butene-1-thiol, and so-called "skunky" or "sunstruck" flavour.

3 Analytical Methods for Determining Volatile Sulphur Compounds in Beer

3.1 General Remarks

Techniques which have been used to determine volatile sulphur compounds in beer include amperometric, colorimetric, fluorometric, gas chromatographic, potentiometric and titrimetric methods (Garza-Ulloa 1980). However, from the standpoint of sensitivity, specificity and reliability, there is no doubt that the current method of choice is gas chromatography (GC) with flame photometric detection, using one of the various procedures described below in Sect. 3.3. Before 1966, however, when Drews et al. first applied GC to the problem of sulphury flavours in beer, the only sulphur compounds determined selectively were hydrogen sulphide and sulphur dioxide. Other sulphur compounds were measured either as the total level of organosulphur compounds, or as groups of compounds of common functionality such as thiols and reducible disulphides.

In addition to problems associated with insufficiently specific methods of analysis, early workers also faced a number of other difficulties. Clearly, in analyzing beer for very low levels of sulphur compounds, great attention must be paid to the possibility of artefact formation. Thus the use of heat, excessive exposure to air and significant changes in pH should be avoided at all stages of analysis. In many instances these precautions were not taken, leading to erroneous results. Moreover, losses of trace sulphur compounds by irreversible adsorption in some of the early gas chromatographic systems, or by reaction with metal impurities in reagents or on glassware, also contributed to misleading data regarding the nature and levels of sulphur compounds present in beer.

The discussion of individual methods for determining sulphur compounds in beer has been divided into those based on chromatographic procedures (Sect. 3.3) and those based on procedures other than chromatographic (Sect. 3.2). The greater depth of the discussion of chromatographic methods reflects the relative importance of the techniques currently used to measure sulphur compounds in beer. In either case, however, the amount of space available here precludes all but the most important methods from being described in detail.

3.2 Non-Chromatographic Methods

For reasons of non-specificity, already mentioned above in Sect. 3.1, the only volatile sulphur compounds commonly measured in beer using non-chromatographic methods are sulphur dioxide and, to a much lesser extent, hydrogen sulphide.

3.2.1 Sulphur Dioxide

Sulphur dioxide is formed naturally by yeast during fermentation, and there is considerable evidence that a significant part of it survives into finished beer, where it contributes to product flavour stability (Nordlöv 1985). In some countries (e.g. Great Britain and the USA) it is permitted to add *limited* quantities of sulphur dioxide (usually in the form of an inorganic salt) in order to further enhance flavour stability, whilst in certain others, only naturally formed sulphur dioxide is allowed. Thus there is a clear need for sensitive and reliable methods of measuring sulphur dioxide in beer. One particular complication encountered with such analyses is that not all of the sulphur dioxide in beer is free, part at least exists in bound form, for example: (1) as bisulphite ions, and (2) combined with carbonyl compounds such as acetaldehyde. The most commonly used non-chromatographic methods are summarized below in Sect. 3.2.1.1 through 3.2.1.3.

3.2.1.1 The Monier-Williams Distillation/Titrimetric Procedure

This widely used technique is amongst those recommended by bodies such as the Institute of Brewing in Great Britain (IOB 1982), and the European Brewery Convention (EBC 1975). Typically, the procedure, which measures total (i.e. free + bound) sulphur dioxide, is carried out as follows.

Dilute phosphoric acid (50 ml; 5% v/v) is placed in a distillation flask (fitted with a purge gas inlet, a dropping funnel and a reflux condenser/delivery tube) and boiled for 2–3 min in a current of carbon dioxide to remove all air from the apparatus. An equal volume of beer is then added and the mixture boiled for 15 min. The sulphur dioxide evolved is carried over by means of a slow stream of carbon dioxide into a receiving vessel containing hydrogen peroxide solution (10 ml; 10 vol). The sulphuric acid produced by oxidation of the sulphur dioxide is then titrated with a dilute sodium hydroxide solution (0.02 M), using bromophenol blue as indicator, and the concentration of sulphur dioxide originally present in the beer obtained from a simple calculation.

The IOB/EBC modification of the Monier-Williams procedure was shown to give a recovery of approx. 100%, and a coefficient of variation of approx. 9%, for levels of sulphur dioxide in beer of ca. 5 ppm (Buckee and Hargitt 1976).

3.2.1.2 The Para-Rosaniline Colorimetric Procedure

This technique is based upon measuring the extent to which the colour (measured spectrophotometrically at 550 nm) of a solution of (acid-decolourised) p-rosaniline hydrochloride is restored, as a result of the action of sulphur dioxide present in beer. The colour development is carried out directly in beer, so no distillation step is required. The level of sulphur dioxide present is calculated from the difference between the absorbance of the developed sample and a reagent blank, multiplied by a conversion factor obtained from a suitable calibration plot. Different modifications of the technique have been designed to measure either free or total sulphur dioxide in beer. The method, which is rather more sensitive than the Monier-Williams procedure, has also been recommended by the EBC (1975), as well as by the American Society of Brewing Chemists (ASBC 1976). More recently, the former body has adopted the alternative DTNB method (see Sect. 3.2.1.3 below). However, this decision was based, at least in part, on the alleged carcinogenic properties of p-rosaniline.

An interesting variation of the p-rosaniline method, employing flow injection analysis and gas diffusion across a PTFE membrane, has also been reported (Tecator 1983). The method was claimed to give a detection limit of 0.3 ppm.

3.2.1.3 The DTNB Method

As mentioned above, the EBC has recently adopted the so-called DTNB method for determining sulphur dioxide in beer. In addition, the method is also currently under investigation by the ASBC (1985). In this alternative procedure, sulphur dioxide (free + bound) is distilled from acidified beer and flushed by means of a flow of nitrogen into a buffered solution of DTNB (5,5′-dithiobis[nitrobenzoic acid]). The coloured reaction product is then measured spectrophotometrically at 415 nm, and the concentration of sulphur dioxide in beer read directly from a prepared calibration curve. Initial results from a collaborative trial involving comparison of the DTNB and p-rosaniline procedures (ASBC 1985) indicated, in both cases, acceptable within-laboratory errors, but high between-laboratory errors. However, comments from collaborators suggested that the DTNB procedure was time-consuming and not preferred for routine use.

3.2.2 Hydrogen Sulphide

Hydrogen sulphide, which in brewing can arise from a variety of sources, is probably the most common cause of sulphury off-flavours in beer. It is therefore important to be able to measure this sulphur compound. A number of methods have been used, of which the most significant, other than those based on GC (see Sect. 3.3), are mentioned briefly below. As with sulphur dioxide, the measurement of hydrogen sulphide in beer is complicated by the fact that it exists in both free and bound forms. In the case of hydrogen sulphide, however, the binding mechanism involves, at least in part, trace amounts of certain metal ions which are present in beer.

The fluorometric titration procedure reported by Brenner and Khan (1974) allows simultaneous determination of both organic thiols and free hydrogen sulphide in beer. The method was claimed to have a coefficient of variation of approx. 6% when measuring levels in the range 0.3–2.4 ppb. Other methods for the determination of hydrogen sulphide in beer include a colorimetric procedure based on the formation of methylene blue (Sinclair et al. 1969), and direct measurement using an ion specific electrode (Owades et al. 1967). However, none of the non-chromatographic methods devised to date are able to compete with those based on GC, from the standpoint of sensitivity, specificity, reproducibility and the speed and ease of analysis.

3.3 Chromatographic Methods

At present, neither high performance liquid chromatography (HPLC) nor supercritical fluid chromatography (SFC) have made any impact on analysis of the sulphur-containing flavour constituents of beer. However, it is possible to use SFC with sulphur-specific flame photometric detection, and as this relatively new chromatographic technique matures, it will more than likely be applied to the analysis of sulphur compounds in beer. The particular attraction of SFC lies in its ability to separate involatile and/or relatively polar materials, whilst operating at much lower temperatures than are usual with GC (Chester 1986). Thus it is possible that the application of SFC could reveal important sulphur-containing flavour compounds in beer which are not detected by GC.

Numerous gas chromatographic procedures have been reported for the determination of volatile sulphur compounds in beer. These are discussed below in terms of the methods of analyte detection, chromatographic separation and sample preparation/introduction used. Chen and Morrison (1983) have reviewed the application of GC to brewing research and quality control in general.

3.3.1 Analyte Detection

Gas chromatographic analyses of volatile sulphur compounds in beer have utilized several kinds of detector, including the microcoulometric, thermal conductivity, flame ionization, mass spectrometric and flame photometric detectors (Garza-Ulloa 1980). Of these, the latter (when used in conjunction with a 394-nm optical filter) is generally the first choice, by virtue of its impressive specificity towards sulphur compounds.

Use of the flame photometric detector (FPD) is not, however, without problems (Perry and Carter 1971). Briefly, these include a critical dependence of the detector sensitivity on factors such as the flow rates of carrier gas and flame gases. Also, the FPD response to sulphur compounds generally does not obey the theoretical square law relationship, exponent values usually lying somewhere between 1.5 and 2. Finally, and probably most importantly, the FPD response to sulphur compounds is reduced or quenched by co-eluting compounds, such as hydrocarbons, which do not contain sulphur. In some cases, quenching can lead to the complete absence of any visible chromatographic peak from an eluting sulphur-containing component. As demonstrated by Lee and Siebert (1986), quenching can lead to substantial errors in quantitation, when sulphur compounds such as DMS are estimated with the aid of calibration curves.

A promising alternative to the FPD, at least in the case of reduced sulphur compounds, is the fluorine-induced chemiluminescence detector (Nelson et al. 1983). This detector monitors chemiluminescence resulting from the reaction of GC effluent with molecular fluorine at reduced pressures. For reduced sulphur compounds, the detector's sensitivity is generally similar to that of the FPD, but it has the advantages of a linear response over at least three orders of magnitude and a greater degree of selectivity. However, it detects neither hydrogen sulphide nor sulphur dioxide.

Gas chromatography coupled with mass spectrometry (GC-MS) has been invaluable in identifying trace sulphur-containing constituents of beer (Peppard 1985a). Furthermore, mass spectrometric detection, particularly when using specific-ion monitoring, can provide a high degree of sensitivity. However, because sulphur compounds do not yield useful common fragment ions, this method is generally limited to the analysis of specific or "targetted" substances.

3.3.2 Chromatographic Separation

Since the technique of GC was first applied to measuring volatile sulphur compounds in beer, there have been a number of marked improvements in the materials used for manufacture and treatment of chromatography columns. In the case of packed columns, these are now mostly constructed from glass, Teflon or FEP (fluorinated ethylene/propylene copolymer) rather than, as formerly, stainless steel, aluminium or copper, which can cause serious losses of sulphur compounds through adsorption and surface catalysis. Advances in the techniques of deactivating columns and solid support materials have also led to fewer problems resulting from adsorption of sulphur compounds (which can in turn lead to peak tailing or even the complete disappearance of peaks from chromatograms). A number of different stationary phases have been successfully used for chromatographing the volatile sulphur components of beer. Two of the most popular phases have been: (1) 12% polyphenyl ether (5 ring) + 0.5% phosphoric acid, and (2) 10% Triton X 305 + 0.5% phosphoric acid. These have usually been coated onto conventional support materials, such as the Chromosorb-type products. In some instances, however, they have been coated instead onto more inert materials such as powdered Teflon. Alternatively, a number of workers have carried out GC of sulphur compounds using columns packed with porous polymers such as Porapak Q.

Within the last few years, an increasing number of researchers have switched to the use of capillary columns for separating mixtures of volatile compounds derived from beer. This has been particularly noticeable since the introduction of polyimide-coated fused silica capillary columns (Dandeneau and Zerenner 1979). Compared to glass capillary columns, these have the advantage of being highly inert; in addition, they are more robust and, at the same time, are tremendously flexible. However, one of the main drawbacks of using capillary columns for separating the trace quantities of sulphur compounds isolated from beer is related to the methods of sample introduction often employed (see Sect. 3.3.3 below). Briefly, a number of these involve introduction of the sample onto the chromatographic column dispersed in a relatively large volume of carrier gas, whilst high resolution chromatography requires introduction of the sample as a narrow band. Capillary columns coated (or in some cases chemically bonded) with a number of different stationary phases have been used for separating the volatile sulphur constituents of beer. These include standard phases such as OV-101 and Carbowax 20 *M*, as well as special purpose phases such as CP Wax 57CB (see Sect. 3.3.3.2 below).

3.3.3 Sample Preparation/Introduction into Gas Chromatograph

The simplest analytical methods for determining the volatile sulphur constituents of beer employ headspace sampling followed by direct injection of the vapours into a gas chromatograph. In this way, highly volatile sulphur compounds, such as DMS, can readily be analyzed on a routine basis at levels below 10 ppb (see Sect. 3.3.3.1 below). Unfortunately, most flavour-active sulphur compounds are not amenable to this type of analysis, as they are insufficiently volatile or are not present in beer at high enough concentrations. Then methods of sample enrichment must be employed, such as those described in Sect. 3.3.3.2 and 3.3.3.3.

3.3.3.1 Static Headspace Sampling

The only sulphur compounds routinely measured in beer using a static headspace sampling technique are hydrogen sulphide, sulphur dioxide and DMS (e.g. White and Wainwright 1977; Nagami et al. 1980; Lee and Siebert 1986). In most cases a simple gas-tight syringe is used, although some researchers have employed a multi-port gas sampling valve and loop system. A procedure for obtaining a headspace sample of beer, using a gas-tight syringe, is described below (White and Wainwright 1977).

Cold beer (20 ml) is pipetted into a 125-ml flat-sided bottle (previously flushed with nitrogen), and 1 ml of a freshly prepared solution of 1 ppm ethyl methyl sulphide (EMS) in absolute ethanol is added to provide an internal standard. A serum cap is fitted, and the bottle pressurized by injection of 30 ml of nitrogen. It is then shaken thoroughly, before being totally immersed for 30 min in a water bath held at 30 °C. Finally, a 2 ml gas phase sample is removed using a gas-tight syringe (maintained at 40 °C between injections in order to prevent sample condensation) and injected into a gas chromatograph equipped with a flame photometric detector. For use with a 5.5 m × 6 mm i.d. glass chromatography column packed with phosphoric acid treated 10% Triton X-305 on 80–100 mesh Chromosorb G AW DMCS (or similar), an isothermal oven temperature of 70 °C is recommended. Concentrations of individual sulphur compounds (e.g. DMS) in beer are determined from standard calibration curves. These are constructed by making additions of the authentic compounds in 3% (v/v) aqueous ethanol to beer, at several different concentrations covering the range of interest. Concentrations of individual components are then plotted against the ratios of their peak height (or area) to that of the internal standard.

3.3.3.2 Dynamic Headspace Sampling (Purge and Trap Methods)

To achieve markedly improved sensitivity in the headspace analysis of volatile sulphur compounds, the trend in recent years has been towards the use of dynamic headspace sampling and so-called purge/trap/desorb (PTD) systems. Such methods generally involve purging volatiles from beer, with an inert gas such as helium or nitrogen, into a cold trap or trap packed with adsorbent material (e.g. charcoal or one of various porous polymers). After a suitable collection period, volatiles are desorbed from the trap by rapid heating in a stream of inert gas, and are transfered directly into a gas chromatograph for analysis. In systems which employ adsorbents, the desorption process can last up to 20 min. In such cases, the volatiles are either re-collected in an intermediate cold trap, or are "thermally focussed" at the front end of the chromatographic column, which usually requires cooling to well below ambient temperature. Alternatively, the contents of the initial collection trap may be extracted with an organic solvent, and the extract (in some cases after a concentration step) injected into the gas chromatograph.

Of various methodologies involving dynamic headspace sampling, that employing the flushing of beer headspace vapours into a cold trap followed by extraction of the trap with organic solvent is probably the most straightforward. Such a method, which requires very little by way of additional specialized equipment, was used successfully by Suggett et al. (1979) to measure a number of hop-derived sulphur compounds in beer at levels well below 1 ppb (e.g. a detection limit of 0.01 ppb was achieved for dimethyl disulphide). Nowadays, however, the most popular dynamic headspace sampling technique is that which involves the use of an adsorbent trap, desorption of the volatile sulphur compounds by a rapid heating process and thermal focussing of these substances at the front end of a chromatographic column cooled to below ambient temperature (usually in the range −30 to −65 °C). Such methods have been described by Leppänen et al. (1981), Narziss et al. (1983) and Peppard (1985b), where detection limits were reported to be well below 1 ppb for most of the sulphur compounds studied.

In designing an analytical procedure based on PTD methodology, it is necessary to optimize numerous experimental parameters. Where a series of different analytes is to be measured in one analysis, the optimum conditions are likely to vary from one compound to the next, and the experimenter will usually be forced to compromise. For procedures employing collection of volatile sulphur constituents in an adsorbent trap, clearly one of the most important factors which must be considered is the choice of the adsorbent material itself. Possibilities include charcoal, and numerous porous polymers such as Tenax, various Porapaks and the Chromosorb series of products. Important factors to consider are the trapping efficiencies of the adsorbents (characterized for individual compounds by their so-called "break-through volume"), the ease with which analytes can subsequently be removed from the adsorbent during the desorption step, and the degree to which the adsorbent material gives rise to artefacts during analyses (e.g. volatile thermal degradation products). Thus Leppänen et al. (1981) used both Chromosorb 101 and Chromosorb 102, whilst Narziss et al. (1983) employed specifically the latter. On the other hand, Peppard (1985 b) tested Porapak Q, Tenax GC, and Chromosorb 102, but chose the first-named of these on the basis of superior recoveries achieved for the particular sulphur compounds of interest. Clearly, other factors connected with sampling which require investigation include the amount of adsorbent material to use, the duration of the headspace purging step (together with the purge gas flow rate), and the temperature at which to hold the sample during purging. Once the analytes are trapped on the adsorbent material in the most efficient manner, parameters which must be optimized include the temperature and duration of the desorption step, together with the carrier gas flow rate, and the temperature at which thermal focussing should be carried out on the chromatographic column. Finally, of course, it is necessary to select optimum chromatographic conditions, including temperatures, hold periods, rates of temperature increase, etc. The choice of stationary phase is also particularly important, in view of the very low temperatures employed in the thermal focussing step (the stationary phase should ideally remain liquid), and the presence of potentially damaging traces of water which are inevitably introduced into the column along with the volatiles. To this end, Peppard (1985 b) chose a capillary column chemically bonded with the stationary phase CP Wax 57CB. In order to avoid the phenomenon of peak-splitting, which tends to occur with sub-ambient use of a capillary column, the latter was placed in a collar made from aluminium foil and asbestos paper.

The principal disadvantage of the PTD method, as outlined briefly above, is that it requires the purchase and/or construction of specialized equipment. This includes items such as a rapid heating unit for desorbing sulphur volatiles from adsorbent traps (see the example shown diagrammatically in Fig. 1), and the means for operating a gas chromatographic oven at sub-ambient temperatures. However, a number of manufacturers now offer complete systems designed specifically for PTD operation and the measurement of trace organic constituents in complex media. The advantages of using a PTD system are manyfold. The technique (with optimization) combines high sensitivity with satisfactory reproducibility. Furthermore, it is a technique which lends itself to automation and a (rel-

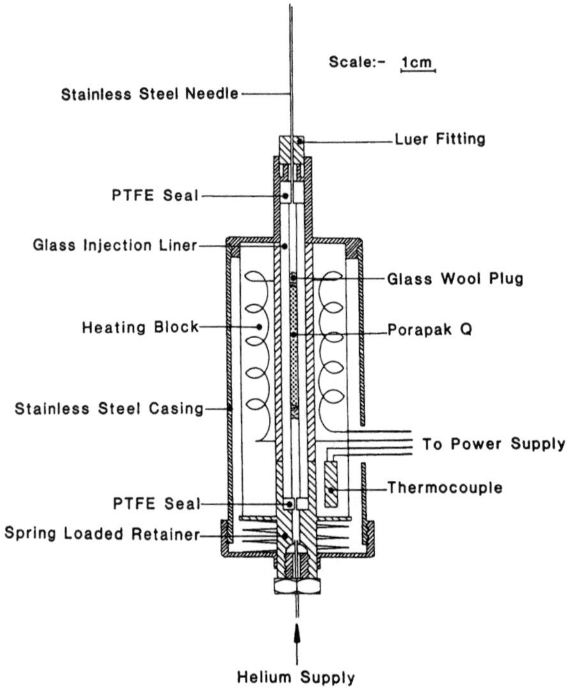

Fig. 1. A rapid heating unit, for the desorption of volatile flavour compounds in purge/trap/desorb procedures. J Inst Brew, 1985, p 368

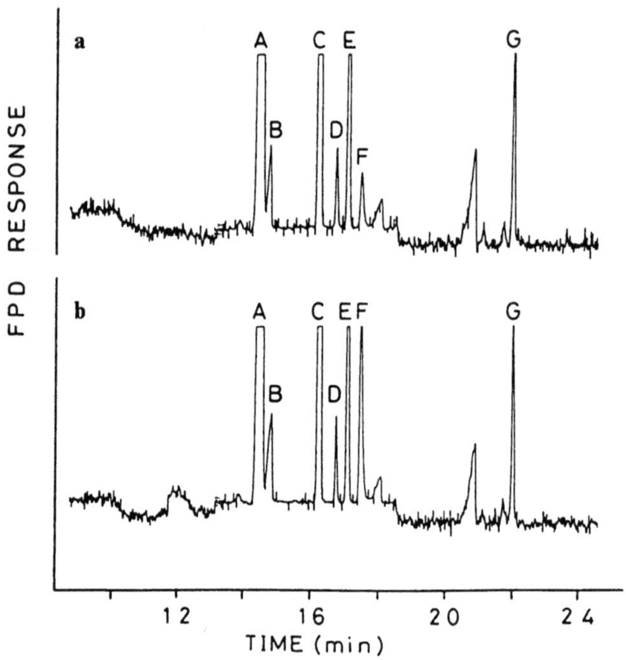

Fig. 2 a, b. Sections of gas chromatograms obtained using a purge/trap/desorb procedure: **a** bottled lager, as purchased from retail outlet; **b** bottled lager, as above but further exposed (1 h) to diffuse daylight. *A* diisopropyl sulphide (internal standard); *B* ethanol; *C* methyl thioacetate; *D* dimethyl disulphide; *E* ethyl thioacetate; *F* 3-methyl-2-butene-1-thiol; *G* unknown. J Inst Brew, 1985, p 367

atively) high rate of sample throughput. In addition, PTD can be used over a surprisingly wide range of volatilities, although careful method optimization is usually necessary for specific compounds, especially those having high volatility and a low break-through volume. For example, methanethiol, a trace sulphur constituent of very high volatility (b. pt. ca. 6 °C), was readily measured in beer using a PTD procedure (Leppänen et al. 1981; Narziss et al. 1983), whereas this is generally not possible using static headspace sampling or solvent extraction procedures.

An example of the use of PTD methodology is given in Fig. 2. Figure 2a shows part of the FPD chromatogram obtained by analysis of a commercial lager (bottled in green glass–see Sect. 2 above) as purchased from a retail outlet. Figure 2b shows the marked increase in level of 3-methyl-2-butene-1-thiol, occurring after an additional 1-h exposure of the beer to diffuse daylight. Both samples of beer had a pronounced "skunky" flavour.

Instead of flushing with a stream of inert gas as described above, an alternative procedure for removing volatile sulphur compounds from beer employs vacuum stripping. Pickett et al. (1976a, b) described the use of such a technique for concentrating about a dozen volatile sulphur-containing components of beer. However, the method was very laborious and required the use of an elaborate all-glass vacuum system.

3.3.3.3 Methods for Concentrating Sulphur Compounds of Lower Volatility

Clearly, the analysis of sulphur compounds of relatively high volatility is most readily carried out using a headspace sampling technique. However, many of the volatile sulphur constituents present in beer have boiling points well in excess of 100 °C; such compounds of relatively low volatility are probably best measured using sampling procedures based on solvent extraction or direct extraction of beer with adsorbent materials such as porous polymer resins or chemically bonded silicas. Thus, Pickett et al. (1976a, b) used vacuum steam distillation of beer, followed by solvent extraction of the aqueous distillate with diethyl ether, to prepare beer flavour concentrates in which about a dozen sulphur compounds could be detected by GC/FPD. A similar method, employing dichloromethane in the solvent extraction step, was later used by Narziss et al. (1983). Peppard and Douse (1979) also used vacuum steam distillation, but instead passed the aqueous distillate down a column of Amberlite XAD-2 porous polymer resin. Extraction of the resin with diethyl ether yielded a concentrate in which more than 25 sulphur compounds could be detected, including a number at or below their flavour threshold levels.

Whilst the types of methods mentioned above are fairly straightforward to carry out, they do have a number of drawbacks. First, both methods are biased in favour of compounds which are relatively non-polar in nature, polar compounds being rather more difficult to remove from an aqueous system. As far as solvent extraction is concerned, the concentration of large volumes of organic solvent (e.g. 1 l) down to microlitre amounts, clearly necessitates the use of highly purified solvents in order to avoid concentration of many trace solvent impurities in addition to the flavour components of interest. Similarly, the use of porous polymer resins, such as Amberlite XAD-2, requires extensive clean-up procedures prior to use. When solvent extraction is used directly with beer, rather than with a vacuum steam distillate of beer, serious problems are encountered due to the formation of emulsions which often persist even in the presence of added anti-foam and/or inorganic salts. Porous polymer resins, on the other hand, can be used directly with beer, although it may be necessary to degas the beer prior to extraction.

4 Conclusions

Deleterious sulphury flavours can arise in beer from a number of sources. The compounds responsible are highly flavour-active, and are usually not present at levels above ca. 1 ppb. Highly sensitive, specific and reliable methods for the analysis of sulphur compounds are therefore a prerequisite to understanding and controlling sulphury flavours in brewing. Current methods are almost entirely based on the use of gas chromatography with flame photometric detection, often combined with a procedure for carrying out sample enrichment. However, adsorption losses and the introduction of artefacts are serious problems with current methods, and the flame photometric detector is not completely reliable in the presence of non-sulphur volatiles. The continued development of new and improved analytical techniques is therefore clearly necessary whilst so many of the questions relating to sulphury flavours in beer remain unanswered.

Acknowledgement. The author wishes to thank the Stroh Brewery Co. for permission to publish this chapter.

References

Arkima V, Jounela-Eriksson P, Leppänen O (1982) The influence of volatile sulphur compounds on beer flavour. European Brewery Convention Monograph VII, Proceedings of Flavour Symposium. Brauwelt, Nürnberg, p 89

ASBC (1976) American Society of Brewing Chemists, Methods of analysis, 7th edn. ASBC, St. Paul, MN, Beer-21

ASBC Subcommittee on Sulphur dioxide (1985) Sulphur dioxide in beer by DTNB. J Am Soc Brew Chem 43(3):167–168

Brenner MW, Khan AA (1974) Simultaneous fluorometric microdetermination of hydrogen sulphide and volatile thiols in beer. J Inst Brew 80:544–548

Buckee GK, Hargitt R (1976) Comparison of two distillation-titrimetric procedures for the estimation of sulphur dioxide in beer. J Inst Brew 82:290–291

Chen EC-H, Morrison NM (1983) Application of gas chromatography to brewing research and quality control: a review. J Am Soc Brew Chem 41(1):14–19

Chester TL (1986) The role of supercritical fluid chromatography in analytical chemistry. J Chromatogr Sci 24:226–229

Dandeneau RD, Zerenner EH (1979) An investigation of glasses for capillary chromatography. J High Resolut Chromatogr Chromatogr Commun 2:351–355

Drews B, Specht H, Kühl E-D (1966) Über die Schwefelverbindungen des Bieres. Monatsschr Brau 19:239–255

EBC (1975) European Brewery Convention, Analytica-EBC III, 3rd edn (with Errata, 1977), E75–E78 & Supplement Method 7.12.1 DTNB

Frew WJ (1898) The causes of "stench" in beer. J Soc Chem Ind 17:561–562

Garza-Ulloa H (1980) Analytical control of sulfur compounds in beer. A review. Brew Dig 55(1):20–26

IOB (1982) Institute of Brewing, Recommended methods of analysis (Revision), p 40

Lee SS, Siebert KJ (1986) A new calibration procedure for the determination of dimethyl sulphide in beer using gas chromatography with a flame photometric detector. J Am Soc Brew Chem 44(2):57–66

Leppänen O, Denslow J, Koivisto T, Ronkainen P (1981) A convenient method for determining volatile sulphur compounds in beer. In: Schreier P (ed) Flavour '81. Proceedings of the 3rd Weurman symposium. De Gruyter, Berlin, p 361

Nagami K, Takahashi T, Nakatani K, Kumada J (1980) Hydrogen sulphide in brewing. MBAA Tech Q 17(2):64–68

Narziss L, Miedaner H, Kattein U (1983) Bestimmung flüchtiger Schwefelverbindungen in Würze und Bier im ppb-Bereich. Monatsschr Brauwiss 36(1):13–17

Narziss L, Miedaner H, Kattein U (1985) Über das Verhalten von schwefelhaltigen Aromakomponenten beim Brauprozeß. Monatsschr Brauwiss 38(11):439–445

Nelson JK, Getty RH, Birks JW (1983) Fluorine-induced chemiluminescence detector for reduced sulphur compounds. Anal Chem 55:1767–1770

Nordlöv H (1985) Formation of sulphur dioxide during beer fermentation. Proc 20th Eur Brew Conv Congr, Helsinki, p 291

Nykänen L, Suomalainen H (1983) Aroma of beer, wine and distilled alcoholic beverages. Akademie, Berlin, p 233

Owades JL, Blick SR, Owades SH (1967) Specific ion determination of hydrogen sulphide in beer. Proc Am Soc Brew Chem 75–78

Peppard TL (1984) Sulphur compounds in beer. In: Nykänen L, Lehtonen P (eds) Flavour research of alcoholic beverages–instrumental and sensory analysis. Found Biotech Ind Ferment Res 3:61–70

Peppard TL (1985a) The application of mass spectrometry in beer flavour studies. J Inst Brew 91:16–24

Peppard TL (1985b) Development of routine investigational tools for the study of sulphury flavours in beer. J Inst Brew 91:364–369

Peppard TL, Douse JMF (1979) Simple and sensitive technique for detecting trace flavour components in beer. J Chromatogr 176:444–447

Perry SG, Carter FWG (1971) The chromatography of sulphur compounds. Gas Chromatography, Proc Int Symp 8:381–398

Pickett JA, Coates J, Peppard TL, Sharpe FR (1976a) Chemical characterisation of differences between ales and lagers. J Inst Brew 82:233–238

Pickett JA, Coates J, Sharpe FR (1976b) Procedure for non-destructive concentration of flavour-active components of beer. J Inst Brew 82:228–233

Sinclair A, Hall RD, Thornburn Burns D (1969) The analysis of volatile sulphur compounds in beer. Proc 12th Brew Conv Congr, Interlaken, p 427

Suggett A, Moir M, Seaton JC (1979) The role of sulphur compounds in hop flavour. Proc 17th Eur Brew Conv Congr, Berlin (West), p 79

Tecator (1983) Determination of total sulphite in beer by flow injection analysis and gas diffusion. Application Notes AN 61/83 and AN 61–24/83

Wainwright T (1972) Sulfur tastes and smells in beer. Brew Dig 47(7).78–84

White FH, Wainwright T (1977) The presence of two dimethyl sulphide precursors in malt, their control by malt kilning conditions, and their effect on beer DMS levels. J Inst Brew 83:224–230

Short Introduction to Headspace Analysis

J. STRATING

1 Introduction

If a liquid or solid sample is placed in a septum-closed vial at constant temperature with a free gas phase over the sample, one might after a certain equilibration time take a volume of vapour from the gas phase over the sample to analyze it. The statement above is a general description of a headspace sampling procedure. The free gas phase over the sample is referred as "headspace".

The term "headspace analysis" means in most cases a gas chromatographic separation of the volatile compounds in the headspace sample to achieve qualitative and quantitative data of the components in the liquid or solid sample via the vapour phase in equilibrium with that liquid or solid sample. This type of headspace analysis allowing the organic volatiles to equilibrate between the sample and the gas phase is named the "static headspace method".

If the static headspace method does not provide a sufficient concentration of the volatile components of interest with respect to the sensitivity of the gas chromatographic detector, the "dynamic headspace method" is required to increase the amount of sample available for analysis.

One of the first investigators to use static headspace analysis was Weurman (1961) in his study of the enzymatic formation of volatile components in raspberries.

The advantages of headspace analysis (HSA) compared with the conventional manipulation of samples requiring extensive preliminary procedures, i.e. solvent extraction, result from the fact that no overloading or contamination of the chromatographic column with undesirable high-boiling compounds or non-volatile materials occurs.

Success or failure of a gas chromatographic HSA depends on whether or not sufficient concentrations of volatiles have been injected from the headspace sample into the gas chromatographic system, and on the sensitivity and type of detector used for the HSA.

The static-HSA has almost become the established practice for the investigation of problems involving food odours, and the resulting chromatogram is sometimes described as an "aromagram", suggesting a resemblance with the organoleptic impression of the human nose.

This is rather inaccurate, in that the specificity of the human nose is not always the same as that of the gas chromatographic detector employed.

Naturally, the flavour of a food or beverage bears upon the distribution of the odorous volatiles in the product, and HSA can be of great importance, but it is not a conclusive procedure. One will have to realize that there are other aspects to food odours which cannot be solved by HSA.

As quoted by Weurman (1969), one has to deal with:

a) "The qualitative and quantitative composition of all the odorous volatiles in the food itself.

This aspect, the total volatile composition, which cannot be solved merely by HSA, forms the basis of all our understanding of the odours of a product".

b) "The composition of the odorous vapours over, and in equilibrium with the food as it is observed by our sense of smell".

c) "The physical state and distribution of the odorous volatile compounds in the food. This aspect relates the composition of the vaporous odours over the product to the overall composition of the volatiles in the product.

So the basic complex mixture of total volatiles in two different food items might be the same, both qualitatively and quantitatively, but the effect might be two different odours. The difference then is caused by the various degrees to which the volatiles are dissolved in the lipoidic or aqueous phase of the food, or by the degree to which their effective concentration is reduced by adsorption onto the solids of the product".

These matrix effects of the food sample are very important with relation to HSA, as will be dealt with later on.

2 The Static Headspace Method

The samples to be analyzed by static headspace techniques can be homogeneous liquid solutions, homogeneous solid solution or non-homogeneous materials.

If a liquid sample is placed in a closed vial with a free gas phase over the sample, then each sample component having a vapour pressure at the temperature of the vial will also be present in the vapour phase.

According to Henry's law, the partial vapour pressure P'_B of the analyte B above the solution depends on the mole fraction, X_B, of the analyte B and the saturated vapour pressure, P^0_B, of the pure compound B at a given temperature, corrected for any deviation from ideality by the activity coefficiënt γ_B:

$$P'_B = \gamma_B \cdot X_B \cdot P^0_B. \tag{1}$$

In terms of partial pressure, one can also state, the chromatographic peak area (A_B) of compound B in the vapour phase above a liquid is proportional to its partial vapour pressure, P'_B, giving the relation:

$$A_B = C_B\, P'_B. \tag{2}$$

A_B = chromatographic peak area response of B in the gas phase.
C_B = compound-dependent proportionality constant.
P'_B = partial vapour pressure of B.

Combining Eqs. (1) and (2) and eliminating P'_B yields an expression relating X_B to the vapour phase composition of the sample:

$$X_B = \frac{A_B}{C_B\, \gamma_B\, P^o_B}. \tag{3}$$

The activity coefficient γ is of importance for gas chromatographic headspace analysis. For practical application of gas chromatographic HSA, the determination of low concentrations of analytes in a mixture, one has to deal in most cases with so-called ideal dilute solutions. Under these circumstances $\gamma=\text{const}$ (ideal dilute solution). If it is necessary to deal with samples in which relative high concentrations of analytes are present, then $\gamma=f(\text{concentration})$. In the case of ideal mixtures $\gamma=1$.

Under the conditions of an ideal dilute mixture ($\gamma=\text{const}$) there is a linear relationship between the concentration of the trace component in the liquid phase and its concentration in the vapour phase. The latter is directly measurable from the headspace chromatogram. However, the activity coefficient is not only a characteristic of the compound of interest, but also depend upon the matrix in which the compound is found. The calibration must therefore be performed in the same, or at least a similar, matrix. This is possible if the matrix is known.

To make a quantitative approach with the static headspace technique, a truly thermodynamic equilibrium has to be established when the headspace sample is withdrawn. The sample in the vial has to reach a constant temperature. The time needed to reach equilibrium depends on the sample under investigation, so a standard value cannot be given a priori.

3 The Dynamic Headspace Method

According to Weurman (1974), the concentration of the main components in food odours are generally in the order of 10^{-3}–10^{-6} g/l of air, while those of the trace components may amount to 10^{-7}–10^{-10} g/l. Since the odour threshold values of many compounds are 10^{-8} g/l or even lower it is important to analyze the trace components as well because they can contribute considerably to the aroma of a product. In g.c. headspace analysis covering the area of wide bore capillary columns to packed columns, volumes in the order of 0.5 ml–10 ml of air can be analyzed directly, which means that only the more abundant and more volatile compounds will exist at detectable levels in real headspace analysis of vapours existing in equilibrium with a sample held in a closed system. For example, 1 ml of headspace may contain 10^{-6}–10^{-9} g of the main components and 10^{-10}–10^{-13} g of the trace components.

Considering the limit of detection for a flame ionization detector at 10^{-10}–10^{-11} g/s, it will be clear that often the amount of trace components in the vapour sample is not enough for detection.

It means that the static headspace method may not provide a sufficient concentration of the volatile components of interest with respect to the sensitivity of the gas chromatographic detector. In this case, a prechromatographic concentration step must be included before injection onto the column.

This can be achieved by switching from static to dynamic headspace analysis. In the latter case, the headspace chamber is purged by an adequate gas flow, which strips the volatile components from the condensed phase and carries them to a trap, where they are concentrated. Various approaches have been postulated for trapping the stripped volatiles, including adsorption on a specific solid surface, immobilization on a gas-liquid stationary phase, or cryogenic trapping on a short piece of fused silica capillary, as published recently by Badings et al. (1985).

Many papers devoted to these "purge and trap" methods and the corresponding instrumental apparatus have been published to date, but they are too numerous to be referenced completely in this short introduction. Publications dealing with the static headspace methods as well with the dynamic headspace methods which give a good survey of the possibilities, including a theoretical approach, are those of Hachenberg and Schmidt (1977), Kolb (1977), Poole and Schuette (1984), and Ioffe and Vitenberg (1984).

4 Sampling Methods

Injection of the vapour phase of a product is one of the most ideal sampling methods, although some precautions have to be taken to perform reproducible analyses with the use of gas-tight syringes.

The advantages of vapour sampling are evident: the analyzed vapours are the same as those observed by smelling, artefact formation is not likely to occur because the pretreatment of the sample is minimal, and quantitative ratios of constituents of a vapour remain unchanged as long as static headspace conditions are valid. With direct vapour analysis a product is contained in an infusion flask closed with a septum secured by a screw cap. The flask is filled to one quarter with the sample and is placed in a thermostatted water bath for 30 min. In most cases half an hour is long enough to establish an equilibration between product and vapour. Vapour samples are taken with a gas-tight syringe by piercing the septum with the syringe needle.

The syringe must be at the same or a slightly higher temperature than the sample to prevent condensation in the syringe and the syringe must be grease-free. The syringe also has to be rinsed three to four times with the vapour sample before the final injection of the sample into the column. This procedure is necessary to minimize losses resulting from adsorption on the inner surfaces of the syringe (Kepner et al. 1964).

Losses by adsorption of volatiles in the septum can be prevented by placing a piece of aluminium foil under the septum to avoid contact of vapour with the septum.

The back pressure in the g.c. injection port should be lower than the pressure of the gas sample and the syringe must be carefully cleaned after each injection by vacuum at higher temperature (60 °C).

For routine analyses the manual sampling procedure has been replaced nowadays by automated headspace analyzers with an electropneumatic dosing system.

Kolb et al. (1979) has described such a system in detail and these automated headspace samplers are commercially available. Vapour injection into packed columns is possible, let us say optimal with sample sizes from 2–10 ml. On capillary columns, sample sizes of 0.5–1 ml are usable for so-called wide bore columns (internal diameter 0.5–0.75 mm). The static headspace sampling may not provide a sufficient concentration of the volatile components of interest with respect to the sensitivity of the gas chromatographic system. In this case a prechromatographic concentration step must be included before injection onto the column. This can be achieved by switching from static to dynamic sampling methods.

In the latter case, the headspace chamber is purged by an adequate gasflow, which strips the volatile components from the liquid or solid samples and carries them to a trap, where they are concentrated.

Various approaches have been postulated for trapping the stripped volatiles, including adsorption on a specific solid surface (Tenax, Porapak materials, Chromosorb 105, activated charcoal, etc.), immobilization on a gas-liquid stationary phase, or cryogenic trapping.

This preconcentration step is usually followed by thermal desorption.

The amount of analyte stripped will depend primarily on the substance-specific partition coefficient and experimental variables such as flow rate, time, and total volume of stripping gas passed through the solution. Only in those instances where the organic compound has low water solubility ($<2\%$ w/w) and is relatively volatile ($\approx$b.p. <200 °C) can quantitative extraction be expected, according to Bellar and Lichtenberg (1974).

Gas phase stripping (sampling) with adsorption of the volatiles on a small charcoal trap in a closed loop system was described by Grob and Zürcher (1976) for determining very low levels of organic volatiles (ng/l) in drinking water. The trapped organic volatiles are removed from the miniature charcoal trap by elution with a small volume of pure carbon disulphide and analyzed directly by capillary column gas chromatography.

5 Factors Affecting Choice of Equilibrium Conditions

The sensitivity of the headspace sampling method can be improved in some instances by adjusting the pH, raising or lowering of the temperature, addition of an inorganic salt. Kepner et al. (1964) investigated some of these factors with samples of beer and rum. The temperature at which samples are equilibrated affects the system in several ways. Raising the equilibration temperature increases the possibility of artefact formation and may do so more for beer than for rum because of differences in content of thermolabile compounds. On the other hand, such a rise will result in an increase of the vapour pressure in the headspace, thus increasing the possibility of measuring the concentration of trace components in the sample.

The increase of vapour pressure is different, however, for compounds of different physical properties. When chromatographic parameters are such that a high boiling ester appears on the ethanol "tailing curve", lowering of the equilibration temperature can give better results. At the lower temperature the vapour pressure of the ethanol is decreased by a much greater factor than that of the high

Table 1. Effect of dilution on vapor pressures of certain volatile constituents in rum

Compound	Peak height,[a] mm, in rum with ethanol content			
	75% (undiluted)	50%	15%	7.5%
Acetaldehyde	175	380	205	135
Ethyl formate	3280	4800	2980	1700
Isobutanol	17	50	43	24
Isoamyl alcohols and 2-methyl-1 butanol	9	56	73	50
Ethyl caproate	10	169	318	227

[a] 6-ml headspace samples from rum equilibrated at 36 °C for 45 min.

boiling ester, thus permitting a much more accurate measurement of the concentration of the ester.

The measurement of the concentration of a trace component in solution may also be facilitated by saturation of the sample with an inorganic salt carefully chosen for the purpose. For example separate headspace analyses of a 15 ppm isoamylalcohol solution in 5% aqueous ethanol showed four- and sevenfold increases in peak heights when saturated with sodium chloride and ammonium sulphate, respectively.

Instead of salts, addition of sugar sometimes might be an alternative to improve results.

Ethanol, which is the main component of fermentation products has a great influence on the vapour pressures of the other volatiles in the sample.

The data in Table 1 illustrate the effect of diluting a rum, which contains 75% ethanol, to 50, 15, and 7,5% ethanol content by the addition of water.

The undiluted and the three diluted rum samples were analyzed by headspace techniques and the vapour pressures of the components measured in terms of peak heights. Essentially, all components showed a marked increase in peak height for the sample containing 50% ethanol, while the higher boiling components continued to show increases in peak heights for the sample containing 15% ethanol.

Without any detailed consideration of the complex interactions (hydrogen bond formation, changing of the activity coefficients) contributing to the results, it is feasible to use dilution as a technique to increase concentrations of volatile components in the vapour phase of alcholic beverages. The most favourable degree of dilution for a given type of fermentation product must be determined experimentally to facilitate the desired analysis.

Adjusting the pH of the sample is feasible if the concentration of the analyte in the gas phase is proportional to the concentration of the undissociated part of the analyte in solution. Adjusting the solution pH so that the analyte exists mainly in the non-dissociated form will improve sensitivity.

Functional group analysis can also be performed by headspace analysis. The normal sample headspace analysis is run, a derivatizing reagent capable of forming an involatile derivative with compounds containing a particular functional group is then added to the liquid phase, and a second headspace sample is run. The difference between the two chromatograms indicates which volatile components contain the functional group.

6 Quantitative Headspace Analysis

As mentioned in Sect. 2 (static headspace method) the equation

$$X_B = \frac{A_B}{C_B \gamma_B P_B^o}$$

is the fundamental relationships for headspace analysis.

In headspace analysis the chromatographic peak area of a substance in the gas phase, is an indirect measure of the concentration of that same substance in the original liquid phase. The liquid and gas phase concentration are related to each other by the activity coefficient and an experimentally derived proportionality constant. The activity coefficient is matrix dependent and unknown for most analyses; it must therefore be accounted for by calibrating the sampling system. To improve the reproducibility of headspace analyses, especially when manually operated gas syringes are used, an internal standard is added to the sample being analyzed. Poole and Schuette (1983) have summarized the properties desired of an internal standard (see Table 2).

Table 2. Properties desired in an ideal internal standard

a) Should be separable chromatographically from all other components in the sample.
b) Should have an elution time close to that of the analyte to be determined.
c) Should respond to the detector in a similar way to the analyte being determined.
d) Concentration should be similar to that of the analyte.
e) Should not occur in the original sample.
f) Should act chemically like the analyte being measured.
g) Should be in equilibrium with the matrix which contains the analyte.

Rarely will an internal standard for improving the reproducibility and reliability of an (headspace) analysis meet all of these requirements but properties a, e, and g are the most obvious and important ones for a multicomponent sample headspace analysis.

If the sample matrix can be obtained in a pure form, then calibration is simply performed by adding known amounts of the analyte to the matrix and analyzing the calibration standards in a manner identical to the original samples (Kepner et al. 1964).

In Table 3 suitable calibration methods are summarized (Poole and Schuette 1983; Kolb 1982).

Table 3. Calibration methods for quantitative static headspace analysis

Sample type	Calibration method	Principle	Examples
Homogeneous solutions	Model system	Matrix available in pure form. Known amounts of substance added to pure matrix and determined by the same experimental procedure as samples	Volatile organic compounds in drinking water, beverages, vegetable oils, mineral oils, etc.
Inhomogeneous samples (liquid + solid)	Model system, standard addition	Model system may be used if a pure sample matrix is available (e.g., blood, milk) in the standard addition method, the sample is analyzed twice, the second time after the addition of a known amount of the substance to be determined. The sample must be re-equilibrated after addition of the standard	Residual monomers in polymer dispersions. Alcohol and toxic substances in blood. Olfactory substances in milk
Soluble solids	Model system, standards additions	For solid samples forming homogeneous solutions the model system may be used if pure sample matrix materials are available otherwise the standard additions method is used	Inorganic, organic, and polymeric solids and salts. Monomers in polymers
Insoluble solids	Multiple headspace extraction	MHE can be used for substances of high volatility with a small partition coefficient. Method is based on a stepwise gas extraction at equal time intervals. Normal headspace chromatogram is run, a fraction of the gas phase exhausted, a second headspace chromatogram is run. The difference in peak areas provides a measure of the total peak area of the analyte	Volatiles and monomers in insoluble polymers. Aroma volatiles from foodstuffs fruits, spices, tobacco, etc. Residual solvents in pharmaceuticals and printed films

7 Final Remarks

The origin of the headspace sampling and analysis was at the early sixties and published results of investigations covered the use of packed columns (Weurman 1961; Kepner et al. 1963, 1964; Kunitake 1965). Later wide bore capillary columns 50–100 mm length were developed and 0.5–1.0 ml headspace samples were analyzed by direct injection without splitting (Slotema 1978; Kolb 1984) and finally with the introduction of fused silica capillary columns in combination with cryotrapping techniques, gradually the dynamic headspace method was used in purge and trap systems (Bellar and Lichtenberg 1974; Pankow and Rosen 1984; Badings et al. 1985). The applications of headspace analysis are to numerous to dicuss in detail in this short introduction to headspace analysis. Two recent references dealing with plant volatiles will be mentioned. A dynamic headspace enrichment system for volatiles released by plants has been published by Termonia and Alaerts (1985) and a direct headspace analysis in quality control of aromatic herbs by Chialva et al. (1983).

Much literature concerning headspace gas chromatography has been published, including books (Hachenberg and Schmidt 1977) and contributions devoted to various topics such as for example: residual monomers in polymers (di Pasquale et al. 1978), pollutants in water (McNally and Grob 1984), volatiles in biological material (Mendenhall et al. 1980). The headspace sampling and analysis have advantages and disadvantages just as any other sampling and separation technique.

Positive features are: the procedure performs a very simple and efficient "pre-chromatographic clean up" of the sample; no involatile material is deposited within the injection port or at the column inlet. It should also be noted that the headspace procedure only modifies the mode of injection. Packed or capillary columns may be used and the full range of gas chromatographic detectors also remains available when increased selectivity or sensitivity toward a particular compound is desired. There is also very little possibility of artefact formation and reproducible ratio changes of the components in the vapour can be used to facilitate a given analysis.

Disadvantages are that components with low volatility (higher boiling components) are less easier to analyse and sometimes not at all by headspace techniques and secondly that calibration can be more difficult and time-consuming.

References

Badings HT, de Jong C, Dooper RPM (1985) Automatic system for rapid analysis of volatile compounds by purge-and-cold-trapping capillary gas chromatography. In: Sandra P, Bertsch W (eds) Proc of the 6th Int Symp on Capill. Chromatogr, Riva del Garda. Hüthig, Heidelberg, pp 666–678

Bellar TA, Lichtenberg JJ (1974) Determining volatile organics at microgram-per-liter levels by gas chromatography. J Am Water Works Assoc 66:739–744

Chialva F, Doglia G, Gabri G, Ulian F (1983) Direct headspace GC analysis with glass capillary columns in quality control of aromatic herbs. In: Rijks J (ed) Proc of the 5th Int Symp on Capill. Chromatogr, Riva del Garda. Elsevier Scientific, Amsterdam, pp 430–440

Di Pasquale G, di Iorio G, Cappaccioli T (1978) GC headspace determination of residual acrylonitrile in acrylonitrile-butadiene-styrene resins and migration into a simulated fatty foodstuffs liquid. J Chromatogr 160:133–136

Grob K, Zürcher F (1976) Stripping of trace organic substances from water. Equipment and procedures. J Chromatogr 117:285–294

Hachenberg H, Schmidt AP (1977) Gas chromatographic headspace analysis. Heyden, London

Ioffe BV, Vitenberg AG (1984) Headspace analysis and related methods in gas chromatography. Wiley, New York
Kepner RE, Strating J, Weurman C (1963) Quantitative determination of esters in beer by gas chromatographic analysis of headspace vapours. J Inst Brew 62:399–405
Kepner RE, Maarse H, Strating J (1964) Gas chromatographic headspace techniques for the quantitative determination of volatile components in multicomponent aqueous solutions. Anal Chem 36:77–82
Kolb B (1977) Applied headspace gas chromatography. Heyden, London
Kolb B (1982) Multiple headspace extraction–a procedure for eliminating the influence of the sample matrix in quantitative headspace gas chromatography. Chromatographia 15:587–593
Kolb B (1984) Analysis of food contaminants by headspace gas chromatography. In: Gilbert J (ed) Analysis of food contaminants. Elsevier Applied Science, London, pp 117–156
Kolb B, Pospisil P, Borath T, Auer A (1979) Headspace gas chromatography with glass capillaries using an automatic electropneumatic dosing system. HRC & CC 2:283–278
Kunitake N (1965) Gas chromatographic determination of low boiling volatiles in beer without any concentration steps. Bull Brew Sci 11:1–9
McNally ME, Grob RL (1984) Headspace determination of solubility limits of the base neutral and volatile components from the Environmental Protection Agency's list of priority pollutants. J Chromatogr 284:105–116
Mendenhall ChL, Mac Gee J, Green ES (1980) Simple rapid and sensitive method for the simultaneous quantitation of ethanol and acetaldehyde in biological materials using headspace gas chromatography. J Chromatogr 190:197–200
Pankow JF, Rosen ME (1984) The analysis of volatile compounds by purge and trap with whole column cryotrapping on a fused silica capillary column. HRC & CC 7:504–508
Poole CF, Schuette SA (1983) Isolation and Concentration techniques for capillary column gas chromatographic analysis. HRC & CC 6:526–549
Poole CF, Schuette SA (1984) Contemporary practice of chromatography. Elsevier, Amsterdam, pp 463–473
Slotema FP (1978) Applications of chromatography in the brewery. Brew Dig 53:48–52
Termonia M, Allaerts G (1985) Dynamic headspace enrichment system for CGC analysis of volatiles released by plants. In: Sandra P, Bertsch W (eds) Proc of the 6th Int Symp on Capill. Chromatogr, Riva del Garda. Hüthig, Heidelberg, pp 538–545
Weurman C (1961) The enzymatic formation of volatile components in raspberries. Food Technol 15:531–536
Weurman C (1969) Isolation and concentration of volatiles in food odor research. J Agric Food Chem 17:370–384
Weurman C (1974) Sampling in air-borne odorant analysis. In: Johnston JW, Moulton DG, Turk A (eds) Human responses to environmental odors. Academic, New York, pp 263–328

The Use of Principal Components Analysis in Monitoring the Quality of Beer

T. Peppard

1 Introduction

In recent years, brewing scientists have made increasing use of multivariate statistical techniques in attempts to gain a better understanding of the problems associated with beer and the raw materials of brewing. Principally, this has come about for two reasons. First, a growing awareness that many such problems are genuinely multivariate in nature, and secondly, the increasing availability of relatively low-cost microcomputers and suitable software packages. The area in which these techniques have had most impact has been that of beer flavour, in which the task of multivariate statistics is essentially to correlate flavour judgements (descriptive or hedonic) and/or physicochemical measurements, collected for a (frequently large) number of samples. It cannot be overstressed, however, that the techniques themselves, like any based on statistics, merely establish *correlations* and do not prove *causal relationships*. The latter can come about only through further experimentation, although statistical correlation does at least provide a good starting point for such studies.

Multivariate techniques are applicable to sets of products or samples for which multivariate data (i.e. data comprising measurements of many variables) have been collected. They have been used extensively in various branches of pure and applied science, as well as in fields such as psychology and the social sciences (Marriott 1974; Harris 1975). An excellent overview of the scope and limitations of multivariate techniques, as relating to flavour studies, has been published by Powers (1982). Frequently used methods include Cluster Analysis, Discriminant Analysis, and Multiple Regression Analysis, whilst more recently, techniques such as Procrustes Analysis and Partial Least Squares Regression Analysis have come to the fore (e.g. Williams et al. 1984). However, one of the most powerful multivariate procedures employed is Principal Components Analysis (PCA) (Marriott 1974; Harris 1975); the use of this technique in flavour studies has been described in detail by Vuataz (1976/77). PCA is related to another multivariate technique known as factor analysis, but authors appear to differ in whether they consider PCA to be a type of factor analysis or a distinct technique.

In the past, the application of multivariate techniques such as PCA has required access to mainframe computers and sophisticated statistical software packages such as BMDP (1985) and SAS (1985). Nowadays, however, it is possible to purchase software enabling PCA to be run on laboratory microcomputers; the technique is therefore potentially available to most researchers. It is the purpose of this chapter to describe the general principles behind PCA, and to demonstrate its use in the field of brewing science.

2 The Principles of PCA

2.1 The Objectives

Principal Components Analysis is based on Euclidean geometry and matrix algebra. It is used to reduce the effective number of dimensions in which a multivariate problem must be considered, by concentrating most of the variance in a data set into a much smaller number of *new* variables (termed principal components). Orthogonal least-squares are used to determine the first principal component axis, with subsequent axes following from successive orthogonal least-squares solutions. Being orthogonal, the axes are perpendicular and hence independent of one another. When the results of PCA are plotted on Cartesian co-ordinates, the first axis represents the maximum variance in the data configuration, the second axis the second largest variance, and so on. PCA therefore enables classification of the individual members of a set of samples according to similarity/dissimilarity, when the results are represented graphically by one or more plots on two-dimensional co-ordinates. Furthermore, the technique allows investigation of the underlying relationships which exist between variables, and enables interpretation of the classification achieved in terms of these variables.

The method of PCA is perfectly general; it involves few assumptions about the original variables, no hypothesis that can be tested and no underlying model (Marriott 1974). It is simply a different and more convenient way of expressing the same set of results.

2.2 Conceptualizing the Multivariate Problem

Consider a set of products or samples, e.g. six beers (labelled A–F), which has been characterized by a small number of measurements (say, three) on each. These measurements may be derived from instrumental analysis, sensory analysis

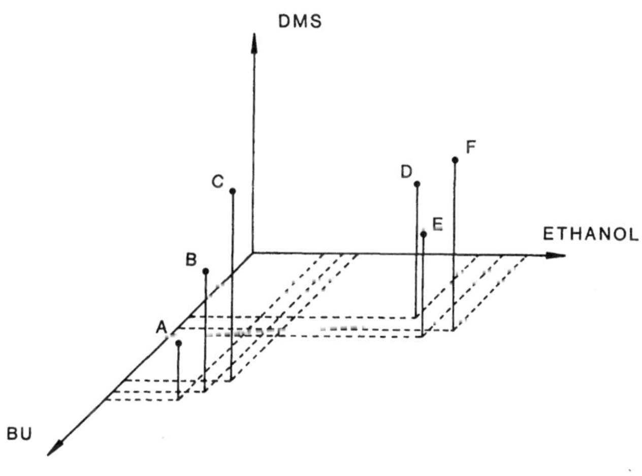

Fig. 1. Plotting beers on three-dimensional Cartesian co-ordinates. (J Inst Brew, 1985, p 326)

or a mixture of both. For the sake of argument, however, let them be analytical bitterness (BU) as measured spectrophotometrically, and dimethyl sulphide (DMS) and ethanol concentrations as measured instrumentally by gas chromatography. It would be possible to investigate the degree of similarity/dissimilarity between members of the set, and therefore to determine the existence of any natural groupings, by plotting the six beers on three-dimensional co-ordinates, as illustrated in Fig. 1. Clearly, in this simple example, the beers tend to fall into two groups, namely A–C and D–F. Furthermore, it is possible to interpret this classification in terms of the measured variables. Thus members of the first group (A–C) tend to have higher BU values and lower concentrations of ethanol than those of the second group (D–F). On the other hand, the concentration of DMS appears to have very little bearing on the observed grouping. The method illustrated here involves just six samples and three variables. In practice, both the number of samples to be compared and the number of variables to be measured are likely to be considerably higher. The use of 20 variables, for example, would require the use of 20-dimensional co-ordinates, making the above approach obviously impracticable. However, this is exactly the type of problem to which PCA can be usefully applied.

2.3 A Stepwise Explanation of PCA

The various stages involved in carrying out PCA are described in this section, together with an explanation of some of the terminology used. However, it should be remembered that with the use of commercially available computer programmes, many of the steps described are, in practice, carried out automatically and do not require input from the operator.

The results of PCA are not independent of scale; multiplying one of the original variables by a constant, for example, will change the result. Frequently the variables are observations of quite different types, recorded in different units or on different scales. Therefore one of the first steps in PCA involves putting the data into a "standardized" form, e.g. all measurements for each variable in turn may be replaced by the term:

$$(x_{ij}-\bar{x}_j)/\sigma_j,$$

where x_{ij} is the measurement of the jth variable on the ith sample, whilst $\bar{x}_j$ and σ_j are the mean and the standard deviation, respectively, of all the measurements made on the jth variable. In this standardized form, data for each variable have a mean value of zero, and unit standard deviation. Thus, measurements on all variables will be expressed in comparable units of standard deviation.

The next stage of the analysis is to establish a correlation matrix involving all variables, i.e. a table indicating the correlation coefficients between all possible pairs of variables. Those variables, which are highly intercorrelated, are identified, and the common variance within this group of variables replaced by a derived variable known as the first principal component. The aspect of the behaviour of the original variables that has been accounted for by the first principal component is removed from the data, and the matrix of residual correlation calculated. From this, a second group of variables is sought, and their common variance replaced by a second principal component which is orthogonal (i.e. not correlated) to the first component. The effects attributed to the second component are then removed from the data, and the above processes repeated until all of the variance in the original data set has been accommodated. The end result of this stage of the analysis is transformation of the original set of variables into the *same number* of new variables (components), all of which are orthogonal.

For each sample, therefore, the standardized measurements on the original variables (e.g. BU, DMS concentration, etc., in the example given in Sect. 2.2) are transformed into values on the new set of component axes. For any one sample, one of the original measurements can be expressed as a linear combination of values on the component axes. Thus, for a data set based on measurements of nine variables:

$$V_1 = [L_{1,1} \cdot C_1] + [L_{1,2} \cdot C_2] + \ldots + [L_{1,9} \cdot C_9]$$
$$V_2 = [L_{2,1} \cdot C_1] + [L_{2,2} \cdot C_2] + \ldots + [L_{2,9} \cdot C_9]$$
$$\ldots$$
$$\ldots$$
$$V_9 = [L_{9,1} \cdot C_1] + [L_{9,2} \cdot C_2] + \ldots + [L_{9,9} \cdot C_9],$$

where V_1, V_2, ... V_9 are standardized measurements on the original nine variables, and C_1, C_2, ... C_9 are values on the nine component axes. $L_{1,1}$, $L_{1,2}$, ... $L_{9,9}$ measure the degree of correlation between original variables and new components, and are known as the component loadings.

At this point in the analysis, a number of components (equal to the number of original variables) will have been computed. However, the major part of the total variance within a data set is usually highly concentrated within the first few components. Therefore, a relatively small number of the components will be selected on the basis of the minimum acceptable proportion of total variance to be accounted for. This will be determined either by the operator, or by the computer using criteria built into the program. In either case, however, the decision as to how many principal components to select is likely to be based on consideration of the so-called eigenvalues and communalities.

To illustrate the use of eigenvalues and communalities, consider the data shown in Table 1 (the origin of this data set will be discussed later in Sect. 3.1). For the sake of simplicity, only the component loadings of the first three principal components are given, as these explain approx. 66% of the variance in the data set and were the ones ultimately used in the analysis. The relationship between the first principal component (C-1) and, for example, the variable colour, is described by the component loading of 0.7143; this is actually the correlation coefficient between C-1 and colour, whilst the square of this value (ca. 0.51) represents the amount of shared variance. The extent to which a given component explains the variance in the original data can be assessed by calculating the sum of the squared component loadings for that component across the entire set of original variables. These sums, which are known as eigenvalues, are computed for each component, as indicated in Table 1. The sum of the eigenvalues across all of the components is, by necessity, equal to the number of original variables; the average

Table 1. Component loading matrix and associated eigenvalues, communalities and variances[a]

Original variables	Components								Communality
	C-1	C-2	C-3	C-4	C-5	C-6	C-7	C-8	
1. pH	−0.4808	0.5016	0.3512						0.61
2. Colour	0.7143	0.3042	0.4246						0.78
3. Gravity	0.1185	0.7639	0.1341						0.62
4. TSN	−0.1237	−0.2509	0.8028						0.72
5. FAN	−0.7878	−0.2656	−0.0948						0.70
6. BU	−0.6853	−0.2219	0.3412						0.64
7. Foam	−0.6472	0.4434	0.1113						0.63
8. CO_2	−0.3888	0.4427	−0.4633						0.56
Eigenvalue	2.43	1.50	1.32	0.90	0.60	0.51	0.45	0.29	
Variance (%)	30.39	18.79	16.48	11.23	7.54	6.35	5.61	3.62	
Cumulative variance (%)	30.39	49.18	65.66	76.89	84.42	90.77	96.38	100.00	

Computations: Communality for jth variable = sum of squares of entries in row j.
Eigenvalue for kth component = sum of squares in column k.
Variance explained by kth component = eigenvalue for kth component, divided by number of original variables [=8] (expressed as %).

[a] Permission to publish the data contained within Table 1 kindly provided by the Director of the Brewing Research Foundation.

eigenvalue is therefore equal to unity. Consequently, an eigenvalue of greater than 1.0 indicates that the component accounts for more variance than the average variable, and this is frequently used as the cut-off point for deciding how many components to utilize in the analysis. The variance explained by a particular component is calculated simply by dividing its eigenvalue by the number of original variables, and expressing the result as a percentage. As the complete set of components would be required to account for all of the variance in each variable, the selection of just a few components will lead to a portion of the variance being left unaccounted for. The communality of a variable is the proportion of that variable's variance which is accounted for by the chosen components; it is computed from the sum of the squares of the variable's loadings on those selected components. Clearly, it is desirable to have high values for the communalities of as many variables as possible.

The last stage of PCA is to calculate for all samples, values on each of the selected principal component axes. These values, which are known as the component scores, together with the component loadings, represent the main result of the analysis. Since most of the variance in a data set is generally concentrated in the first few components, a series of plots of standardized component scores on two-dimensional co-ordinates enables classification of the members of the set according to similarity/dissimilarity. Furthermore, superimposition of the component loadings onto such plots allows interpretation of the classification achieved in terms of the original variables. These points, and many of the others made in this section, are probably best illustrated by reference to the example discussed below in Sect. 3.1.

Principal Components Analysis can be carried out as described above. However, there are a number of so-called rotation procedures which may be used in order to make the results of PCA more readily interpretable. In the most well-known of these procedures, Varimax rotation, the solution is mathematically rotated in order to obtain a pattern of component loadings which comprise the maximum number of high and low values (i.e. values near unity or zero) and the minimum number of intermediate values. The aim is to simplify the interpretation of component axes in terms of the original variables, so that components and original variables are either very highly or very poorly correlated. Rotation may be thought of as the movement of the investigator from one perspective to another in a search for the most meaningful representation of the data!

3 The Use of PCA in Brewing Science

During the last decade, Principal Components Analysis has been used by a number of investigators as an aid to solving problems in the field of brewing science. The technique has mostly been used to probe degrees of similarity/dissimilarity existing within sets of distinctly different samples, the members of which vary to a greater or lesser extent. Some of these applications of PCA are described in Sect. 3.2. However, PCA can also be used to deal with situations involving subtle changes in the nature of one particular product. This can, for example, involve monitoring the consistency of specific product lines from a brewery, or following changes in a particular batch of product during storage. In either case, the advantage of using PCA over the univariate alternative, i.e. plotting changes in individual variables with time, is that the former more easily reveals product drift due to concerted shifts in a number of linked variables.

Peppard (1985a), working at the Brewing Research Foundation (B.R.F.) in Great Britain, recently demonstrated the application of PCA in monitoring the consistency of beer quality, by making both between-brand and within-brand comparisons based on the results of instrumental analysis. Section 3.1 gives a detailed account of the use of within-brand product consistency tracking by PCA, for beers originating from a pilot-scale brewery. An important function of this section is to highlight some of the concepts described in Sect. 2.3.

3.1 A Detailed Example of the Use of PCA in Brewing Science

At the time the work was carried out, the experimental pilot-scale brewery at the B.R.F. produced annually more than 400 beer samples, all of which were routinely analyzed for a number of different parameters related to various facets of beer quality. Many of the beers were brewed as part of a broad-based R&D programme, involving many aspects of raw materials and processing. However, a regular supply of control or "standard" beer was required for these investigations, and it was routine analytical data from this single "product line" which was used in the studies described.

Table 2 gives the analytical data which were obtained for 28 standard beers during the course of 1 year. The eight analytical parameters measured were pH, colour, gravity, total soluble nitrogen (TSN), free amino nitrogen (FAN), bitterness units (BU), foam and carbon dioxide level; these were mostly determined using methods recommended by the Institute of Brewing (1982). The need for standardization is clearly seen in this data set, where the measurements of foam, for example, are approximately an order of magnitude greater than are those for, say, colour.

Principal Components Analysis, carried out on the data set presented in Table 2, yielded the component loading matrix and associated eigenvalues, communalities and variances previously given in Table 1. As noted earlier (see Sect. 2.3), only the component loadings of the first three principal components (having eigenvalues greater than unity) are given. This three-component solution accounted for approx. 66% of the variance in the data set, although of course this still leaves approx. 34% unaccounted for. Moreover, communalities of the original variables ranged from 0.56 to 0.78. In other words, for some of the variables (e.g. carbon dioxide level), a little over half of the original variance was incorporated into the three-component solution. However, despite these caveats, the results of PCA were still able to provide a useful insight into the inter-relationships existing within this set of analytical data. It should be noted here that rotation techniques, such as Varimax rotation mentioned above (Sect. 2.3), were not used in the analysis.

Table 2. Routine analytical data for 28 beers on 8 parameters[a]

Beer no.	Variables							
	pH	Colour	Gravity	TSN	FAN	BU	Foam	CO_2
1	3.90	11.6	10.0	49.4	4.9	24	117.8	2.82
2	3.94	10.6	9.5	48.4	4.1	24	102.3	2.52
3	3.78	14.0	10.0	46.2	3.7	19	96.9	2.35
4	3.87	13.3	9.1	49.4	3.8	22	102.0	2.30
5	3.82	12.6	8.6	48.4	3.9	24	97.2	2.21
6	3.80	11.8	9.3	50.7	4.7	26	102.9	1.83
7	3.80	14.4	8.6	52.8	4.5	23	94.5	2.25
8	3.83	14.0	10.1	52.7	3.7	29	97.2	2.31
9	3.86	18.8	9.3	52.3	3.7	23	96.9	2.29
10	3.84	13.7	10.0	50.8	4.0	24	107.0	1.81
11	3.89	14.4	9.8	53.2	3.6	22	97.0	2.25
12	3.77	12.8	10.4	51.6	3.8	22	101.0	2.12
13	3.82	11.6	9.2	53.6	4.2	24	102.0	2.46
14	3.81	12.0	9.1	52.3	3.4	24	100.0	2.36
15	3.94	10.9	10.1	50.1	5.0	24	101.7	2.68
16	3.74	12.9	9.9	44.2	3.7	20	88.6	2.29
17	3.70	11.7	8.8	50.3	3.9	22	96.4	2.36
18	3.87	13.5	8.5	49.8	4.0	24	102.3	2.39
19	3.85	12.1	9.7	51.7	4.6	26	101.3	2.03
20	3.82	13.9	8.7	50.7	4.6	22	94.2	2.36
21	3.79	13.7	10.0	44.6	3.6	24	101.0	2.58
22	3.83	26.8	10.2	52.7	2.9	19	96.0	2.16
23	3.87	13.4	10.0	51.2	4.1	25	102.0	2.46
24	3.91	14.5	10.1	51.5	3.7	22	95.0	2.27
25	3.78	12.6	10.1	51.9	3.9	23	107.0	2.37
26	3.80	11.1	8.7	53.9	4.6	24	93.0	2.28
27	3.82	11.7	9.7	53.8	4.0	22	101.0	2.44
28	3.82	11.8	9.1	52.3	3.9	23	104.0	2.52
Mean	3.83	13.4	9.5	50.7	4.0	23	99.9	2.32
Std. dev.	0.05	3.0	0.6	2.5	0.5	2	5.4	0.21

[a] Permission to publish the data contained within Table 2 kindly provided by the Director of the Brewing Research Foundation.

Figure 2 shows the component loadings of the original variables and the standardized component scores for all 28 beers on the first two principal component axes (which account for approx. 49% of the variance in the data set). It can be seen that whilst the majority of beers are clustered together quite closely, a few are distributed away from this main group. It turns out that of these, beers 1, 2, 3, 15, and 16 were brewed using a different supply of malt compared with other beers. On the other hand, beer 22 had actually been included in the data-base by mistake; it was in fact not from a standard brew, having received a post-fermentation addition of synthetic melanoidin. This is the dark brown material largely responsible for the colour of beer; therefore the relative position of beer 22 is readily interpreted, as might be expected, in terms of the component loadings for the variable colour.

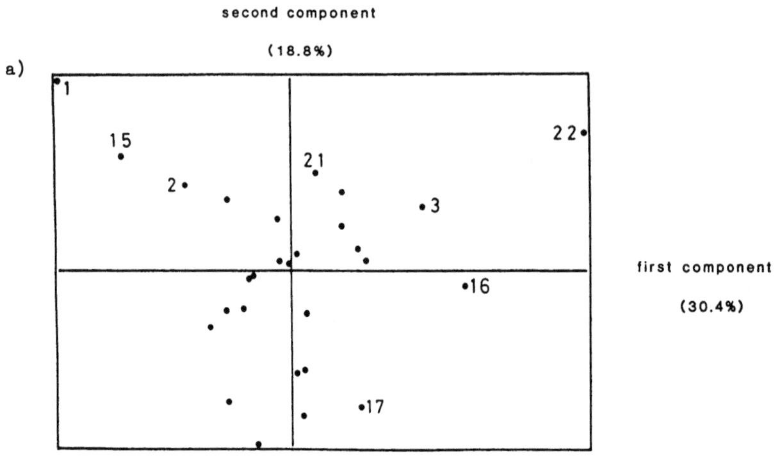

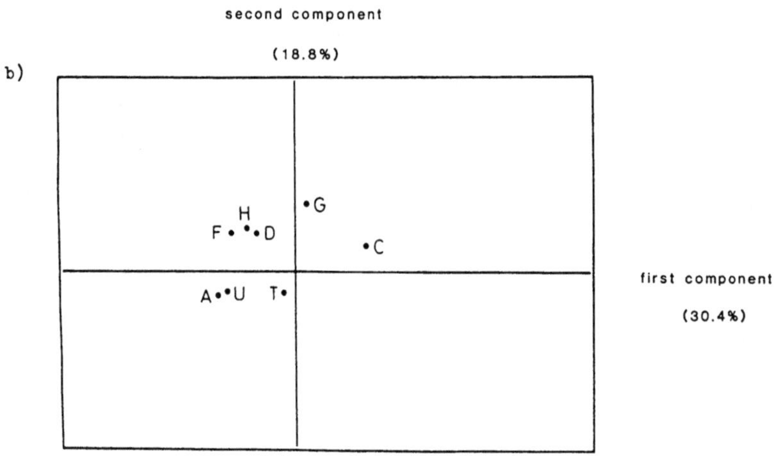

Fig. 2 a, b. Result of PCA on analytical data from 28 pilot brewery scale beers. **a** Standardized component scores. **b** Component loadings, on the 1st and 2nd principal component axes. Identification of variables: *A* FAN (free amino nitrogen); *C* colour; *D* CO_2; *F* foam; *G* gravity; *H* pH; *T* TSN (total soluble nitrogen); *U* BU. (J Inst Brew, 1985, p 327)

For many data sets, examination of the first two principal components leaves a relatively large proportion of variance still unaccounted for. In such cases it can be worthwhile investigating further components. This can be seen in the present example, where the first two components account for only about 49% of the original variance, whilst the third component, for example, explains a further ca. 16% of the variance. Figure 3 shows the component loadings of the original variables and the standardized component scores for all 28 beers on the first and third principal component axes. As in Fig. 2, the majority of beers are clustered together, notable exceptions again include beers 1, 2, 3, 15, 16, and 22, but additionally beers 17 and 21. The relative positions of the latter two beers (i.e. as outliers) may again be interpreted by consideration of the component loadings of the original

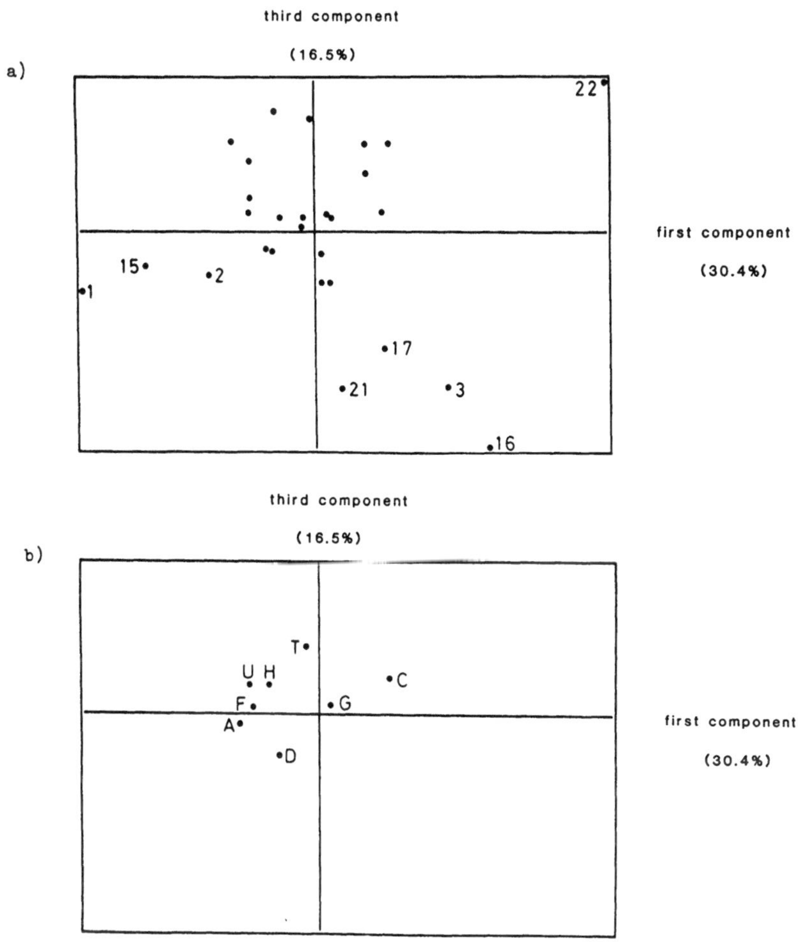

Fig. 3 a, b. Result of PCA on analytical data from 28 pilot brewery scale beers. **a** Standardized component scores. **b** Component loadings, on the 1st and 3rd principal component axes. Identification of variables: *A* FAN (free amino nitrogen); *C* colour; *D* CO_2; *F* foam; *G* gravity; *H* pH; *T* TSN (total soluble nitrogen); *U* BU. (J Inst Brew, 1985, p 328)

variables. Beer 21 has a rather low TSN measurement (44.6 cf. the mean value of 50.7 and the standard deviation of 2.5), whilst beer 17 has a rather low pH (3.70 cf. the mean value of 3.83 and the standard deviation of 0.05).

For product consistency checks based on within-brand comparisons, a database of analytical results built up gradually over an extended period of time can be subjected to PCA after the addition of each new sample. With plots of standardized component scores on principal component axes, a 95% confidence ellipse may be superimposed, as demonstrated in Fig. 4. This indicates the approximate area into which there is a 95% probability that the next sample will fall. It is derived from values of the mean and standard deviation on each principal component axis plotted, after the exclusion of beers designated as outliers (e.g. beers

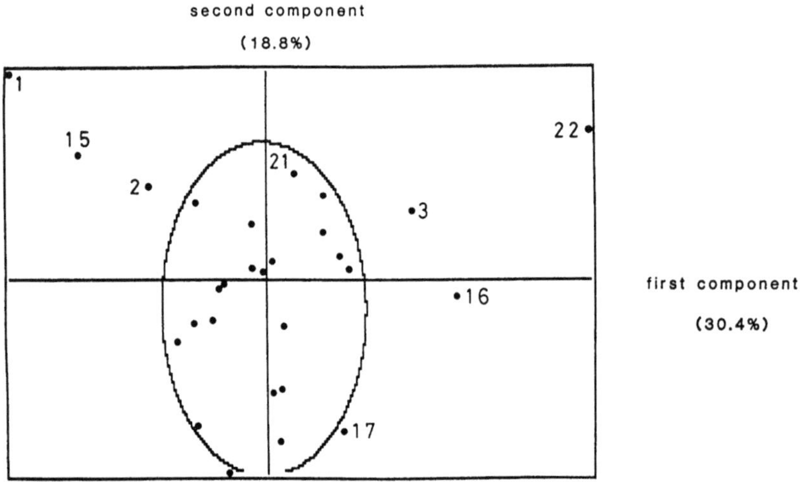

Fig. 4. Result of PCA on analytical data from 28 pilot brewery scale beers. Standardized component scores on the 1st and 2nd principal component axes, showing the 95% confidence ellipse. (J Inst Brew, 1985, p 328)

1, 2, 3, 15, 16, and 22 in Fig. 4). The elliptical nature of the 95% confidence perimeter is due simply to the different scales used on the two principal component axes. It should be noted that the results of PCA are strongly influenced by the presence of outliers (e.g. MacFie 1983). Thus the spatial arrangement obtained for a series of beers will be significantly perturbed by the addition (or deletion) of an outlier. It is therefore recommended that beers designated as outliers (i.e. by inspection, rather than on the basis of a formal statistical test) are not retained amongst those in the data-base to be included in future Principal Components Analyses.

3.2 Other Examples of the Use of PCA in Brewing Science

In brewing science, PCA has been most widely used in the area of flavour, where it is generally employed in one of three ways. In the first of these, the technique is used to classify the individual members of a set of samples (according to similarity(dissimilarity) on the basis of sensory data alone. The second use relies instead on the results of instrumental and/or wet chemical analyses, whilst in the third, results of both sensory and objective tests are employed. Some examples of each of these uses of PCA are discussed below.

Of the various types of sensory testing which are commonly applied to beer, that known as flavour profiling or descriptive flavour analysis (Hudson 1982) produces the most voluminous data. Moreover, the data set is multivariate, as it comprises scores for each beer on a number of different descriptive flavour terms (or attributes), and as such it is suitable for treatment by PCA. Typically, a flavour profile test might involve tasters assessing beer for 6–25 individual flavour attributes.

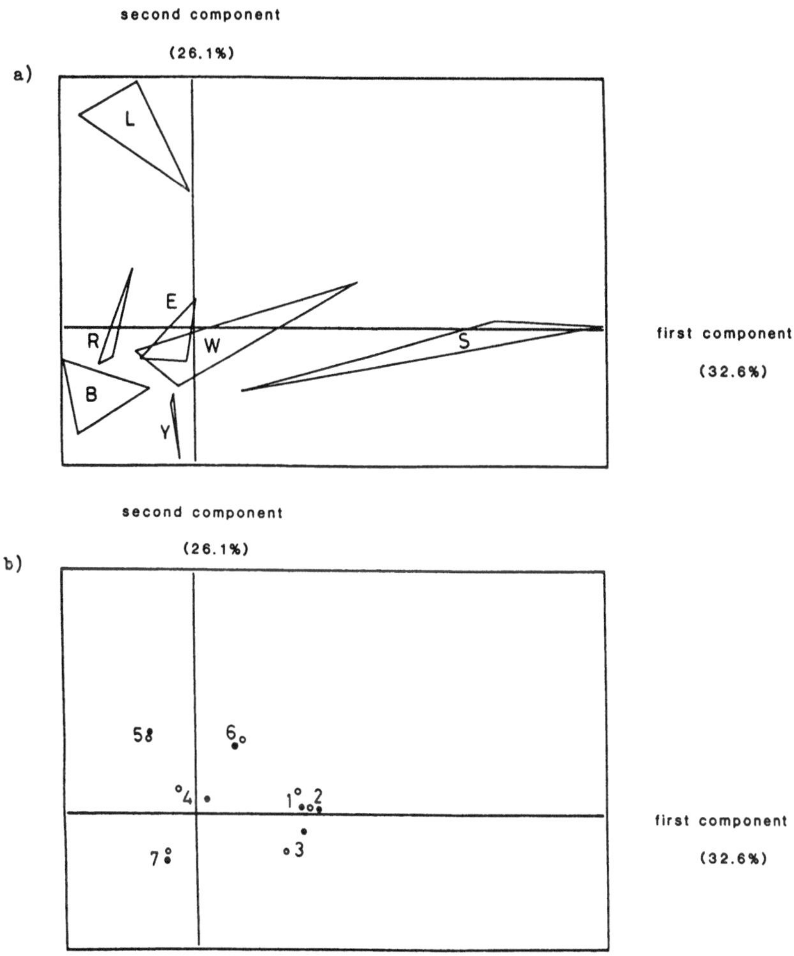

Fig. 5 a, b. Result of PCA on descriptive data from 7 beers. **a** Standardized component scores. **b** Component loadings, on the 1st and 2nd Principal component axes. ● = aroma, ○ = taste; *1* light-struck; *2* sulphidic (H_2S); *3* rubbery; *4* cabbagy; *5* DMS; *6* onion-like; *7* meaty. (J Inst Brew, 1985, p 366)

Figure 5 illustrates the use of PCA for determining the ability of a flavour profile panel to discriminate between the sulphury characteristics exhibited by a series of seven ales (Peppard 1985 b). All beers were tasted on three occasions, in accordance with an incomplete block design which involved seven tasting sessions with three beers assessed in each, so that ultimately each beer was compared against every other. Therefore the different beers are represented in Fig. 5 by seven triangles with the three replicate analyses as vertices. Clearly, seven relatively small, well-separated triangles would indicate the panel's ability to discriminate successfully between all beers. On the other hand, large over-lapping triangles would be indicative of poor discrimination. Therefore, in the example shown in Fig. 5, the panel has been able to discriminate to some extent between all beers except W and E. Beers L and S are particularly well separated from the remainder (and each other), although one replicate of beer S showed rather poor agreement with the other two. As judged by the positions of the beers relative to the original variables (cf. Fig. 5a with b), beer L was characterized by DMS flavour, whilst beer S was characterized somewhat by sulphidic, light-struck and rubbery flavour notes.

Clapperton (1979) (see also Clapperton and Piggott 1979a) was able to differentiate between the flavours of ales and lagers, each of two different alcoholic strengths, by subjecting the results of descriptive flavour analysis to PCA. The flavour panel tasted 32 beer samples, comprising two samples of each of 16 brands of beer selected to represent the four different types. Plots of results on the first two principal component axes showed resolution of the four groups of beers, and the close proximity of the majority of duplicate samples. The first component separated the two strengths of lager beer and the two strengths of ale; sensory parameters associated with the first component were those known to be directly linked with the original gravity of beer, such as the flavour attributes body, high-gravity fullness, estery and viscous. Both of the first two principal components contributed to the separation of ales from lagers. The flavour notes responsible, toffee-like, caramel and burnt for ales, and sulphury, cooked vegetable and DMS in the case of lagers, are related to the different types of malt that are used in brewing these two types of beer.

In separate studies, Clapperton and Piggott (1979b) used the technique of PCA to compare the results of descriptive flavour analysis, obtained using panels of trained and untrained assessors. They showed that training and experience improved reproducibility and discrimination of tasters, without distorting the general pattern of flavour terms reported. Further, they claimed to be able to identify assessors as either novices, those with some experience or experts, on the basis of their performance.

Arkima et al. (1982) investigated the influence of volatile sulphur compounds on beer flavour using Principal Components Analysis. They treated three Finnish lagers with a series of volatile sulphur compounds, one compound at a time. Each spiked beer was then analyzed by descriptive flavour analysis, and the results obtained for each of the three series of spiked lagers in turn were subjected to PCA. They were able to elucidate relationships between the various (sulphury and non-sulphury related) flavour attributes used, and to relate specific flavour descriptors to the addition of individual sulphur compounds. Interestingly, these effects were found to differ somewhat between the three series of lagers.

Beer is an exceedingly complex product, whose quality is determined by many factors which include those physical, chemical and biological in nature. Therefore, numerous parameters related to various aspects of beer quality have been measured in order to provide a means of ensuring a sound and consistent product. These include parameters determined by microbiological methods, and procedures involving instrumental and/or wet chemical analysis. A number of analyses, e.g. foam, pH and BU, were mentioned briefly in connection with the use of PCA described above in Sect. 3.1. However, the type of analysis which yields the most voluminous data is that based on a chromatographic method such as high performance liquid chromatography (HPLC), and particularly gas chromatography (GC). Beer is reported to contain more than 800 components (Meilgaard 1981), and it is not uncommon, when using a technique such as capillary GC, to be able to resolve between 100 and 200 of these. Such analyses therefore readily yield large volumes of multivariate data which are suitable for treatment by PCA.

In attempting to monitor the consistency of beer quality, Peppard (1985a) made both between- and within-brand comparisons, based on subjecting the results of physicochemical analyses to PCA. The within-brand comparisons have already been discussed in detail in Sect. 3.1. For the between-brand comparisons, BU, colour, pH and the levels of seven volatile flavour constituents (as measured by GC) were determined for a series of commercial beer samples. A plot of the component scores on the first two principal component axes mapped the degrees of similarity/dissimilarity amongst beers. It was described how this type of plot could be a useful means of regularly monitoring the various products of a brewery, at least where a sufficient number of product lines are involved. Assuming that the majority of products within a given set of regularly produced beers are consistently "sound" (i.e. within target specification), a marked shift in the position of one beer, relative to the others, would act as an indication that that particular batch is not sound.

Moll and co-workers (1981a) and Moll and Flayeux (1982) have used PCA and other multivariate statistical methods in attempts to predict the organoleptic quality of beer from results of physicochemical measurements made on the corresponding malts and barleys. Beers were classified as good, average or poor, as regards flavour quality, on the basis of PCA using sensory data obtained by descriptive flavour analysis. The groupings obtained were then compared with those achieved using PCA (and factorial discriminant analysis) of malt or barley data alone. The raw material parameters measured included proteins, polyphenols, minerals, enzymes and, in the case of malt, total nitrogen, diastatic power and fermentable extract. The reliability of flavour prediction in these studies was found to be quite good, although malt analytical data performed rather better than did data from barley. In addition, a significant dependence of the results on barley crop year was noted (using either malt or barley data).

In separate studies, Moll et al. (1981b) used PCA in combination with other techniques to predict changes in the organoleptic quality of beer during storage. They collected sensory and analytical data from several series of beers, each comprising products of different original gravities from different breweries. Prior to tasting or analysis, beers were subjected to four storage treatments: 6 days at 0 °C, 6 days at 40 °C (accelerated ageing test conditions), 6 months at 0 °C or 6 months at 25 °C (likely storage conditions in trade). The sensory data consisted of scores for 34 flavour attributes; the analytical data included measurements of polyphenols, so-called sensitive proteins, metal ions, bitter substances, original gravity, DMS, free amino nitrogen and flavor volatiles extracted by carbon disulphide. From subjecting the analytical data alone to PCA, it was concluded that the accelerated ageing technique did not allow the prediction of beer flavour stability, often giving results more similar to those obtained from the treatment using 6 days at 0 °C than to those using 6 months at 25 °C. When sensory data from the 6 days/0 °C and 6 months/25 °C beers were subjected to PCA, the results could readily be rationalized in terms of the positive flavour attributes and the flavour defects making up the 34 original terms. The results of this analysis were then used, in combination with those of hierarchical ascending classification, to calculate a "stability index" for each beer. Accordingly, beers were classified as being of poor, medium or good flavour stability. Finally, Discriminant Factorial Analysis of the analytical data was used to select a set of 12 analytical parameters which were found to be capable of reproducing the same classification, and therefore of predicting the organoleptic stability of the beers during 6 months storage at 25 °C.

As mentioned in the Introduction, Factor Analysis is a multivariate statistical technique somewhat related to Principal Components Analysis. In the field of brewing science, it has been applied to problems of a similar nature to those discussed above. Notable uses, discussion of which is beyond the scope of the present chapter, have been reported by Reiner (1971), Narziss and Reiner (1972), Reiner and Piendl (1974), Jacobsen et al. (1979), Jacobsen and Volden (1981), and Narziss et al. (1985).

4 Conclusions

Principal Components Analysis enables classification of the members of a set of samples (for which multivariate data have been collected) according to similarity/dissimilarity; the results are generally represented graphically by one or more plots on two-dimensional Cartesian co-ordinates. Furthermore, the technique allows investigation of the underlying relationships which exist between variables, and enables interpretation of the classification achieved in terms of these variables.

PCA is potentially available to all researchers having access to a laboratory microcomputer. In the field of brewing science, the technique has been applied to data obtained by combinations of sensory, instrumental and wet chemical analysis, where it has been used both in investigational work, as well as in monitoring the consistency of beer quality.

Acknowledgement. The author wishes to thank the Stroh Brewery Co. for permission to publish this chapter.

References

Arkima V, Jounela-Eriksson P, Leppänen O (1982) The influence of volatile sulphur compounds on beer flavour. Eur Brew Conv Monograph VII, Proc Flavour Symp. Brauwelt, Nürnberg, p 89

BMDP (1985) BMDP statistical software. University of California Press, Berkeley, CA

Clapperton JF (1979) Sensory characterisation of the flavour of beer. In: Land DG, Nursten HE (eds) Progress in flavour research. Applied Science, London, p 1

Clapperton JF, Piggott JR (1979a) Differentiation of ale and lager flavours by principal components analysis of flavour characterization data. J Inst Brew 85:271–274
Clapperton JF, Piggott JR (1979b) Flavour characterization by trained and untrained assessors. J Inst Brew 85:275–277
Harris RJ (1975) A primer of multivariate statistics. Academic, New York
Hudson JR (1982) Profile taste assessment. Eur Brew Conv Monograph VII, Proc Flavour Symp. Brauwelt, Nürnberg, p 17
IOB (1982) Institute of Brewing. Recommended methods of analysis (revision)
Jacobsen T, Volden R (1981) Variations in trace elements in malt as shown by factor analysis. MBAA Tech Q 18(3):122–125
Jacobsen T, Volden R, Engan S, Aubert O (1979) A chemometric study of some beer flavour components. J Inst Brew 85:265–270
MacFie HJH (1983) Some useful multivariate techniques relevant to classifying and identifying microbes. In: Martens H, Russwurm H (eds) Food research and data analysis. Applied Science, London, p 215
Marriott FHC (1974) The interpretation of multiple observations. Academic, London
Meilgaard MC (1981) Beer flavour. Thesis, Technical University of Denmark, Copenhagen
Moll M, Flayeux R (1982) Practical Studies of the role of raw materials in beer flavour. Eur Brew Conv Monograph. VII. Proc Flavour Symp. Brauwelt, Nürnberg, p 45
Moll M, Flayeux R, Muller P (1981a) Relationship between beer flavour and the composition of barley and malt. MBAA Tech Q 18(1):31–35
Moll M, Vinh T, Flayeux R, Muller P, Monnez J-M (1981b) Prediction of the organoleptic quality of beer. In: Charalambous G, Inglett G (eds) The quality of foods and beverages. Academic, New York, p 147 (Chemistry & technology, vol 1)
Narziss L, Reiner L (1972) Zusammenhänge zwischen Malz- und Bierqualität, dargestellt mit Hilfe der Faktoranalyse. Brauwissenschaft 25:149–153
Narziss L, Miedaner H, Gresser A (1985) Hopfenölverbindungen und ihr Beitrag zum Hopfenaroma im Bier. Monatsschr Brauwiss 38:406–409
Peppard TL (1985a) Principal components analysis–its application in monitoring the consistency of beer quality. J Inst Brew 91:325–328
Peppard TL (1985b) Development of routine investigational tools for the study of sulphury flavours in beer. J Inst Brew 91:364–369
Powers JJ (1982) Techniques of analysis of flavours. Integration of sensory and instrumental methods. In: Morton ID, Macleod AJ (eds) Food flavours, Part A. Introduction. Elsevier, Amsterdam, p 121
Reiner L (1971) Die Faktorenanalyse in der brautechnologischen Forschung. Brauwissenschaft 24:403–412
Reiner L, Piendl A (1974) Zur Gruppierung von Biereigenschaften mittels der Faktorenanalyse. Brauwissenschaft 27:1–10
SAS (1985) SAS user's guide: statistics, version 5 edn. SAS Institute Inc., Cary, NC
Vuataz L (1976/77) Some points of methodology in multidimensional data analysis as applied to sensory evaluation. Nestle research News, pp 57–71
Williams AA, Rogers C, Noble AC (1984) Characterisation of flavour in alcoholic beverages. In: Nykänen L, Lehtonen P (eds) Flavour research of alcoholic beverages. Instrumental and sensory analysis. Found Biotech Ind Ferment Res 3:235–253

German Beer Purity Law and its Influences on the Properties and Analysis of Beer

S. Donhauser

1 The German Beer Purity Law

During the past centuries, the changing composition of the raw materials and production methods have led to an abundance of different beers. These beers differ not only in appearance, but also considerably in flavour, aroma and bitterness. The condition and quality of the raw materials have a decisive influence upon the quality and character of the beer. Malted barley is preferably used in the classical production methods for producing a fermentable extract.

On April 24. 1516 in Ingolstadt, the German purity law written by the Bavarian Dukes Wilhelm IV. and Ludwig X. was enacted with regards to the methods of beer production. It states that only barley, hops and water may be used for the production of bottom fermented beer. In other regions of Bavaria, for example in Munich, since 1487, there had already existed a similar beer purity law.

This ordinance, established solely to protect the consumer, prevented all forms of aduleration from the often questionable intentions of food chemistry, and thus established beer as the number one beverage of choice among the population. The worldwide reputation of the Bavarian and German beers came into existence through the establishment of the beer purity law. The Bavarian brewer has maintained these standards even during economically bad times. Until now, the consumer could rest assured that beer would be produced unadulterated by the nutritionally inferior food additives that are often found in other food products. The removal of the beer purity law has caused the beer drinker of the Federal Republic of Germany to fear for the reputation of their favourite beverage. With the passage of time, other German states accepted the purity law until 1906, when it was passed as a law for the entire German Empire. In its present form, the purity law is included as part of the German beer tax law. The relevant sections of the beer tax law with regards to beer production guidelines are as follow:

Beer Production

§ 9

1. Only malted barley, hops, yeast and water may be used in brewing bottom fermented beer. Exceptions are mentioned in Sects. 4 through 6.
2. The production of top fermented beer must follow the same standards but with the following additions: it is also permissible to use other malted grains, technically pure cane, beet, or invert sugars and starches and colouring agents produced from sugar.
3. Malt is considered to be all artificially germinated grains.
6. Permissible clarifying agents for sweet wort and beer are those which function using mechanical or adsorption principles. Residual traces are only allowed when technically unavoidable and nutritionally, aromatically and flavour-related harmless.

The exceptions mentioned in § 9 Sect. 2 are not applicable for the states of Bavaria and Baden-Württemberg.

The standards for brewing water quality are described in the German drinking water ordinance (Trinkwasserverordnung TVO).

2 Complaint of the EEC Commission at the European Court of Law Against the Federal Republic of Germany

Previously, the EEC commission filed suit on 12.2.1982 for breach of contract according to article 169 EWG-V against the Federal Republic of Germany because of the beer purity law. The following issues were presented:

- The beer purity law represented a trade barrier within the EEC.
- The widely accepted raw materials for brewing used in the other member countries represent no health risk that justify restrictions of use.
- Listing all consumer-relevant ingredients on the label of the beers would further protect the consumer.

The EEC commission came to the conclusion that, for the above reasons, applying the purity law to imported beer from member countries would be impermissible according to article 30 of the EEC contract.

3 Arguments for Retaining the Beer Purity Law

The German government presented the following reasons for applying the purity law to both domestically produced and imported beer in response to the case of the EEC commission from 12.2.1982 entitled "German Laws Applying to the Production and Distribution of Beer":

- According to article 36 of the EEC-contract, measures taken to protect the consumers' health prevail over those mentioned in article 30 that would contradict these actions. The German legal restrictions of the beer purity law are linked to the distribution of beer for reasons of protecting the health of the consumer and for this reason apply equally towards domestically and foreign produced products.
- The Federal Government stands fast by the traditional restriction of the purity law because a beer of superior quality can be brewed using only the ingredients previously mentioned. Should beer be partially or totally produced from other raw products, numerous other technological substances will have to be used in the production process in order to attain the qualities the consumer desires. According to previous releases of the Federal Government, other member countries have allowed food additives such as acids, enzymes, tannins, hazing stabilizers, antioxidants, foam stabilizers, anti-foam agents, and preservatives for those reasons.
- The importance of the beer purity law is above all health related. Because of the high per capita level of beer consumption in Germany, beer is treated as a basic food. It is, on the average, the most widely consumed food product in the Federal Republic of Germany.
- By removing the distribution restriction of beer not brewed according to the beer purity law, the danger is increased that the German consumer's intake of these food additives increases through the added consumption of this beer.

- It is not necessarily the effect of the isolated substance with respect to preventative health measures that is critical; the actual or preventative health risks lie more in the fact that the summation, cumulation and combination reactions of these substances on people have not been sufficiently researched and are not known. Since the use of food additives always contains an added risk, it cannot be disregarded that a possible summation and chemical interaction of food additives in beer not brewed according to the beer purity law, could produce long-term health effects for the consumer. The use of food additives for food products is regulated by the Food Additive Registration Ordinance in the Federal Republic of Germany in the Food Product and Basic Needs law. In the course of time, a large number of substances that were at one time considered harmless had to be removed from this list of acceptable food additives because modern research techniques have proven them to be carcinogenous or mutagenous, or to produce allergic reactions. An example of this is Diethylester PKE that became popular in the 60's for preserving wine and alcohol-free drinks. Permission to use this substance had to be rescinded in 1973 when it was discovered that during decomposition not only alcohol and CO_2 were produced, but actually other secondary byproducts. Ethanol is able to react and form diethylcarbonate, and ammonia can react to form carcinogenous ethylurethane.
- Based on the high rate of beer consumption in the Federal Republic, the total intake of food additives for various sectors of the population would increase exorbitantly. For this reason, it is inconsiderable for the health reasoning of the beer purity law that food additives allowed for beer production in other member countries are allowed to be used in the Federal Republic of Germany for other food products.
- Last but not least is the contradiction that after the abolishment of the German purity law for beer, the internationally recognized principles of food additive controls will apply. It is thereby understood that a food additive may only be used when harmless to health and when technologically absolutely necessary. The centuries-old brewing tradition of the beer purity law, that is also accepted in countries other than Germany, has demonstrated without question that the production of beer of superior quality without the use of additives is possible.

On 12. 3. 87, the European Court of Law in Luxemburg ruled, without taking note of the health reasoning expressed by the Federal Government, that beers from EEC countries that did not brew according to the German beer purity law would be allowed to be imported into Germany so long as the products were properly labeled.

4 Beers with Adjuncts

It is quite common in almost all beer-producing countries to use adjuncts for both economical and technological reasons. Additionally, many of these countries use other additives, such as enzymes, coagulants, antioxidants and foam preservatives to improve or rather maintain the physicochemical and biological stability of the beer.

The most widely used adjuncts are primarily corn and rice, with quantities of around 20 and 30% being frequently used. Proportions of these amounts can be used without problem when brewed in conjunction with well-modified malt. Higher percentages, even with the use of unmalted barley, hinder sugar conversion during mashing, reduce the lautering velocity, and often negatively influence the beer quality. The enzymatic potential of the malt no longer suffices to break down the larger molecules in the mash. In such instances it is necessary to add enzyme complex preparations made from fungi and bacteria with high α-amylase, glucanase, and proteinase activity.

Should corn and rice be used as adjuncts for beer production, the amount of nitrogen is reduced as the concentration of these cereals containing a smaller amount of soluble nitrogen increases. Corn- and rice-brewed beers also prove to have smaller amounts of formol- and α-amino nitrogen along with anthocyanogens and tannins. The same applies to prolin and phosphate values. These values, however, cannot be used for making a general statement about brewing with adjuncts for the total and formol nitrogen are not just dependent upon the protein content of the malt, but also the level of modification. With adjunct brewed beers there is a slight tendency towards lower nitrogen, prolin, phosphate, and tannin contents. When adjuncts are used in conjunction with well-modified malt, these differences are eliminated, since more nitrogen, phosphate and prolin go into solution. The same applies to malts that have been treated with gibberellic acid. Due to the great variety of malting and brewing processes with the varying technologies and the variety of generated data, it is inappropriate to make a statement about the general use of adjuncts.

5 Analytical Analysis of Additives and Harmful Substances

The quality control of beer, whether brewed according to the German beer purity law or not, is carried on everywhere using much the same analytical methods, the analytical measures of the Middle European Brewing Technological Analysis Commission (MEBAK) or the methods of the American Society of Brewing Chemists (ASBC). With this in mind, no further discussion of analytical methods is necessary.

Additional methods are only necessary when substances have to be analyzed that are not allowed according to the beer purity law.

All attempts to identify adjuncts and enzyme additives using chemical methods, for example total maltose, specific gravity difference and the total nitrogen of the beer, the amount of α-amino nitrogen and carbohydrate isotope identification have failed up to now. Biological material usually consists of a mixture of macromolecules that demonstrate analogue or similar physicochemical properties, so that it is often difficult to differentiate between them. Additionally, most functional tests cover a whole group of factors.

For the detection of adjuncts, enzyme and hormonal additives in such complex matrices as malt and beer it is necessary to use very sensitive and specific methods. These prerequisites have been achieved only by immunochemical methods. These methods are not only important for the medical, biological and biochemical sciences, but are an important tool in quality control of food products where it is necessary to detect cheap and nutritionally inferior surrogate substances, difficult to detect using conventional methods. The quality control of food products without the use of immunochemical methods would therefore be incomplete. The primary reason for this is not to detect fraud, but rather to protect the health of the consumer.

Immunochemical methods for the identification of adjuncts and enzyme additives in malt and beer were first applied by Donhauser (1964).

The following methods have been employed for the identification of additives in malt and beer.

5.1 Immunological and Immunochemical Methods

The exact structure of protein is not always known. It is, however, known that proteins are specific for the type of organism. This applies also to the proteins of the various cereals, so that specific differences can be detected. Based on the relationship (the family Graminaceae), many of them are structurally similar so that they create cross-reactions. Careful preparation of the seren, that is immunization with an antigene that causes no cross-reaction, is therefore a mandatory preparatory action when unexpected precipitation is to be avoided.

As previously discussed, it is allowed to stabilize protein in beer using adsorbents in the Federal Republic of Germany, but not with the use of proteolytic enzymes, since this is not acceptable according to the beer purity law. The identification of proteolytic enzymes usually involves analyzing the level of enzymatic activity, and numerous methods exist to determine the higher levels of proteolytic activity. Casein or hemoglobin serve as the substrate for these laboratory methods. Indication of enzyme additives is attained when a certain level of substrate degradation occurs. This does not apply if the beer has been previously pasteurized. Enzymatically stabilized beers are pasteurized to inactivate the enzyme, thus limiting the undesirable effects (for example deterioration of foam retention) and insuring the biological stability of the beer. Using the previously mentioned methods produces inaccurate results, since the proteolytic activity of unstabilized beers and unpasteurized beers is often higher than the pasteurized and illegitimately stabilized beers. The possibility of false positive or false negative results using immunological indicators is not possible since these methods are based on the structure of the enzymes and not on their activity levels.

5.1.1 Qualitative and Half-Quantitative Indication Methods

5.1.1.1 Double Diffusion in Agar

This method is used often for the identification of adjuncts and enzyme additives in beer. The principle of this qualitative-half quantitative method, also known as the Ouchterlonytechnique (Ouchterlony 1949, Donhauser 1972, 1976, 1979a, b) is based on highly specific immunoreactions. While the functional tests encompass large groups of factors, the immunoreactions are highly specific. The antibody reacts with only one area of the corresponding antigen which is characterized by a small number of amino acids and their sequence.

A minor change in this area will cause a change in the immunoreaction. The advantage of this method is the ability to identify antigens in complex mixtures and still be able to determine cross-reactions, that is usually not as easily done using other immunological methods.

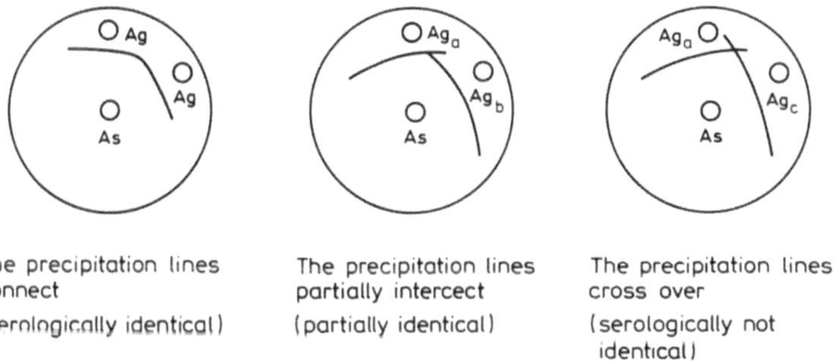

Fig. 1. Schematics of two-dimensional double diffusion for qualitative analysis. *Ag* Antigen; *As* = Antiserum

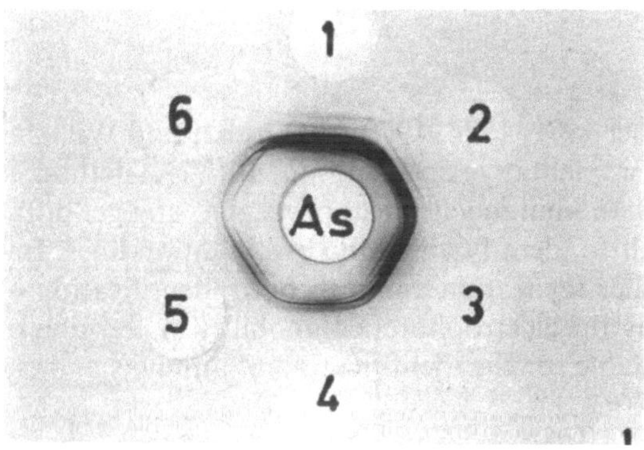

Fig. 2. Agar gel diffusion test (rosette-shaped). Detection of corn adjuncts in beer in serial dilutions of *1* = 10%–*6* = 0.31%. Antiserum: Corn 8

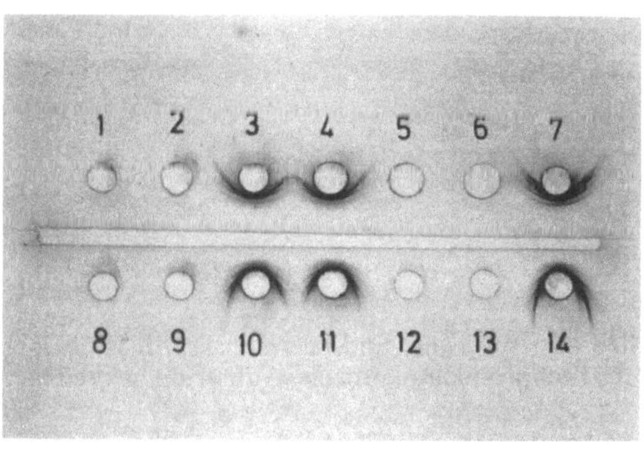

Fig. 3. Agar gel diffusion test (linearly arranged). Testing of 7 beers for rice additives. Beers *1–7* and *8–13* are identical, differing only in application well distance from the serum trough. As seen from the curvature of the antigen-antibody complex (precipitate), the formations between the antigen and antibody are highly influenced by differences in concentration. Antiserum: rice.

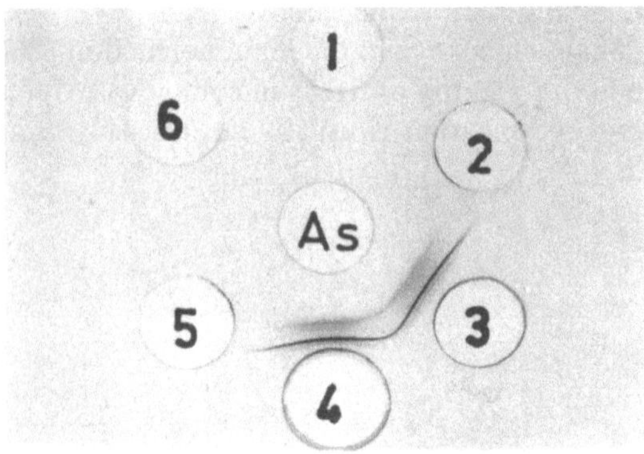

Fig. 4. Agar gel diffusion test. Detection of papain use for preserving beer. *1*, *2* Beer with bromelin added; *3*, *4* beer with papain added; *5*, *6* beer with ficin added. Antiserum: papain (The antiserum is specific and does not react with the other Sulfhydrile proteases bromelin and ficin)

With this method, the antigen and antibody diffuse in a layer of agar, thus building precipitation lines in the areas of optimal concentration. Each antigen/antibody system produces a precipitation line. If the two antigens are serologically identical, the precipitation lines will bend towards each other and connect. If they are partially identical, the precipitation lines move partially towards each other, forming spurs. Should the two antigens be opposites, the corresponding precipitation lines cross over each other (Figs. 1–4).

5.1.1.2 Immunoelectrophoresis

Immunoelectrophoresis (Donhauser 1972, 1976, 1979 a, b; Grabar and Williams 1953; Scheidegger 1955), likewise a qualitative and semi-quantitative method, is the combination of a physical and an immunochemical process. Using gel-diffusion techniques, it is often difficult to identify all of the precipitation lines. Immunoelectrophoresis enables further separation and improved identification of the substance, based on harnessing the electrophoretical mobility in addition of diffusion. This method is also suitable for the identification of adjuncts and enzyme additives (Figs. 5 and 6).

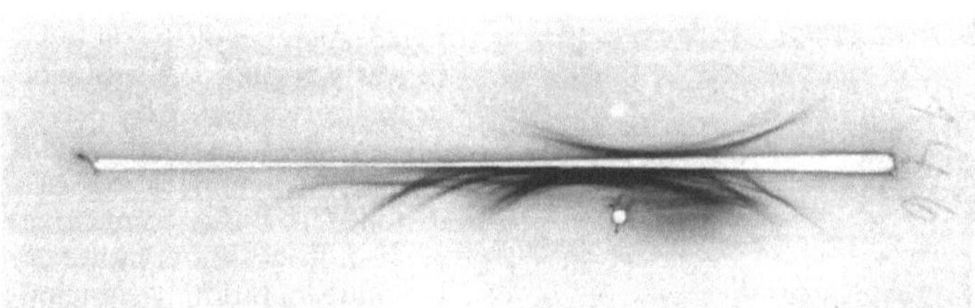

Fig. 5. Immunoelectrophoresis of malted barley *(below)* and beer *(above)*. It is obvious that the basic proteins degrade during the beer production process. Antiserum: malted barley

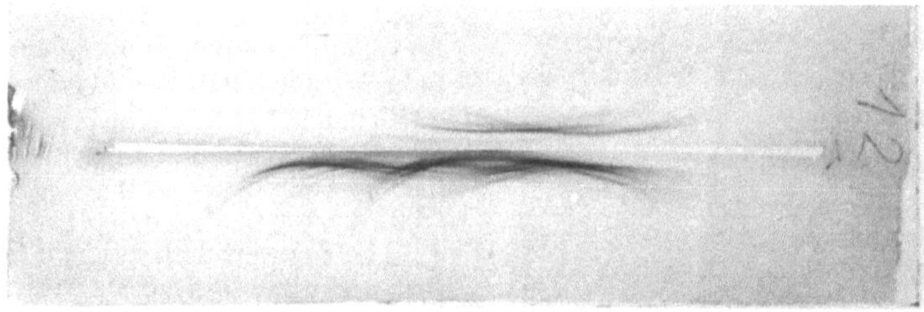

Fig. 6. Immunoelectrophoresis of wheat *below)* and wheat beer *(above)*. The use of wheat (or other cereals) for brewing can be specifically detected. Antiserum: wheat *(Triticum aestivum)*

5.1.1.3 Counter Current Electrophoresis

Counter current electrophoresis (Gocke and Howe 1970) is also a semi-quantitative method, used to identify proteins that are attracted to the anode under specific conditions in an electric field. A precipitation line forms when the antigen meets with the specific antibody in agar gel. This method is suitable for adjunct identification, since adjunct grain proteins are attracted to the anode.

5.1.2 Quantitative Identification Methods

Newer quantitative methods such as radial immunodiffusion, electroimmunodiffusion and various other two-dimensional electrophoresis techniques in gels containing antibodies have enabled the further characterization of antigens. Since the areas enclosed by the precipitation lines are proportional to the antigen/antibody ratio, these methods are also suitable for the immunological and quantitative identification of additives. For the undertaking of quantitative analysis, it is necessary in most instances to use an oligo- or monospecific antiserum.

5.1.2.1 Basic Radial Immunodiffusion

Basic radial immunodiffusion was first applied by Petrie (1932). Ouchterlony later adapted it for half-quantitative applications. Mancini and colleagues finally developed it into a quantitative method. The process involves diffusing an unknown amount of antigen radially from a circular-shaped application tumbler in a period of time that enables all antigens to react. The area of the precipitation rings is proportional to the antigen concentration and inversely proportional to the antibody concentration. The diameter of the precipitation rings is measured and graphed against the logarithmic concentration of the antigen. Unknown concentrations are plotted on a calibrated curve which has been prepared based on standardized concentrations. Radial immunodiffusion is suitable for adjunct and enzyme additive identification (Donhauser 1976, 1979 c) (Fig. 7).

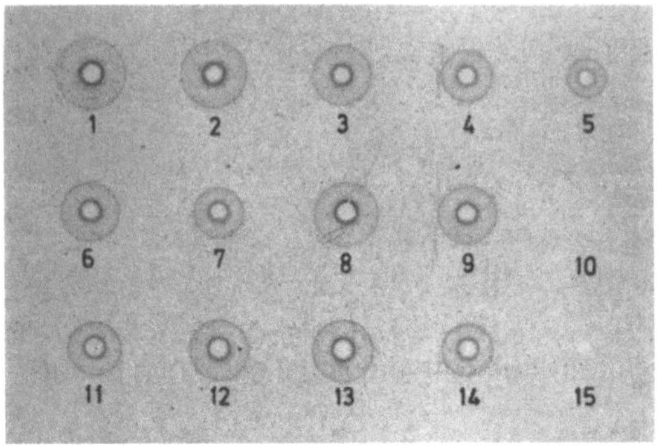

Fig. 7. Radial immunodiffusion for quantitative detection of ficin in beer. *1* Ficin 0.02% = 20 g/hl; *2* = ficin 0.016% = 16 g/hl; *3* = ficin 0.012% = 12 g/hl; *4* = ficin 0.008% = 8 g/hl; *5* = ficin 0.004% = 4 g/hl; *6, 7, 8, 9, 11, 12, 13, 14* = beers with different ficin additives; *10, 15* beers without ficin additives. Antiserum: ficin

5.1.2.2 Electroimmunodiffusion

Electroimmunodiffusion, according to Laurell (1966), deals with quantitative electrophoresis in a gel containing antiserum. An antigene causes an extended immunoprecipitate from its electrophoretic movement in agar gel, when the gel contains an evenly distributed monospecific antiserum. The length of the precipitation is proportional to the concentration of the antigen when applied in a known volume. This method can be used for all proteins whose movements in electrophoresis can be differentiated from the movement of the antibodies distributed in the agar gel. This method is suitable for identification of adjunct grain proteins. Enzymes can be identified only when negatively charged (Donhauser 1976, 1979a). Studies completed on wheat beers produced by southern German breweries without previous knowledge of wheat/barley ratios produced exact results. The results produced with immunoelectrophoresis matched perfectly with beers produced from a Munich brewery using 25, 50, and 60% wheat malt.

Using immunoelectrophoresis techniques in gel containing antiserum, it is possible not only to differentiate between winter and summer barley, but also to quantify the results. It is taken commonly for granted that malt of brewing quality is only to be produced from summer barley. One of the most important questions when purchasing brewing barley is the pureness of the barley variety and that less suited varieties have not been mixed. For the purpose of varietal identification, immunochemical methods have with time proved to be superior to normal electrophoresis methods, based on the speed and ease of the analysis together with precise results.

These techniques are also suitable for the evaluation of malt quality with respect to α- and β-amylase content. These methods prove to be superior when compared to the enzymatic methods. Using the same principles, it is possible to determine the glucanases in malt. The advantages of immunochemical methods are the simple analytical procedures, the minimal amount of time needed to analyze a large number of samples together with multiple applications due to the many corresponding antiserums. Immunochemical methods offer the further possibility of identifying other elements contained in beer (Figs. 8 and 9).

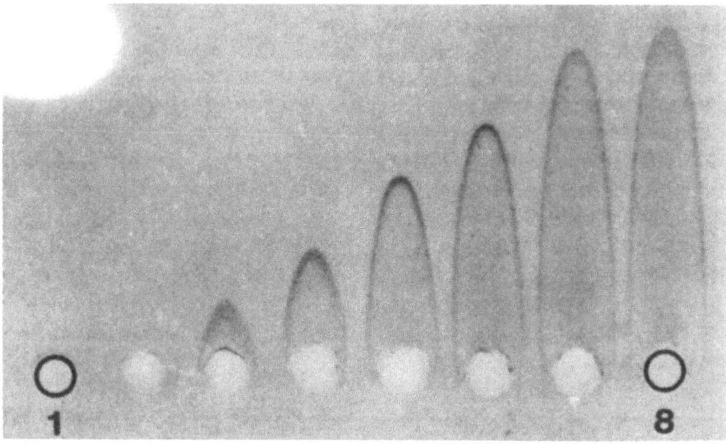

Fig. 8. Electroimmunodiffusion. Development of α-amylase in "green malt" from the 1st–8th germination days. α-amylase does not yet exist in the barley) Antiserum: α-amylase from malted barley

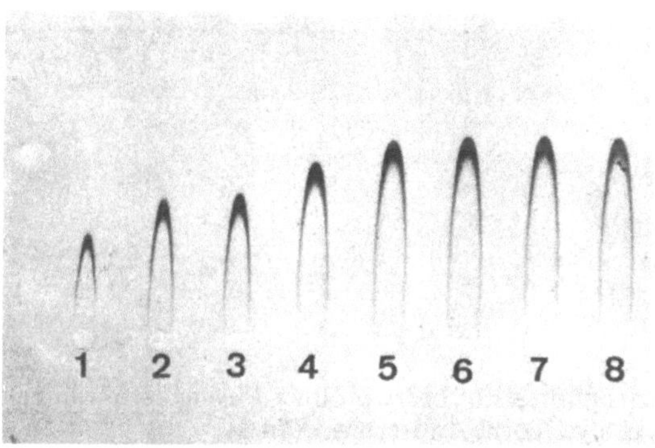

Fig. 9. Electroimmunodiffusion. Development of the β-amylase during the 1st–8th days of germination (pre-existing in barley). Antiserum: β-amylase from malted barley

5.1.2.3 Two-Dimensional Immunoelectrophoresis

The identification of adjunct and enzyme additives can also be completed using two-dimensional electrophoresis (Laurell 1965; Clarke and Freeman 1968). This method combines basic agar gel electrophoresis with electroimmunodiffusion. The first dimension entails electrophoretic separation of the protein mixture in an antibody-free agar gel, where the second dimension quantifies the immunoprecipitation in the agar gel containing the antiserum (Figs. 10 and 11).

5.1.3 Highly Sensitive Immunochemical Methods

Of increasing popularity for identifying small amounts of existing antigens are Radio-Immuno-Assay (RIA) and even more popular, Enzyme-Linked-Immu-

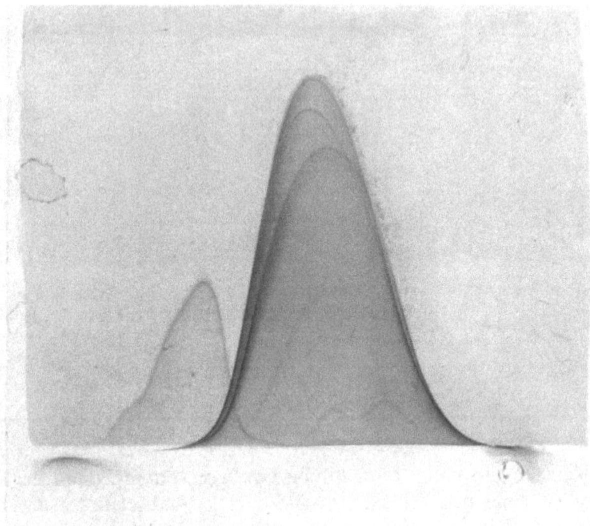

Fig. 10. Two-dimensional immunoelectrophoresis of an unstabilized beer produced from malted barley. Antiserum: beer (produced from malted barley)

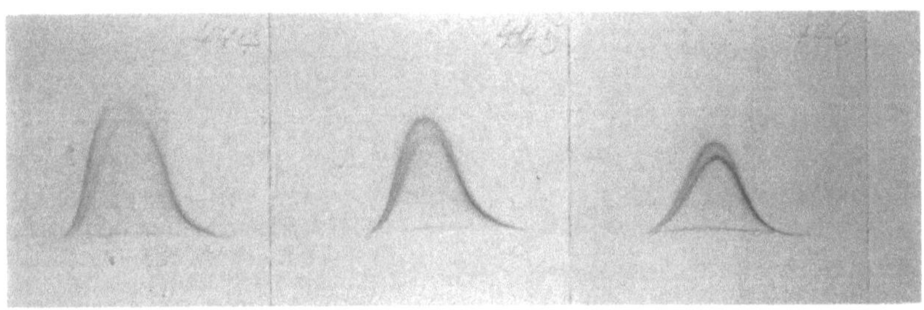

Fig. 11. Two-dimensional immunoelectrophoresis of beers produced using corn adjuncts. *From left to right.* 30% corn, 20% corn, 15% corn. Antiserum: corn

nosorbent-Assay (ELISA). They belong to a group of extremely sensitive identification methods. In general, they are used to identify molecules in the ng to pg range. By using appropriate testing procedures, the sensitivity can be increased even more, as the tests we have developed will demonstrate. Both methods rely on the specific antigen-antibody reaction with the use of marked reactants.

5.1.3.1 Radio Immuno Assay

Radio Immuno Assay (Donhauser 1979d; Yalow and Berson 1960; Eskins 1960; Greenwood et al. 1963) combines the high specifity of the antigen-antibody reaction with the extremely great sensitivity of radioactivity measurement, so that the smallest substance amounts can be identified.

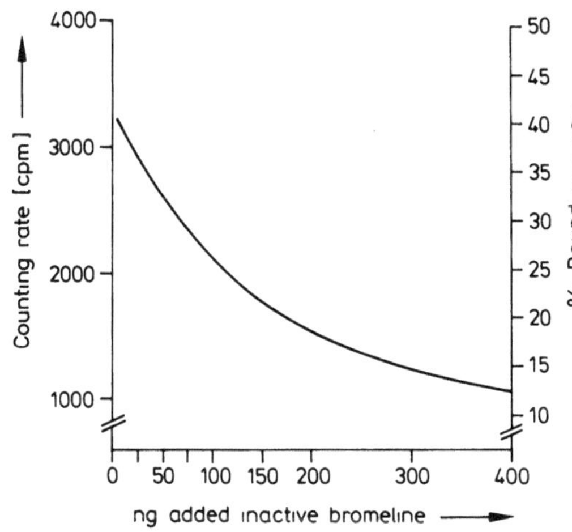

Fig. 12. Calibration curve of radioactively marked bromeline

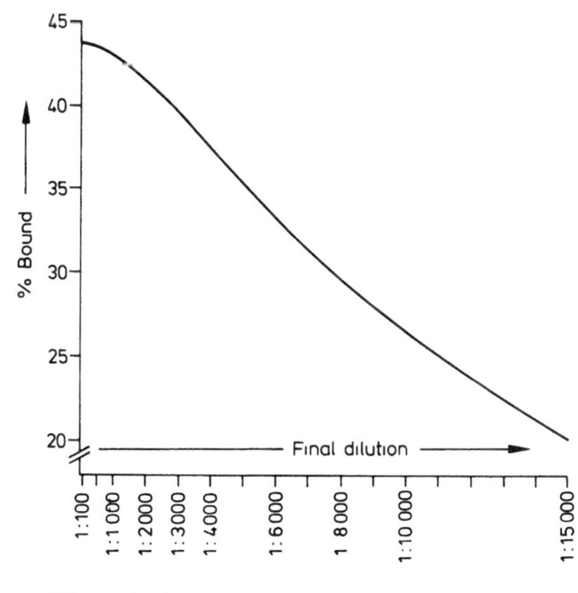

Fig. 13. Antiserum dilution series

The Principle of Radio Immuno Assay

RIA involves coupling a radioactive isotope that functions as a marking substance with a chosen antigene or a substance that reacts with a hormone, thus functioning as a receptor. The principle of RIA is based on the reversible interaction between antigen and antibody quantities.

The prerequisite in developing a RIA test is that sufficient quantities of the respective antigen (protein or protein-coupled substance) are available in a pure form for immunization and especially for radioactive marking and for drawing up a calibration curve. The antigen may not change its immunological identity due to the radioactive marking. In order to carry out RIA, the optimal antibody concentration and the required adequate amounts of radioactive marked antigens must have been previously tested (Figs. 12, 13).

5.1.3.2 ELISA or EIA (Enzyme Linked Immunosorbent Assay)

Enzymes are used for marking in this test (Engvall and Perlman 1971, 1972; Van Weemen and Schuurs 1971; Donhauser et al. 1985; Donhauser and Winnewisser

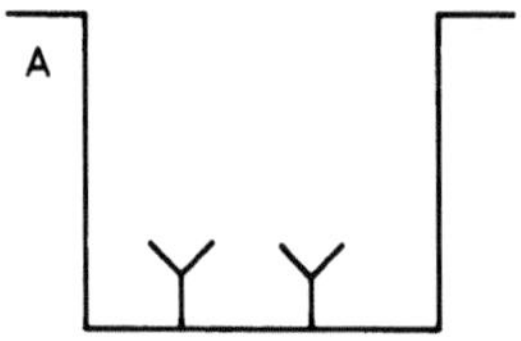

A. Application
The undiluted antibody solution is poured into the container and incubated. A portion of the antibodies is thereby adsorbed by the container inner surface/synthetic contact area. Unattached material is removed by pouring out the contents and repeated washing of the container.

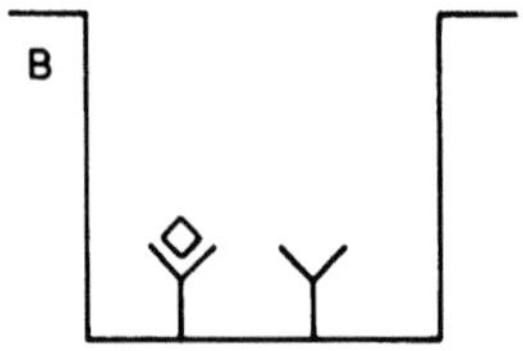

B. Antigene binding (1st immunoreaction)
The sample which is to be analyzed is added to the container and incubated: The antigene is bound with the antibody that is immobilized on the inner surface area. Unbound substances are separated by washing the container.

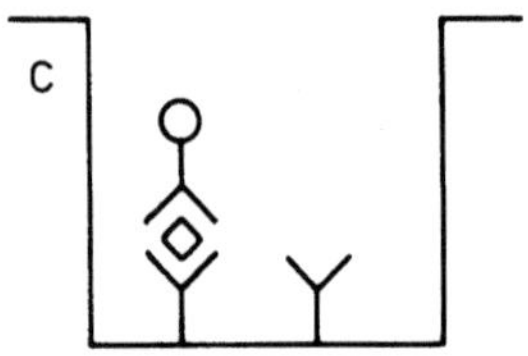

C. Enzyme conjugate binding (2nd immunoreaction)
With this 3rd incubation, antibodies marked with enzymes bind with the previously bound antigen. Extraneous conjugate is removed once again using water.

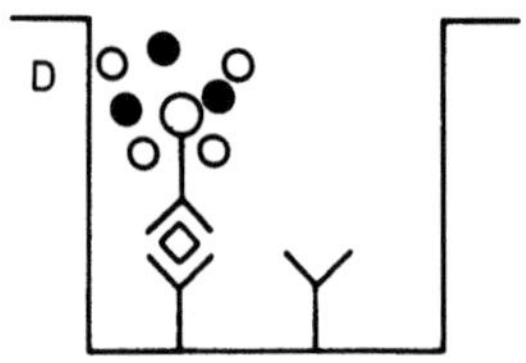

D. Enzyme activity measurement
The amount of immuno-bound enzyme conjugate is measured using the enzyme activity as an indicator, enzyme substrate is added. After a predetermined incubation time, the enzymatic reaction is interrupted and the resulting product is measured, usually photometrically.

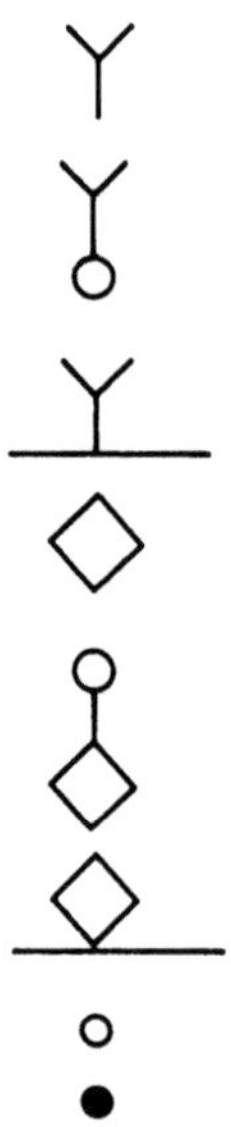

Antibody

Enzyme-marked antibody

Container-bound antibody

Antigen

Enzyme-marked antigen

Container-bound antigen

Enzyme substrate

Fig. 14. Description of ELISA using the "sandwich" principle

1985), with alkaline phosphatase, horseradish peroxidase, glucose-oxidase and β-galactosidase being the most common. Their activities are the most easily determined. The Enzyme Immuno Tests differ in the methods and in being homogenous or heterogenous. Homogenous EIA, the so-called EMIT technique, calls for modulating the activity of the enzyme used for marking using the antigen antibody reaction, usually as an inhibitor. This technique allows the identification of relatively small molecules (haptenes).

Heterogenous EIA, named Enzyme Linked Immunosorbent Assay and shortened to ELISA, has either the antigen or antibody attached to a supporting structure, usually a container wall when working with synthetic containers. There are three appropriate processes for antigen detection, all of which relate back to the variants:

- the so-called Sandwich ELISA
- the competitive ELISA and
- the Immunoenzymetric Test, shortened to IEMA

I would like to briefly discuss the Sandwich ELISA (Donhauser et al. 1985). The principle is described in Fig. 14.

A prerequisite for this process is that the antigen that is to be identified or quantified possesses at least two different antigen determinants, meaning binding positions, for antibodies. A Sandwich ELISA is therefore only suitable for analyzing macromolecules.

Tests for the identification of proteolytical enzymes with plant ancestry and α-amylases and amyloglucosidases from fungi in beer have been developed in our department based on these principles. ELISA is quickly and easily completed. The samples do not need to be cultivated, as is called for with the gel precipitation techniques. Sample preparation involves simply diluting the beer and removing the carbon dioxide. The advantages of ELISA become quite clear when a large number of samples must be analyzed in addition to the extreme sensitivity. Even when the tests are performed manually, it is possible to analyze around 20 beers in one day.

5.2 Gas Chromatography and High Performance Liquid Chromatography (HPLC)

Gas Chromatography and High Performance Liquid Chromatography are needed due to the use of formaldehyd during mashing to improve the chemicophysical shelf-life of beer and the use of preservatives to maintain the biological stability. The identification of formaldehyde is based on a derivational process that is preferably completed using 2,4-dinitrophenylhydrazin.

5.3. Electrophoresis

Electrophoresis is used for the identification of alginate additives which are used in beer as foam stabilizers (Pfenninger et al. 1983; Schur et al. 1981). Electrophoretic separation is performed after alkaline hydrolysis and concentrating the alginate, and followed by colouring with toluidin blue.

5.4 Isotachophoresis

This electrophoresis method has only recently found its way into the area of food analysis (Donhauser et al. 1984). It is therefore worthwhile to give a short introduction to this technique.

The basic theory of isotachophoresis is based on the Kohlrausch function, which states that all other ion concentrations are determined by the self-determined "Leition" concentration.

Isotachophoresis is conducted in a capillary without a supporting structure. A discontinuous electrolyte is used as a separating system. It consists of a "Leition" with higher mobility and a following ion with lower mobility. The pH range in which the separation should take place is regulated by a counter ion. The electron neutrality of the discontinuous electrolyte should be the highest possible buffer capacity of the counter ion.

Preservatives such as sorbinate, benzoic acid and PHB-ester, all of which are not allowed by the beer purity law, are all also identifiable using isotachophoresis.

5.5 Atomic Absorption Spectrometry

Using Atomic Absorption Spectrometry it is possible to detect heavy metals that are used, for example, to improve the foam retention of beer. In the 1960's in North America, many people became sick, with several fatalities, due to the use of cobalt chloride and cobalt sulfide to improve foam retention in beer.

6 Summary

Until the ruling of the European Court of Law from 12. 03. 1987, the German beer purity law from 1516 covered all beers distributed in the Federal Republic of Germany, including imported beer.

Because of this new ruling it is possible in the future for imported beer to be sold in the Federal Republic of Germany under the name "beer" even when it contains malt substitutes (adjuncts) and providing that the type (for example rice, corn grits) and their amount are clearly visible on the beer label. The ruling does not change the continued acceptance of the beer purity law for domestically produced beer. It applies only for beer imported from other member countries of the European Economic Community.

To the question of additives, the Court generally accepted that restrictions based on protecting the health of the consumer are legally valid. Importers must, however, have the opportunity, in the form of a hearing, to petition for the use of additives. Restricting *all* additives for imported beer is excessively strict. Therefore, in the future, it will be necessary for the Federal Republic of Germany to relax its absolute additive restriction for beers imported from other member countries, providing that the use of these additives is clearly noted on the beer label.

The Federal Ministry for Youth, Family and Health has clearly stated that all foreign substances disallowed by the beer purity law contained in the imported beers must be easily recognized on the label. Furthermore, the ministry made it known that customs offices of the state authorities be instructed with regards to the identification of suspicious beer transactions. Also, the Bavarian Parliament unanimously decided that all foreign beers in Bavaria must have their contents analyzed by the Ministry of Health. Additionally, the Bavarian Government will be informed by a legal petition stating the desires of the Parliament, that it must be made easy for the consumer to differentiate between "beer brewed according to the purity law" and adulterated barley juice.

The German lawmakers found themselves in agreement with the Recommendations of the World Health Organization, in that food additives should only be allowed for use only when technically unavoidable. Chemical additives for the food product beer are not allowed in Germany for the three following reasons:

- Because beer intake among men in Germany has the highest *volume percentage* of all the consumed food products.
- Because of the specific health risk added due to the combination of alcohol and additives and the interaction among the additives with each other.
- and last but not least, because the use of additives is technically unnecessary as the quality and varieties of German beer have consistently demonstrated.

The German position in rebuttal to the complaint relied on article 36 of the EEC-contract. It clearly states that member countries may restrict the free movement of goods when done for reasons of health protection. This position was very clearly proved by the Federal Republic of Germany during the hearing.

The Federal Government called upon previous rulings of the European Court and quoted them saying that no member country is required to reduce their level of protective health measures.

Consequently, the European Court of Law refused to acknowledge the complete aspect of health protection of the consumer in the beer purity law in its ruling against the Federal Republic of Germany. According to the German view, health issues should take precedent over purely economic issues. The combination-, cumulation- and summation effects that come into question with regards to additives have been inadequately researched. This is a particularly important aspect considering that the per-capita consumption of beer in the Federal Republic of Germany is around 150 l/year (Bavaria: approx. 240 l/year). Furthermore, there is not the smallest technological necessity to produce beer using the previously mentioned additives. Without using these substances a beer of superior quality can be produced and has been produced during the past centuries.

With the detection methods discussed it is easy to detect the use of adjuncts and chemical additives.

References

Biersteuergesetz der Bundesrepublik Deutschland v. 15. April 1986

Brauwelt (1982) EG-Vertragsverletzungsverfahren gegen Reinheitsgebot 122:384

Brauwelt (1984 b) Die EG-Kommission hat am 9. Juli Klage gegen die BR Deutschland in Sachen Reinheitsgebot für Bier beim EuGH eingereicht 124:1219

Clarke MHG, Freeman T (1968) Quantitative immunelectrophoresis of human serum proteins. Clin Sci 35:403
Donhauser S (1964) Auftrennung und Spezifizierung der bei den technologischen Vorgängen der Bierbereitung auftretenden salzlöslichen Proteine von Gerste und Rohfrucht durch immunologische und physiko-chemische Methoden. Diss TH München
Donhauser S (1972) Nachweisversuche an enzymatisch stabilisierten Bieren mit immunologischen Methoden. Brauwiss 25:189
Donhauser S (1976) Die Anwendung immunologischer Methoden in der Brauerei-Analytik. Brauwelt 116:1560
Donhauser S (1979a) Immunologische Untersuchungen zum Nachweis von proteolytischen Enzymen im Bier. Brauwiss 32:93
Donhauser S (1979b) Immunologische Untersuchungen zum Nachweis von proteolytischen Enzymen im Bier. II. Brauwiss 32:119
Donhauser S (1979c) Immunologische Untersuchungen zum Nachweis von proteolytischen Enzymen im Bier. III. Brauwiss 32:173
Donhauser S (1979d) Radioimmunoassay zum Bromelin- und Ficin-Nachweis im Bier. Brauwiss 32:211
Donhauser S, Winnewisser W (1985) Enzym-Immuntest zum Ficin-Nachweis im Bier. Monatsschr Brauwiss 38:121
Donhauser S, Vogel H, Winnewisser W (1985) Enzym-Immuntest zum Papain-Nachweis im Bier. Monatsschr Brauwiss 38:244
Donhauser S, Glas K, Gruber B (1984) Nachweis von Konservierungsmitteln im Bier mittels HPLC und Isotachophorese. Monatsschr Brauwiss 37:252
Engvall E, Perlman P (1971) Enzyme-linked immunosorbent assay (ELISA) Quantitative assay of immunoglobulin G. Immunochem 8:871
Engvall E, Perlman P (1972) Enzyme-linked immunosorbent assay, ELISA. III. Quantitation of specific antibodies by enzyme-labelled antiimmunoglobulin in antigen-coated tubes. J Immunol 109:129
Eskins RP (1960) The estimation of thyroxine in human plasma by an electrophoretic technique. Clin Chim Acta 5:453
Gocke DJ, Howe C (1970) Detection of Australia antigen by counterimmunoelectrophoresis. J Immunol 104:1031
Grabar P, Williams C (1953) Méthode permettant l'étude conjuguée des propriétés électrophoriques et immunochimiques d'un mélange de protéines. Application au sérum sanguin. Biochim Biophys Acta 10:193–194
Greenwood FC, Hunter WM, Glover JS (1963) The preparation of ^{131}I-labelled human growth hormone of high specific radioactivity. Biochem J 89:114
Laurell CB (1965) Antigen-antibody crossed electrophoresis. Anal Biochem 10:358
Laurell CB (1966) Quantitative estimation of proteins by electrophoresis in agarose gel containing antibodies. Anal Biochem 15:45
Mancini G, Vaerman JP, Carbonara AO, Heremans JF (1964) A single radial diffusion method for the immunologic quantitation of proteins. Protides Biol Fluids. Proc Colloq 11:370
Mancini G, Carbonara AO, Heremans JF (1965) Immunochemical quantitation of antigens by single radial immunodiffusion. Immunochem 2:235
Ouchterlony O, (1949) Acta Pathol Microbiol Scand 23:516
Petrie GF (1932) A specific precipitin reaction associated with the growth on agar plates of Meningococcus, Pneumococcus, and B. Dysenteriæ (Shiga). Br J Exp Path 13:380
Pfenninger HB, Anderegg P, Vykoukal E (1983) Alginate determination in beer. J ASBC 41:40
Scheidegger JJ (1955) Une micro-méthode de l'immuno-électrophorèse. J Int Arch Allergy 7:105
Schur F, Anderegg P, Pfenninger H (1981) Spezifischer Nachweis von Alginat in Bier. Brauerei-Rundschau 92:25
Van Weemen B, Schuurs A (1971) Immunoassay using antigen-enzyme conjugates. FEBS Lett 15:232
Yalow RS, Berson SA (1960) Immunoassay of endogenuous plasma insulin in man. J Clin Invest 39:1157

Subject Index

MIX
Papier aus verantwortungsvollen Quellen
Paper from responsible sources
FSC® C105338

If you have any concerns about our products,
you can contact us on
ProductSafety@springernature.com

In case Publisher is established outside the EU,
the EU authorized representative is:
Springer Nature Customer Service Center GmbH
Europaplatz 3, 69115 Heidelberg, Germany

Printed by Libri Plureos GmbH
in Hamburg, Germany